Beyond History of Science

Professor Robert E. Schofield

Beyond History of Science

Essays in Honor of Robert E. Schofield

Edited by
Elizabeth Garber

Bethlehem: Lehigh University Press
London and Toronto: Associated University Presses

Associated University Presses
440 Forsgate Drive
Cranbury, NJ 08512

Associated University Presses
25 Sicilian Avenue
London WC1A 2QH, England

Associated University Presses
P.O. Box 488, Port Credit
Mississauga, Ontario
Canada L5G 4M2

The paper used in this publication meets the requirements
of the American National Standard for Permanence of Paper
for Printed Library Materials Z39.48-1984.

Library of Congress Cataloging-in-Publication Data

Beyond history of science : essays in honor of Robert E. Schofield /
edited by Elizabeth Garber.
    p. cm.
Includes bibliographical references.
ISBN 0-934223-11-4 (alk. paper)
    1. Science—Historiography. 2. Technology—Historiography.
3. Schofield, Robert E. I. Schofield, Robert E. II. Garber,
Elizabeth.
Q126.9.B49   1990
509—dc20   89-45365
                                      CIP

PRINTED IN THE UNITED STATES OF AMERICA

Contents

Introduction

ELIZABETH GARBER

A collection of essays such as this is traditionally a means of recognizing a scholar's influence in his chosen discipline. This tribute is offered by students who remember Robert E. Schofield's patience as a teacher and colleagues who remember incisive discussions with him on important historiographic issues that materially altered or expanded their own research horizons. Some of these directions were suggested before they became respectable topics of scholarly attention, and others reflect Bob's own broad interests. Relating science and technology to their cultural and social context guided his own research, and Bob has always encouraged others to explore the ties between the two and between each and its social and cultural contexts.[1] Initial explorations preceded the recent reorganization of research in history of science and technology around the issues of contextualism and their theoretical borrowings from the other social sciences. For his students neither science nor technology could be isolated from the politics, economy, social structure, or culture of the era. As a complement to this Bob felt compelled to draw the history of science and technology closer to the broader discipline of history.

Historians of science and technology, if not historians in general, have long accepted that science and technology are integral to the development of Western civilization and society.[2] In the past twenty years they have transcended the historiography that treated the development of modern science and technology as historically accidental.

Current historiography of science and technology is dominated by the assumption that the ideas, language, images, and the concrete productions of scientists and engineers are as much the embodiments of the social structures, values, and cultural resources of modern Western society as are the arts, literature, politics, and the economy. Many historians now organize their research around problems and themes rather than periods or geographically defined nation states. Therefore, they share with others in various subdisciplines of history many presuppositions and methodological problems of research.[3]

The most obvious characteristic of current historiography is the plethora of available explanations. However, the origins of these apparently competing theories of history reside in the social sciences. And in these origins lies a unity behind the facade of diversity. These theories try to integrate science and technology into Western society and culture using socially or culturally derived categories that characterize the larger whole. The same categories label the lives, work, and ideas of the individual; hence the integration is accepted as complete.

Social structure and cultural values are the only keys necessary to understanding scientific and technological ideas and the behavior of scientists and engineers individually and collectively. Understanding the place of science and technology in modern Western society and culture is the deciphering of those values, external to the sciences and technology, that they incorporate and express. They are simply symbolic activities, manifestations of behaviors, values, or beliefs more fundamental in social or cultural terms than the form—science or technology—in which they appear. Science and technology are masks for a more fundamental reality, unknown and unknowable to the participants.

What, precisely, this deeper reality is depends wholly upon the preconceptions of the historian. It can be social place (class), the beliefs of a social group (ideology, interests), cultural values (usually ideology disguised as religious beliefs), and so on. All of these categories are external and assumed as prior to the historical evidence. Imposed on that evidence, they reveal the hidden, underlying reality beyond the empirical facade.

However, more and more misgivings have surfaced in the professional literature in the past five years about the theories and methods so recently and enthusiastically endorsed as the direction for historical research. What to some is a sign of health, interpretive diversity, is to others a sign of a profession without direction, even a sign of serious malaise. Most criticism has centered on the inability of the theories to encompass the complexities and particularities of the historical data.[4] Historians are finding the theoretical models they eagerly adopted from the social sciences are too rigid and simplistic to encompass the fragmentary and discontinuous nature of the evidence. More serious critiques have focussed on the structural weaknesses in current theories of society and culture both as models of society and culture and as explanatory models in history.[5] Such criticisms challenge the monopoly that contextual historians have claimed to examine and interpret science and technology in society and culture. Their theories simply do not address important historical problems, especially those of uniqueness and discontinuity.[6]

Without losing sight of the methods they have learned or the questions that these excursions into other disciplines have allowed them to address, some mainstream historians have abandoned close adherence to the theories of economists, sociologists, and anthropologists.[7] They are returning to problems of particular concern to historians: the kinds of remains historians have to work with, how to integrate these disparate pieces of evidence into an understanding of historical process and the relationships between this kind of available evidence and theory.

An appreciation is emerging of more traditional scholarly approaches in history in general and history of science and technology in particular. Yet this return to tradition cannot be done in the innocence of ignoring the interpretive alternatives now available. The limitations of the older traditions endure and the useful and perceptive nature of closely defined historical questions cannot be denied. We now incorporate as historical problems questions framed within the newer historiographies. These questions were stipulated more specifically, focused by the theories within which they were framed, and they led to closer examinations of the connections between subject and context that were impossible from older perspectives. Nonetheless, there are important historical issues that cannot be addressed by the terms available in current contextual historiography. Complexity is one such issue, as is a growing demand that explanations be in terms that relate to the actors' understanding of the meaning of their lives. The recurrent problem of developing explanatory frameworks that encompass the evidence rather than deny its relevance has resurfaced. However, there is the certainty that narrative is not enough.[8]

Until recently many of the more traditional approaches were pursued apologetically because of serious questions about their validity as history in light of historiographies derived from sociology and anthropology. No genre of history fell under more odium than that of biography. Belief in the priority of socially defined criteria necessarily destroys the importance of individual lives, especially in their peculiarities and eccentricities. Seeing the cognitive realm as derivative of social place, cultural values or political ideology undermines any concern with what makes an individual's contribution to the sciences or to technology critical or crucial. Character and originality were lost to the crowd, suppressed as subjects for legitimate historical investigation.[9] However, studies of individuals are proving invaluable in probing the values, behavior, and social life in complex societies. The idiosyncracies of the subject even help to shed light on the characteristics of the collective.[10]

Biographers are, in fact, faced with two, seemingly incompatible tasks, to examine the individual and his life in all its particularity while connecting that life to the state and conditions in society.[11] That is, to examine how an individual coped with and in some areas triumphed in a particular society. In this the biographer has to treat the details of the life while simultaneously dealing with ways in which the individual functioned in society. And this can be done, even for as enigmatic a personality as Henry Cavendish. Russell McCormmach concentrates on showing how Cavendish's behavior reflects that of his family, and how his role in the community of science reflects that of his social group in the political life of late eighteenth-century Britain. He also ties the workings of scientific societies to the operations of other social institutions of the same era.

Although, the operations of scientific societies may be local, even provincial, ideas and theories transcend political or geographic boundaries. As Martin J. Klein demonstrates, both intellectual and personal relationships can be forged through ideas and ways of doing science. In this case the values embedded in scientific ideas are not tied to a particular cultural form. Josiah

Gibbs's ideas and style spoke directly to Pierre Duhem as an exemplar for physics and as a key to Gibbs's character.

Perhaps one of the more difficult individuals to deal with biographically is the person who on the surface seems successful yet upon closer examination reveals failures that are traceable to personality. Here the judicious use of psychology can be helpful as Michael M. Sokal explains in his examination of James McKeen Cattell's difficult character. Cattell was successful. During the last decade of the nineteenth century, his work was crucial in defining the discipline of psychology, and he was a pivotal figure in organizing it professionally. Life span development psychology helps the biographer to understand some of the changes that Cattell brought to his profession at particular times in his life, and his inability to forge certain kinds of relationships at critical periods that wounded him professionally.

Last, but not least, to write about these individuals as such, historians need to know who they are and the size and social shape of the community of research within which they worked. Clark A. Eliott reminds us how dependent historians are on fragmented, scattered bits of information and people who bring those resources to our attention in a form we can use immediately.

The above studies have demanded consideration of some of the details of what scientists and engineers do on a day-to-day basis. Such concern with the "internal" details of science or technology is still less than fashionable. The assumption is that understanding the cultural and social institutions scientists and engineers have fashioned for themselves does not require an understanding of what they do.[12] What the actors actually do or think they are doing is superfluous. Actual behavior is merely symbolic. However, as yet we have no process that links the evidence that relates to individuals, to its explanation, that resides in a socially based theoretical structure. What then makes the explanation valid?

Sociologists focus on the process of production that for them defines the product. Except, they presuppose a priori the nature of that product. Sociologists have always assumed science to be knowledge of a certain type, usually specified by philosophers. The process of production actually mirrors philosophically prescribed ways of executing this type of knowledge. Their accounts follow this pattern whether the structure of science as knowledge is logical and rational, or arational and socially determined. In the former case scientists act as logical positivists. In the latter they are limited by the social behavior and norms of their society. This regretably, is not a procedure that joins together their productions—be they theories, ideas, buildings, machinery, or processes—and the values and social behaviors they represent. The relations are simply assumed but never examined. Historians cannot even link together, through any general process, the language available in the general culture to their usage in solutions to particular problems. By rooting the analysis outside the work of scientists and engineers, all theorists only establish a series of analogies in the separate domains using categories that come from the social model. However, this remains an analogy because there is no explanation of how the chosen set of social values are transformed into solutions to particular, technical problems. Nevertheless, these connections have to be made, and to do that historians must take into account the activities of scientists and engineers and how they accomplish their ends.

Virginia P. Dawson does just this; using the evidence in the work of scientists themselves, she demonstrates the values they share with other members of a broader culture.[13] In this case the values were taken from the prevailing culture and brought to bear on the solution of the problem of generation. Here is a description of the ways in which religious and metaphysical values shaped theory without implying the existence of meanings hidden to the actors or cultural categories that do not appear from the evidence.

In her argument, Dawson establishes a network of correspondents that made up two of the opposing groups interested in the problem of generation. The theories and ideas of the protagonists reflect both concern with the data and their sense of what constituted a proper scientific argument. Alan J. Rocke focuses on the latter issue and how this changed over the course of the nineteenth century within the discipline of chemistry. The documentary evidence shows no appreciable change in actual practice, but a change in values affected the descriptions of what chemists thought they were doing. As the use of hypotheses became more acceptable, chemists read their own behavior differently. This is another example of how much values guide how scientists describe what they do, but not necessarily how they actually do use empirical evidence and theory. Perhaps this is an object lesson in differentiating rhetoric and practice in historical investigations and the necessity for a sensitivity to the kinds of historical questions that can be asked using these two types of evidence.[14]

At the same time, Rocke considers the values of an international discipline and the way in which members of that discipline redefined it. Studies of disciplines usually trace the establishment and development of institutions, and this is assumed to map the development of a field in science or engineering. Nowhere is this more true than in studies of American science. In this lies the assumption that the key to the uniqueness of American science is in its institutional differences with the sciences of other Western nations. Institutions are not, however, a foolproof guide to the actual existence of any intellectual activity at all and need not indicate the development of new activity either.[15] One of the problems with this institutional approach is of knowing the context of activities and values from which the institution was formed. The need for formal social institutions seems to mark a stage of growth rather than to indicate beginnings. Patsy Gerstner examines the difficult terrain of the development of geology in the United States before the development of any institutional structure. She deals with the ways geologists, or those later called geologists, communicated problems encountered in field work and solutions to those problems to one another before the existence of specialist journals or societies. In this informal network, Samuel G. Morton played a key role, particularly in resolving issues that emerged in the detailed field work of others. Morton seems to have acted as a cross between president of a scientific society, mentor, and consultant to the younger members of a growing community. Such work leaves few obvious traces. The investigation of formal communications and published papers would not uncover this kind of network or community.

And yet detailed analysis of published papers clearly has a function in the history of science. The published paper is the public face of science and the culmination of a facet of research, and scientists spend a great deal of time preparing their work for public scrutiny. In the course of putting findings into a form that is publicly acceptable, the scientist often must confront his own misgivings, failures of understanding, and vaguencess in interpretation. Indeed, the actual composition of the formal paper can be an important act of clarification and interpretation of evidence and ideas; it is part of the creative process.[16] Published papers are more than ritual objects, more than a life-long series of rites of passage in which scientists have lived for three hundred years. Even such a long-lasting ceremony that has spread to all scholarly activities that claim to be "scientific" deserves close scrutiny. In addition, if historians are to understand what a piece of research accomplished, they must carefully consider this ritualized form. Only by examining the actual content of the papers can historians see the difference between what the writer actually accomplished versus what he claims in the rhetoric and the format of the paper.

While trapped within a formal rhetoric, published papers give the historian an immediate insight into what the scientist has done. Separation of rhetoric from content can also reveal a quite different set of goals and problems than has been accepted in the historiography of the discipline.[17] What writers have claimed to be doing has misled historians who have accepted these claims at face value and compounded their problems by translating the phrases wielded by the scientist into contemporary usage. This leads to certain expectations of what the historian ought to find in the paper, most of which are usually fulfilled. In the case of Siméon-Denis Poisson, simply reading his papers for their content reveals problems for the usual interpretation of his work. Elizabeth Garber's essay on Poisson examines these papers and the customary interpretation of them as physical. Reading them without this expectation shows that they are mathematical. But the tradition of mathematics within which they sit is not that of the twentieth but of the eighteenth century and was already under siege during Poisson's lifetime. His papers were an attempt to reestablish a basis for the calculus in physical reality, not simply to explore the physics of heat, static electricity, or fluids. His work also demonstrates that the expansion of the use of mathematics into all domains of physics was neither smooth nor easy, nor was this the obvious direction for the development of theoretical physics.[17] Lastly, this case demonstrates that the mathematization of a problem does not lead automatically to its clarification, to a discussion of the important physical issues, or to a deeper understanding of the relevant physical processes.

That science leads to improved methods in the "arts and manufactures" has been assumed as true since the Renaissance. In the twentieth century philosophers and sociologists of science have elaborated this idea until technology became simply "applied science."[18] Engineers and more recently historians of technology have resented the assumption and tried to assert the historical importance and independence of technology and engineering from science.[19] Yet historians cannot continue to think in terms of the complete independence of either from the other. Both science and technology are inextricably intertwined in ways that are both complex and historically intriguing.[20]

Edwin Layton brings to light the very real difficulties of transforming scientific principles into engineering practice outside the support structure of formal educational institutions for engineers. Engineers and craftsmen had to connect the abstract ideas of Newton's physics to the design criteria and materials limitations. To do this they had first to understand the physical concepts and then transform them into the terms that brought out their engineering usefulness in the specific problem context of their immediate interest. Layton demonstrates that understanding physical concepts is not an easy first step. Understanding entails interpretations that are not obvious, and in the case of Newton's laws were only useful when developed in a sophisticated mathematical language—language only available to the engineer through a formalized and institutionalized education. The unorganized and informally educated were excluded from access to such highly developed interpretations and the mathematical language in which to develop those interpretations. In the United States, for a short time, there existed two engineering traditions—that of the less educated and that of the mathematically sophisticated. By the end of the nineteenth century, both were replaced by the academically educated engineer.

Education is not the only barrier to engineers' use of physical ideas. The concepts must be put into a context and terminology where the engineer can appreciate their utility in solving particular problems of immediate interest in engineering. As Andrew J. Butrica shows, only the appearance of problems that were no longer solvable with normal engineering methods forced engineers to use Kirchhoff's laws, although they had appeared in the engineering literature. These physical laws laid fallow until an engineer demonstrated their use in what was becoming an important problem area, undersea cables. Put into this new problem context in a form that demonstrated their engineering usefulness, they quickly became part of engineering practice.

In both cases the physics required translators who were themselves engineers, who understood the physics, could visualize its use in an engineering context, and could write of it in engineer's terminology. By the time Kirchhoff's laws came to the attention of engineers, engineering itself had become "scientific," and the education of an engineer had become academic. Scientific engineering was the means of educating the new engineer and the reason why his education had become so formal. David F. Channell investigates how the influential teacher and engineer W.J.M. Rankine created a new academic discipline from the disparate traditions of theoretical physics and engineering practice. Engineering science was a deliberate creation. Rankine drew on Common Sense philosophy to distinguish physics and scientific engineering and yet show their connectedness. For Rankine, the difference between engineering science and physics lay in how engineers and physicists constructed and understood the laws of nature. While the laws of physics are necessary to engineers, physicists have license in the ways in which they can construct those laws that were not open to engineers. Rankine drew a strict demarcation between his own research in thermodynamics and his engineering work using the laws of thermodynamics. Yet, both disciplines were partially his own creations.[21] By distinguishing the two, Rankine opened up an intellectual space in which engineering could flourish both as a science and an academic discipline.

Channell's essay illuminates the connections between science and technology just at the time when engineers accepted in practice and in their training the mantle of science. However, in that same era one can find examples where the only visible connection between science and technology lay in their context, a common social matrix. Edward Jay Pershey draws this conclusion from his study of the Cleveland engineering firm of Warner and Swasey. Its work on precision scientific instrument systems was merely another contract, not even used to promote its lines of precision machine tools and other engineering products. Not even a rhetorical gloss exists connecting the two.

Nevertheless, even if the connections between science and technology are more rhetorical than real, Darwin H. Stapleton discloses the depth of the faith and commitment to those connections in the early industrial life of Cleveland. Whether or not there is any intellectual affinity between science and technology, scientists were a resource and research regarded as necessary. The industrial world of Cleveland developed a commitment to research that was quickly encapsulated in institutional form within particular industries. The process of the economic exploitation of research followed a characteristic path of development. This path led to increasing dependence on such research along with the financial commitment to support it. By 1900 the research laboratory, academic consulting, and industrial funding of research were all well entrenched in certain industries.

This range of subject matter demonstrates that the ties scientists have to industry and technology are complex and on several levels. And in such a context it may not be easy to isolate science or technology one from the other. Scientists and engineers retain the same sense of themselves no matter on what kind of problem they are working. In some cases social labels and institutional settings are no good as clues to the kind of problem the individual is actually attacking.[22] At times scientists have to solve formidable engineering problems simply to get on with their research, and engineers in exploring the ramifications of an engineering problem are lead into basic research.

Currently historians of science and technology are exploring these relationships from various sides of the issues. They also share a commitment to exploring the central problem that puts science and technology into their cultural context while joining their disciplines to those of mainstream history. Putting any facet of a society or culture into its broader context brings up the same difficulties no matter what the historian's specialty.

Orville Butler offers an interesting approach to putting science and technology in context by simultaneously exploring a specific change and examining the social matrix within which scientists formed only a small group. Their needs and demands were one small factor in this change. The concerns of the scientific community are literally enmeshed in a cultural change of much broader social significance. Butler's is an elusive subject, the sense of time in everyday life. However, such a sense may be elusive, but its usage, particularly within a society and culture becoming more economically dependent on consumer and industrial demands, leaves definite evidence. But that evidence shows that no one group within society drove Americans into changing their sense of time from local to regional and then national. Scientists' needs as scientists for a rational system of time was only one among many, none of which overwhelmed local custom. Neither the needs of the railroads nor those of business forced the pace of change. Scientists and other groups thus represent the very heterogeneity that is often lacking in many accounts and at the same time becomes integrated into the narrative itself as do other factors in people's normal existence. David Landes's recent work has elevated time measurement into a symbol of the emergence of the modern world. Implicit in Butler's essay is a challenge to that thesis. The acceptance of standard time involved not the mere existence of the technology but a series of complex cultural choices that illustrate changes in a people's sense of the society within which they live.[23]

While Butler's essay integrates science into a description of cultural change by concentrating on the cultural change, Alan I Marcus also shows the richness of cultural resources available to scientists and engineers without resorting to reductionism. In this, Marcus touches upon some of the issues involved in the transformations necessary for cultural values to become meaningful in scientific terms. This is done by rooting his account in the problems addressed in a particular research area and the particularities of the results of those researches. By doing so Marcus also reinterprets a significant era in the early history of biochemistry. The intellectual continuity of Ehrlich and Waksman is shattered through an analysis of the starting point of their respective research and the goals and results of that research. Simultaneously, to emphasize those differences, he explores the relationship of their ideas to their very different cultural contexts.

Marcus leaves moot the criteria used when choosing those aspects of the cultural context reflected in scientific results. On the other hand, Harry J. Eisenman delineates some of the aspects of both science and technology that tie them most directly with other facets of their contemporary culture, in this case, art. Scientists and engineers have not yet faced a duality in their descriptions of what they do. Research papers are laid out as if the research process and product were wholly empirical and rational. And yet the language of aesthetics and emotions predominate when they describe their reactions to theories or experiments, buildings, processes, or technical solutions to engineering problems. This language becomes heightened when they describe their reactions to their own personal creative activity. Two seemingly opposed kinds of discourse coexist—one to describe formally in rational terms what they do and why they accept ideas and theories, another that justifies them on a different cognitive level—and both to the same audience. Artists do not separate these levels of discourse. However casually they have used the language, scientists and engineers describe what they do in exactly the same terms in precisely those areas in which the same terms are used in the arts. Experiences with and in science and technology appeal to areas of conciousness that need to be described in the same language as experiences with and in art or poetry. In particular, describing the act of creativity in any domain of human endeavor evokes the same language; it seems to allow for no other. Analogy gives way to metaphor. Precisely where the language of logical discourse is no longer useful, scientists and engineers invoke beauty, simplicity, symmetry, and similar phrases of aesthetic judgment. And, in describing the act of creativity, those introspective enough to try describe the process in precisely the same terms as creative individuals in other fields.

These aspects of technical work have been taken more seriously by historians of technology than those of science.[24] Using the evidence of the structures built by engineers and the art works of sculptors, Eisenman concentrates on just those characteristics shared by the arts and in this case technology: creativity, aesthetic judgment, craftsmanship, and individual style. Works of the engineer can be works of art; conversely works of art are necessarily works of artifice.

The connections between science and art are equally direct. Both are replete with the marks of craftsmanship as well as creativity and individuality. In studying the parallels between art and science, David Topper demonstrates that in replacing a view of culture that segregates the two, new categories emerge with which historians can categorize both science and art. They must rethink both the connections between them and the ways in which both are described either as science or art. Some works of art are based on and may accurately depict aspects of nature as closely as any scientist could wish. Others are schematic and abstract, their originals in nature barely recognizable in their artistic form.

The essays in this collection have expanded in wider and wider circles from the personal to the discipline and then to the social and cultural. In each of these spheres they have addressed many of the issues that arise from controversies in both the history of science and technology, some of recent origin, others of longer duration. Most of them illustrate the misgivings that have arisen recently within the context of current historiography.

While this volume offers no definite alternative to the confusion of approaches available today, the authors at least share the belief that they know where the solutions do not lie and that the search in more promising directions is well worth the effort. Many of the problems we have tried to address are longstanding, and we have discussed all of them at some time or another with Bob Schofield. He was urging us to investigate the institutional history of science, as he himself did, and pointing to the opportunities it offered as a window to the past before it became current. He was also arguing with colleagues over the relationships between science and technology. For him there was always more to this problem than implied in an easy assumption of science and "applied science." In addition, he assumed, here as elsewhere, that any history involving a technical subject should be grounded in an understanding of the technical details. And he has always insisted on a mastery of the technical details of being a historian and is passionate in his insistence that the resources in the field be researched and published to as broad a public as possible.

Above all else Bob has nurtured a long-term interest in biography and its problems in general and that of Joseph Priestley in particular. In pursuing scholarly interests based on Priestley, Bob has confronted the social and cultural context of science and explored its connections with metaphysics, theology, and early industrialization.

While it is unusual in a volume such as this to treat its contents in detail, it seemed worthwhile in light of the breadth of Schofield's interests and their interconnections with current issues within both the history of science and the history of technology. All the essays in this volume were written especially for it and the editor wishes to thank all the contributors for their patience in dealing with delays and editorial suggestions and to Lehigh University for the opportunity to demonstrate publicly the dimensions of career and colleague so influential to many of us.

Initially and finally we must thank Bob Schofield himself for the encouragement he gave all of us and the guidance he so generously gave some of us to transcend the confines of our own conceptions of research and to venture into more challenging historical problems.

Notes

1. This interest is manifest in his articles on the connections of the Lunar Society and its particular members to industrialization that appeared in the 1950s, articles on the history of scientific societies and in particular the Lunar Society published in 1963, and his papers on the relationship of science to both technology and art. Details of these papers are in the "Bibliography of the Writings of Robert E. Schofield" p. 311.

2. Whether all mainstream historians—political, social, or economic—have come to terms with the implications of this fundamental assumption is yet to be seen. For example see William J. McNeil, "Myth-history, or Truth, Myth, History, and Historians," *American Historical Review* 91 (February 1986): 1–15. See also recent work on postwar foreign policy where technical considerations are crucial but where for historians and political scientists technical issues and personnel remain in the background, i.e., in the footnotes at best.

3. For example, psychoanalytical biography presents the historian with the same problems of linking, say, creativity to psychoanalysis, whether the subject is a scientist such as Newton, or the artist Leonardo da Vinci. Marxist perspectives bring with them the same problems of the use of evidence whether "false consciousness" refers to natural philosophers of the seventeenth century or of industrial workers of nineteenth century France. Many problems in the history of institutions are the same whatever the purpose served by the institution. One historian of science recently used the techniques of family reconstruction to detail the social status, economic situation, and success of a family of eighteenth-century French scientists. See David Sturdy, "Pierre-Jean-Baptiste Chomel (1671–1740)," *British Journal for the History of Science* 11 (1986): 301–22.

4. A symptom of this has been called the "revival of narrative." Historians are casting aside theoretical commitments that do not explain the evidence and can only present the evidence as narrative. See Lawrence Stone, "The Revival of Narrative: Reflections on a New Old History," *Past and Present* 85 (November 1979): 3–25.

5. Fred Weinstein, *The Challenge to History* (Chicago: University of Chicago Press, forthcoming). For the history of science, see Elizabeth Garber and Fred Weinstein, "History of Science as Social History," in *Advances in Psychoanalytic Sociology*, ed. Jerome Rabow, Gerald Platt, and Marion Goldman (Malabar, Fla.: Krieger, 1987), 279–98.

6. These critiques do not arise from a conservative dismissal of such contextual histories as subversive or simply intellectually distasteful. We are not here dealing with Allen Bloom or Gertrude Himmelfarb, but with scholars who have examined the methods and assumptions and found them wanting. See David Knight, "Background and Foreground: Getting Things in Context," *British Journal for the History of Science* 20 (1987): 3–12: Charles Rosenberg, "Science in American Society," *Isis* 74 (1983): 356–67, Richard Westfall, "Marxism and the History of Science," *Isis* 72 (1981): 402–6.

7. For an example of the subtlety of argument that excursions into anthropology—sensitivities to perspectives offered by gender studies and history from the bottom up—can introduce into historical argument while embracing the particular problems of the complexity and the uniqueness of the data, see Natalie Ziman Davis, "On the Lame," *American Historical Review* 93 (1988): 572–603. This article was in reply to a critique of Davis, *The Return of Martin Guerre* (Cambridge: Harvard University Press, 1983). See Robert Finlay, "The Refashioning of Martin Guerre," *American Historical Review* 93 (Winter 1988): 553–71.

8. Carl Degler, "In Pursuit of an American History," *American Historical Review* 92 (1987): 1–12, is one of the latest discussions on the problems of constructing viable theories of history and their relationship to the evidence in the context of American history.

9. Not quite all historians accepted this assessment. See Thomas Hankins, "In Defense of Biography: The Uses of Biography in the History of Science," *History of Science* 17 (1979): 1–16.

10. See David Cassidy, *Werner Heisenberg* (forthcoming). This study is one in understanding how an extremely intelligent and socially privileged young man could have become enmeshed in the moral degradation of Nazi politics. This question is also the focus of Thomas Sheehan, "Heidegger and the Nazis," *New York Review of Books*, 16 June 1988, 38–47.

11. Besides the issue of the connection of evidence to theoretical structure, any historian has to face the fact that ultimately the writing of history is an act of creativity. This is the focus of a continued attack on scholarly history by such fiction writers as Gore Vidal. They do not see the difference between fiction and history. This opinion is even shared by some, particularly cultural, historians. That historians can write credible fiction, based in the discipline of their craft is

demonstrated by Russell McCormmach, *Night Thoughts of a Classical Physicist* (Cambridge: Harvard University Press, 1982).

12. For an example of the creation of a language system and a set of social values through which scientists wished to see themselves viewed, i.e., the creation of a particular meaning for the new term scientist that lead to both the terminology and the moral and social definition of the scientist, see Jack Morrell and Arnold Thackray, *Gentlemen of Science* (Oxford: Clarendon Press, 1982). This study, however, says nothing about the content of the science discussed at the orchestrated meetings of the British Association. For another study of how scientists conduct their business according to the practices and expectations of their society, see Roger Hahn, *The Anatomy of a Scientific Institution: The Paris Academy of Sciences, 1660–1803* (Berkeley: University of California Press, 1971). See also Charles C. Gillispie, *Science and Polity at the End of the Old Regime* (Princeton: Princeton University Press, 1980).

13. Another example of uncovering cultural values through painstaking reanalysis of data is Stephen Jay Gould's study of Samuel G. Morton on cranial capacity. Morton's commitment to an hierarchical view of race was such that he could not see the discrepancies between his data and his theory. See Gould, *The Mismeasure of Man* (New York: W. W. Norton, 1981).

14. The relation between rhetoric and method has been examined also by Bruce S. Eastwood, "Descartes on Refraction: Scientific versus Rhetorical Method," *Isis* 75 (September 1984): 481–502.

15. For a study of the limitations of using the formal establishment of institutions as a measure of intellectual development, see Rachel Laudan, "Ideas and organizations in British geology: A Case Study in Institutional History," *Isis* 68 (1977): 527–38.

16. See Frederic L. Holmes," Scientific Writing and Scientific Discovery," *Isis* 78 (June 1987): 220–35, which relates the process of writing a paper to the creation of a new theory. See also Peter Dear, " 'Totius in verba': Rhetoric and Authority in the early Royal Society," *Isis* 76 (June 1985): 145–61.

17. Histories of the development of theories and disciplines are necessarily convoluted. The participants are often not interested in the technical issues around which the theory or disciplines might eventually organize themselves. Such histories cannot be written from the perspective of the late twentieth century or from within. Yet such histories do have their uses; see Loren Graham et. al., eds., *Functions and Uses of Disciplinary Histories,* (Dordrecht: Reidel, 1983).

18. Sociologists seem particularly fond of picturing technology as "applied science," especially in the 1950s as they struggled to establish the sociology of science as an academic discipline against the opposition of mathematicians and physicists. As a model for themselves, sociologists chose that of the German academic with all the symbolism of "pure" versus "applied."

19. For the inverse, that science is dependent on technology, see Derek Price, "Notes towards a Philosophy of the Science/Technology Interaction," in *The Nature of Technological Knowledge,* ed. Rachel Laudan (Dordrecht: Reidel, 1984), and "Of Sealing Wax and String," *Natural History* 93 (1984): 49–56.

20. The seminal paper on this topic is Edwin T. Layton Jr., "Mirror Image Twins: The Communities of Science and Technology," *Technology and Culture 12* (October 1971): 562–80. Recent discussions of the relationship between science and technology include George Wise, "Science and Technology," *Osiris* 1 (1985): 229–46, in which there is a survey of available historiography. Yakov M. Rabkin, "Technological Innovation in Science The Adoption of Infrared Spectroscopy by Chemists," *Isis* 78 (March 1987): 31–54, concludes that technology drove chemical research after World War II and that the intermediary for that impetus was industry. For scientific research dependent on technology, see also David O. Edge and Michael Mulkay, *Astronomy Transformed: The Emergence of Radio Astronomy in Britain* (New York: Wiley, 1976).

21. Rankine believed that physicists could indulge in making models of how nature operated an indulgence not open to engineers. In his own work in thermodynamics, Rankine moved from a very particular molecular model to the development of energetics that was based on the concept of energy and the measurables of thermodynamics. This was the form of thermodynamics on which he based his engineering discussions of the subject. Common Sense philosophy is also linked to the development of James Clerk Maxwell's ideas in gas theory and electro-magnetism as well. In Rankine's case philosophical principles allowed him to develop a space for a new concept of engineering. On Maxwell, see E. Garber, S. G. Brush, and C. E. F. Everitt, *Maxwell on Gases and Molecules* (Cambridge: MIT Press, 1986), xxii–xxiv.

22. As an example, see Karle Packard's work on the invention and explanation of wave guides,

A History of Guided Electro-Magnetic Waves (M.S. thesis, Polytechnic Institute, New York 1978).

23. David Landes, *Revolution in Time: Clocks and the Making of the Modern World* (Cambridge: Harvard University Press, 1983).

24. See Cyril Stanley Smith, especially in the history of technology. He has argued that aesthetics is an important part of the history of technology because many of the products of engineers lie between science and art in the social landscape, shaping and changing that landscape. More interest has been shown recently in the process of creativity in the history of science, although few historians have consistently examined it either as a general problem or through case studies. Exception's in the history of science include Gerald Holton, *The Scientific Imagination* (Cambridge: Cambridge University Press, 1978); "Finding Favor with the Angel of the Lord: Notes towards the Psychobiographical Study of Scientific Genius," in *Interactions between Science and Philosophy,* ed. Yehuda Elkana (Atlantic Highlands: Humanities Press, 1974), "Mainsprings of scientific discovery," in Owen Gingerich, ed., *The Nature of Scientific Discovery* (Washington D.C.: Smithsonian, 1975), and Howard Gruber, *Darwin on Man,* 2nd ed. (Chicago: University of Chicago Press, 1981), chap. 12. See also, Raymond Aris et. al., eds. *The Springs of Scientific Creativity* (Minneapolis: University of Minnesota Press, 1984).

To Robert E. Schofield: A Personal Tribute

MELVIN KRANZBERG

The second annual meeting (1959) of the fledgling Society for the History of Technology (SHOT) was a milestone in the development of that organization. For one thing, the first issue of *Technology and Culture* had just arrived from the printer and was available for distribution at the meeting. This meant that the new organization had acquired that basic element of a learned society: a scholarly journal. A second favorable sign was the large number (one hundred to two hundred people) who attended each of SHOT's program sessions. Because the meeting was held in conjunction with two great and large organizations—the Association for the Advancement of Science (AAAS) and the American Historical Association (AHA)—notable scholars from many different fields participated in the SHOT program, and that too was an augury of future growth.

Another—less public—event at that Chicago meeting was also to have great impact upon future scholarship in the history of technology—and the history of science. This was a personal encounter at the close of the SHOT-AHA program session dealing with industrial revolutions. When the unusually large and receptive audience had left the auditorium, one person remained behind and spoke to me just as I was leaving.

"I understand you're looking for a historian of science at Case," he said.

"Yes, we are," I responded.

"Have you thought about me?" he asked.

"No, I didn't think you'd be interested. Are you?"

"I certainly am," he answered.

He was Robert E. Schofield. That brief conversation marked the beginning of a fruitful collaboration and a lifetime friendship. More than that, it was to provide a major impetus to a new kind of scholarship in the history of science and technology that has gathered momentum in the quarter century since then.

*   *   *

Although today's historians of science regard the history of technology as a sister discipline possessing full scholarly credentials, that was not the case some three decades ago. There was no solid body of critical scholarship in the history of technology; most of the writings in the field were by popularizers utilizing a "Gee Whiz!" approach, indicating how technological developments had created wonderful things for mankind. Few historical scholars took cognizance of technological developments, and those who did concentrated on the technical devices themselves, with little reference to their sociocultural context.

On the other hand, the history of science had achieved the status of a full-fledged academic discipline: it was taught at major universities, had produced many excellent scholarly publications, and possessed an active learned organization, the History of Science Society. Yet it too was "internalistic" in approach; historians of science studied "pure" science, unsullied by the practical and sociocultural problems besetting its technological "applications" (technology was regarded as "applied science"). In part this was because the founder of the history of science as an academic discipline, George Sarton, had indicated his interest in "thinkers, not tinkerers." Then, under the influence of Alexander Koyré and his followers, the history of science had become devoted almost entirely to the development of scientific ideas themselves, without much reference to the social, economic, political, and cultural context in which they occurred.

The faculty had already interviewed several historians of science for the opening at Case, but those interviewed so far were not particularly interested in developing a program that called for the history of science *and* technology. Also they did not like the idea of teaching in an institution where there were neither majors nor graduate students in history, or even any of the liberal arts. Schofield had no such hangups. Even though he had been recently promoted to an associate professorship at the University of Kansas, he had no compunctions about moving—at the same academic rank—to an engineering institution or about helping develop a program that regarded the history of technology as seriously as the history of science.

Perhaps Schofield's interest was due to his own education and experience before entering into doctoral work in the history of science. Completing a bachelor's program in physics at Princeton in 1944, Schofield was immediately caught up in the war effort, doing research at laboratories in Oak Ridge. After this taste of "applied physics," Schofield returned to the academic study of physics, being awarded his master's degree in that subject from the University of Minnesota in 1948. This too received immediate application when he became a research associate at the Knolls Atomic Power Laboratory of General Electric.

Schofield's background in both the theory and practice of physics was evident when in 1951 he left GE to pursue a doctorate in the history of science at Harvard, a degree he received in 1955. His dissertation topic, "Founding of the Lunar Society of Birmingham (1760–1780): Organization of Industrial Research in Eighteenth Century England," demonstrated that Schofield was fully cognizant of the technological dimensions of science and the scientific parameters of technology.

Schofield's early publications, particularly his article, "The Industrial

Orientation of Science in the Lunar Society of Birmingham," (*Isis* 48 (1957): 408–15, reprinted in A. E. Musson, *Science, Technology and Economic Growth in the Eighteenth Century* (London: Metheun, 1972) 136–97), attracted my attention to him as a possible participant in the Society for the History of Technology, which was still in the process of being born. Here was a genuine historian of science who also seemed interested in the history of technology! Not surprisingly, I asked him to serve on the Advisory Council of the Society for the History of Technology, and he agreed to do so.

Is it any wonder that I was enthusiastic when Schofield approached me at that SHOT meeting and expressed an interest in the opening at Case? The case for Schofield was made even stronger when it was learned that he had been a Fulbright Fellow at University College, London in 1953–54, and that in 1959–60—as we were conducting our search—he had received a Guggenheim Fellowship. In the fall of 1960, Bob Schofield took up his post as Associate Professor of the History of Science at Case Institute of Technology, in Cleveland, Ohio. This marked the beginning of what was to become a major center for the training of historians of science and of a new, allied field, the history of technology.

*   *   *

For the next dozen years Schofield and I worked closely together in developing a graduate program in the history of science and technology and then extending it to the undergraduate level. Development of that program stimulated our colleagues in related areas of the humanities and social sciences to develop congruent courses and graduate programs in their fields, and soon doctoral degrees were offered in the emerging area of science, technology, and public policy and in anthropology (stressing the sociocultural elements of tool-making and tool-using). The science-technology thrust in Schofield's historical studies had a synergistic effect, and an outstanding program in American studies took root, fueled in part by the recognition of the role that technology had played in the making of American society.

Merely to list some of the faculty involved in the history of science and technology indicates the quality that was achieved in the Case program: Martin Klein, Carroll Pursell, Edwin T. Layton, and Reese Jenkins. Some outstanding foreign scholars also came as visiting faculty. These included David G. C. Allen, librarian of the Royal Society of Arts, with whom Schofield collaborated on *Stephen Hales: Scientist-Philanthropist* (London: Scolar Press, 1980), and Alex Keller of the University of Leicester, who filled in when Schofield was awarded his second Guggenheim Fellowship in 1967–68 and spent the year at the Institute for Advanced Study in Princeton.

Concomitant with the establishment of the graduate program, Schofield pressed toward establishment of the Case Institute Archive of Contemporary Science and Technology to provide research materials for graduate students and faculty. As a "reward" for his efforts, Schofield was made curator of the archive, which was housed in the Case Library. To this was added the Williams Memorial Collection, and Schofield encouraged certain Cleveland industries to deposit their records in the archive, thereby providing exellent documentation for theses. Neighboring manuscript and archival collections at Western Reserve Historical Society, the Cleveland Natural History Museum, the Allen Memorial Medical Library, and the Cleveland Art Museum also provided a richness of research material for both students and faculty.

*   *   *

While busy developing the program at Case, Schofield did not neglect his own research. His book *The Lunar Society of Birmingham: A Social History of Provincial Science and Technology in Eighteenth-Century England* (Oxford: Clarendon Press, 1963) was awarded the Pfizer Prize at the 1964 meeting of the History of Science Society. A few years later *A Scientific Autobiography of Joseph Priestley (1733–1804)* (Cambridge: MIT Press, 1966) appeared, consisting of selected scientific correspondence with Schofield's commentary. *Mechanism and Materialism: British Natural Philosophy in an Age of Reason* (Princeton: Princeton University Press, 1970) was a third major publication, which was greeted like his earlier work with high praise for its combination of scholarly erudition and meaningful interpretation of a host of data.

Not suprisingly, in 1972 Schofield was named Lynn Thorndike Professor of the History of Science at Case Western Reserve University. From 1969 to 1979 Schofield served on the Editorial Board of Historical Studies in the Physical Sciences, and during 1978–80 he was a Sigma-Xi National Lecturer. Reports on his Sigma Xi lectures were unanimous in pointing out Schofield's ability to entertain and enthrall an audience while at the same time informing its members and making them think.

Bob Schofield was not the only member of his family to achieve recognition for scholarly work: Mary Peale "Perry" Schofield won equal acclaim for her book on the landmark architecture of Cleveland. At a time when that great city was suffering from an inferiority complex as a result of serious economic woes and outside criticism, Mrs. Schofield's book stressed those elements of cultural history in which Clevelanders could deservedly take pride. Thus, when the Schofield family moved from Cleveland to Ames, Iowa, in 1979, it was not only a loss for Case Western Reserve University but for the city of Cleveland as well.

*   *   *

At Iowa State University Bob Schofield has carried on the great work begun during his tenure at Case Institute (and Case Western Reserve University). Developing the program in the history of technology and science, he has gathered a staff of excellent faculty in those fields. To enlarge publishing opportunities for scholarly monographs in the history of technology and science, Schofield has established and taken on the editorship of the ISU Press Series in the history of technology and science. The series is designed to "increase public understanding of the nature of technological and scientific creativity and their relationship to one another and to the social, cultural, political, and economic context in which they arise."

It is clear that the same devotion to scholarship that has animated Bob Schofield's career still characterizes his work at Iowa State. His renewed enthusiasm—after serious illness and heart bypass surgery had temporarily sapped his strength—is evidence of his continuing commitment to knowledge. Bob Schofield remains the same great teacher, the same warm friend, the same great scholar, the same wonderful human being that he has shown himself to be throughout his long and distinguished career.

This volume, containing papers by former students and colleagues—for colleagues too were influenced by Schofield's example and instruction—gives full evidence of the quality and stimulation afforded by Schofield's teaching and example. The authors represented in this book have become authorities in their particular fields and occupy posts at major universities. They are testimony to the great and wonderful influence of Robert E. Schofield in teaching and training new generations of scholars. In dedicating this volume to him, they show that they are following in the footsteps of their great friend and mentor.

Contributors

ORVILLE BUTLER is currently at the University Center for the Historical Studies of Technology and Science, Iowa State University, Ames, Iowa. His interest in time standardization grew out of his research into the development of astrophysics in the United States, work he is continuing under the direction of Bob Schofield.

ANDREW J. BUTRICA, a visiting Fellow, Beckman Center for the History of Chemistry, was the first Ph.D. (1986) from Iowa State University in the Program in the History of Science and Technology. Written under the direction of Bob Schofield, his dissertation was on the genesis of electrical engineering in France in the nineteenth century. He is currently in Paris exploiting the archives of the Society for the Encouragement of National Industry for a study on the organization of industrial research between 1891 and the First World War.

DAVID F. CHANNELL is an Associate Professor in the History Department, the University of Texas at Dallas, Richardson, Texas. His Ph.D. dissertation (Case, 1975) was on William J. Macquorne Rankine around whom his research is still organized. This recently culminated in a biography in the Scottish Men of Science series, an annotated bibliography on *The History of Engineering Science,* and a study of the interactions between technology and organic life.

VIRGINIA P. DAWSON is adjunct Assistant Professor at Cleveland State University, Cleveland, Ohio. Her Ph.D (Case, 1983) dissertation was on the problem of the polyp in eighteenth century French biology. Expanded and extended, it was recently published by the American Philosophical Society as *The Problem of the Polyp in the Letters of Bonnet, Trembley and Reaumur.*

HARRY J. EISENMAN is now Professor, Department of History and Political Science, University of Missouri-Rolla. His Ph.D. (Case, 1966) dissertation was on Charles Francis Brush and the early electrical engineering industry in Cleveland. His interest quickly turned and has remained in exploring the connections between art, technology, and science.

CLARK A. ELLIOTT is an Associate Curator, Harvard University Archives, Cambridge, Massachusetts. He has always specialized in the archival problems associated with American history and the history of science. He participated in the meeting founding the Forum in American Science of the History of Science Society and established and edited the first issue of *History of Science in America: News and Views,* which later became the newsletter of the Forum. Author of *Biographical Dictionary of American Science* and coauthor and editor of *Progress as Process: Documentation of the History of Post-War Science and Technology in the U. S.,* he recently contributed to the *Osiris* volume on American science. His continuing interest in documenting the lives, origins and careers of American scientists has expanded into coediting (with Margaret Rossiter) *Harvard University as Context for Science.*

ELIZABETH GARBER is Associate Professor, History Department, State University of New York at Stony Brook, is coeditor of *Maxwell on Saturn's Rings* and *Maxwell on Gases and Molecules.* Her current research is on the relationship between physics and mathematics in eighteenth- and nineteenth-century Europe.

PATSY GERSTNER is Chief Curator, Historical Division, Cleveland Health Sciences Library, Cleveland, Ohio. Her research centers on the development of science in America, especially of geology, anthropology, and medicine.

MARTIN J. KLEIN is Eugene Higgins Professor in the History of Physics, Yale University, New Haven, Connecticut. He is author of the biography of Paul Ehrenfest and recently has worked on the early career of Albert Einstein and the physics of Josiah W. Gibbs.

MELVIN KRANZBERG is emeritus Callaway Professor of the History of Technology, Georgia Institute of Technology, Atlanta. He was a major force in establishing the Society for the History of Technology and was its past president and long-time editor of its journal, *Technology and Culture.* He was recently coeditor of *Bridge to the Future,* the centennial volume on the Brooklyn Bridge, and is currently concentrating on innovation and technology and human values.

EDWIN T. LAYTON, JR., is Professor of the History of Science and Technology, Department of Mechanical Engineering, University of Minnesota, Minneapolis. He is past president of the Society for the History of Technology, author of *The Revolt of the Engineers,* which won the Dexter Prize and *Technology and Social Change in America.*

ALAN I MARCUS is Professor of History and Director of the Center for Historical Studies of Technology and Science, Iowa State University. He recently published *Agricultural Science and the Quest for Legitimacy: Farmers, Agricultural Colleges and Experimental Stations, 1870–1890* and coauthored (with P. Segal) *Technology in America: A Brief History*. His current research explores the use of DES as a cattle growth promoter and the controversy surrounding its use.

RUSSELL MCCORMMACH of Eugene, Oregon, established and edited for the first decade *Historical Studies in the Physical Sciences* and is author of *Night Thoughts of a Classical Physicist*. He recently coauthored (with Christa Jungnickel) the 2-volume *Intellectual Mastery of Nature: Theoretical Physics from Ohm to Einstein,* which won the 1987 Pfizer Prize of the History of Science Society. He is currently working with Christa Jungnickel on a study of the practice of science in eighteenth-century Britain.

EDWARD JAY PERSHEY is Director, Tsongas Industrial History Center, Lowell, Massachusetts. The Tsongas Center was established to encourage the teaching of the history of science, technology, and industry in elementary and secondary schools. In addition he has worked on the problems of the preservation of all types of historical materials and the problems of bringing the history of industrial America to as wide an audience as possible. His own research has focussed on the engineering firm of Warner and Swasey in Cleveland.

ALAN J. ROCKE is Associate Professor, Program in History of Science and Technology, Case Western Reserve University, Cleveland, Ohio. He is author of *Chemical Atomism in the Nineteenth Century: From Dalton to Cannizzaro*. He won the university's 1988 Carl F. Wittke Award for excellence in teaching.

MICHAEL M. SOKAL is Professor, Department of Humanities, Worcester Polytechnic Institute, Worcester, Massachusetts. He is currently also Executive Secretary of the History of Science Society and won the 1988 Trustees Award for Outstanding Creative Scholarship from Worcester Polytechnic. He has edited James McKeen Cattell's *Journal and Letters from Europe 1880–1888* and *Psychological Testing and American Society, 1890–1930*. He is currently preparing a full-length biography of Cattell.

DARWIN H. STAPLETON is Director, Rockefeller Foundation Archives, Pocantico Hills, Tarrytown, New York. His research has focused on the career and interest of Benjamin Latrobe, including *The Transfer of Early Industrial Technologies to America; The Engineering Drawings of Benjamin Latrobe,* which he edited; and a bibliography on the *History of Civil Engineering since 1600*. He is currently investigating the role of the Rockefeller Foundation in the funding of research throughout the world in the decades before the Second World War.

DAVID TOPPER is Professor, Department of History, University of Winnipeg,

Winnipeg, Manitoba. He is coeditor of *Leonardo: Journal of the International Society for the Arts, Sciences and Technology.* His own research interests revolve around exploring many aspects of the relationship between science and art. He has also won awards for excellence in teaching including a 1987 3M Teaching Fellowship.

Beyond History of Science

Part I
Biography

Henry Cavendish on the Proper Method of Rectifying Abuses

RUSSELL McCORMMACH

In his massive *History of the Royal Society,* Charles Richard Weld wrote that it was "painful" for him to turn to the events of 1783 and 1784. He would rather have passed over them in "silence," but duty forbade it. He then proceeded to give what he regarded as an impartial account of the events, the so-called dissentions, which "turned the hall of science into an arena of angry debate, to the great and manifest detriment of the Society."[1]

The origin of the dissentions, Weld explained, was a widespread resentment of the conduct of Joseph Banks, since the end of 1778 the elected president of the Royal Society. Certain of the fellows especially resented Banks's conduct in the elections of new members to the society, which took the following form. Fellows wishing to elect a new member usually brought him to one of Banks's Thursday morning breakfasts. If Banks approved of him, the candidate would then be invited as a guest to a dinner of the society's club, at which Banks also presided and where the candidate would meet influential members. But if Banks disapproved of the candidate, he would urge individual members to blackball him at balloting time.[2]

For the good of the society, Banks believed, the members should bring in two kinds of persons: men of science and men of either wealth or rank or both. Rarely, as in the case of Henry Cavendish, they were one and the same. Like the membership at large, the ruling council of the society contained men of both kinds, and here again in the elections Banks made clear his likes and dislikes, exposing himself to the charge of packing the council with pliant friends. The results of Bank's forceful interference in elections revealed a pattern, or so certain members thought, which was a bias against men of science, particlarly men of the mathematical sciences, and in favor of men of rank. Their dissatisfaction with Banks came to a head in, as Weld termed it, the "violent dissentions, foreign to matters of science," of 1783 and 1784.[3]

In Weld's account and in other historical accounts of the dissentions, Henry Cavendish receives only one brief mention, if any at all. While passages from violent speeches are quoted at greater or lesser length, Cavendish is recalled only for his seconding of a motion of approval of Banks as president of the society.[4] This, to be sure, was the one time Cavendish entered the public record of the dissentions, and even this one seemingly small public act would appear exceptional coming from someone as private as Cavendish. But there was much more to Cavendish's involvement than this, as there almost had to be given the stakes and given the crucial importance of the motion and the eminence of Cavendish in the society. Charles Blagden, who at the time was both a kind of scientific assistant to Cavendish and personal assistant to Banks, wrote daily letters to Banks at the height of the dissentions. These afford us a detailed account of Cavendish's developing thoughts about the dissentions and of the actions he took on the basis of them.[5] The picture of Cavendish as an all-but-inaccessible recluse is wrong. To understand Cavendish's behavior during the dissentions, it is useful to place him within his family, that is to say, beside the other, political Cavendishes.

Of the Cavendishes—that clannish family, wealthy, proud, and powerful— Edmund Burke observed in 1771: "No wise king of Great Britain would think it for his credit to let it go abroad that he considered himself, or was considered by others, as personally at variance with ... the families of the Cavendishes."[6] The historian Richard Pares writes of the Cavendishes:

> Much was heard of the "great Revolution families"—of whom some of the proudest, as Sir Lewis Namier has pointed out, were in fact descended from Charles II's bastards. These families—above all, perhaps, the Cavendishes—could not forget that their ancestors had, as it were, conferred the crown upon the king's ancestors, and they did not mean to let him forget it either, for they alluded to it in season and out of season. They looked upon themselves as his creators rather than his creation: one would almost say they had forgotten that the dukedom of Devonshire itself had been established, less than a century earlier, by the merely human agency of a king.[7]

Into this family the natural philosopher Henry Cavendish was born in 1731. He was grandson of William Cavendish, the second duke of Devonshire; nephew of the third duke; first cousin of the fourth duke; and cousin of various other politically influential Cavendishes.[8]

The political exertions of the first duke of Devonshire in conferring the crown were extended by the second duke, who above all was concerned with consolidating the power of the Whigs. The second duke worked hard behind the scenes, offering his absolute loyalty to the shrewd Robert Walpole, who governed the kingdom in the interest of the aristocracy. It was to advance the cause of the Whig ascendancy that the sons and sons-in-law of the second duke were placed in Parliament. The second duke's younger son Lord Charles was elected to the House of Commons three times before he retired from politics into his other pursuits, especially his scientific ones, at which time his oldest son, Henry, the future natural philosopher, was eleven. Nearly all of the adult Cavendish males around young Henry were in politics. But these later Cavendish politicians were not moved by the same urgency as their predecessors had been, as there was no need for them to be. The Cavendish family stood at the head of the peerage, increasingly confident and wealthy.

The third and fourth dukes pursued their political careers out of a sense of duty, as if above the fray, and the fifth duke had a positive aversion to politics. Henry Cavendish's adult life concided with the fourth and fifth dukedoms.[9]

Devonshire House, the Picadilly mansion of the dukes of Devonshire, was the London headquarters of the Whigs.[10] The Whigs of the 1780s, the so-called New Whigs, were libertarian, passionately opposed to the king's policy on the American colonies, and admiring of Charles James Fox, the most implacable of George III's personal enemies.[11] This Whig leader and his king were in fundamental disagreement about power: Fox believed that power was properly exercised only through the king's ministers, whereas George III believed that his ministers were bound by loyalty to uphold his policy. George III found unintelligible Fox's doctrine that the king was to enjoy no personal power, that he was merely to sit on the throne, not to rule from it. In the ensuing constitutional struggle between George III and Fox and his allies, the government of the kingdom was brought to a standstill. The person of George III was *the* political issue, as John Dunning's famous resolution of 1780, which was favored by a parliamentary majority, asserted: "That the influence of the Crown has increased, is increasing, and ought to be diminished."[12] The years 1783–84, it has been argued, witnessed the greatest political crisis in Britain since the Revolution of 1688.[13]

Recall that it was these same years, 1783–84, that witnessed the dissentions of the Royal Society, which challenged the personal power of the president, Joseph Banks, and brought the regular business of that society to a standstill. While Henry Cavendish's relatives, above all his first cousin and chancellor of the Exchequer Lord John Cavendish, were actively concerned with the constitutional crisis, Henry himself was actively concerned with the crisis in the Royal Society. Henry Cavendish was, according to a relative who was in a position to know, "very proud of his family name,"[14] and the nature of his activity in the political affairs of the Royal Society was characteristic of a Cavendish. Just what this means I will take up later after first discussing the dissentions of the Royal Society and Cavendish's place in them.

The political crisis, the dissentions and debates, in the Royal Society started over a disagreement between the president and his council on the one hand and the foreign secretary, the mathematician Dr. Charles Hutton, on the other. Unlike the two regular secretaries of the society, the foreign secretary was not necessarily on the ruling council. When Hutton was elected to his office in 1779, he happened also to be an elected member of the council, but after 1780, when the dissentions occurred, he was no longer. The first indication of the disagreement was recorded at a meeting of the council on 24 January 1782, at which time Hutton's responsibility and performance were taken up. The one was judged onerous, the other inadequate: Hutton, it was decided, had not dealt punctually with the foreign correspondence, his first obligation; he was also overworked and underpaid, which seemed a likely reason for the tardiness. The council resolved that in the future, Hutton should not be expected also to translate foreign articles and extracts from foreign books, and in return he was not to fall behind in the foreign correspondence. Hutton agreed to continue on as foreign secretary with this new understanding. Nothing more was heard of the matter publicly until nearly two years later when, at a meeting of the council on 20 November 1783, it was resolved that the foreign secretary of the society had to live permanently in London. Hutton was professor of mathematics at the Royal Military Academy of Woolwich and so could not live in London. Two members of the council, the Astronomer Royal Nevil Maskelyne and one of the regular secretaries of the society Paul Maty, dissented from this move, which was obviously directed against Hutton. Hutton promptly resigned. At the ordinary meeting of the society on 11 December 1783, it was moved that Hutton be formally thanked for his services as secretary for foreign correspondence. Banks opposed the motion, which was vigorously debated. The motion passed by a narrow margin, and Banks duly thanked Hutton. At the following meeting, on 18 December, Hutton delivered, and a secretary read aloud, a written defense of his handling of the foreign correspondence. Afterwards, a motion was made and carried that Hutton had justified himself, which again was attended by a vigorous debate. The mathematician Dr. Samuel Horsley attacked Banks, accusing him of infringing upon the chartered rights of the society. Horsley said he knew of enough wrongs to keep the society "in debate the whole winter ... perhaps beyond the winter."[15]

The prospect of a winter or longer spent in acrimonious debate was abhorrent to Henry Cavendish, who regarded the serious scientific purpose of the society as inviolable. At this point he became actively—if invisibly to all but a handful of members—engaged in shaping the outcome of the dissentions. His activity is reported in letters Blagden wrote daily from London to Banks at his country house.

It quickly became apparent that the person of Joseph Banks was *the* issue. The debates, highly personal in tone, turned on a scientific judgment. The question the members had to answer was, Had the society been seriously damaged scientifically by its president, Banks? To inform Banks, Blagden delicately inquired into Cavendish's position on the question. Naturally, Banks needed to know where the society's *scientifically* most eminent member stood.

Four days after the stormy meeting of the Royal Society, after dining at their scientific club, Cavendish went with Blagden to his home, where they discussed the troubles of the society.[16] That morning Cavendish had gone to see Dr. William Heberden, a distinguished physician and highly respected older member of the society, and the two of them had arrived at a common position. Blagden reported that Cavendish and Heberden would support Banks, but "just." While Cavendish did not "absolutely refuse a vote of approbation" of Banks, he would absolutely reject any resolution that, by its wording, would seem to pass censure on Horsley and his friends for what they had done in the past. They had given no evidence of acting out of any motive other than the good of the society, Cavendish said. Furthermore, the good of the society required just such vigilant watch by its members over their president and council. But Cavendish did not mean for this watch to take the form of debates during regular meetings, which disrupted the scientific business of the society. To put a stop to the debates without denying the members their rights, Cavendish proposed a resolution, which he believed would be passed by a very large majority. From dictation Blagden wrote it down and then read it back, making sure of the wording. The resolution read:

> That the proper method of rectifying any abuses which may arise in the society is, by choosing into the council such persons as it is supposed will exert themselves in removing the abuses and not by interrupting the ordinary meetings of the society with debates.

Blagden did not think that this resolution would have the result Cavendish expected of it. Horsley would agree that it was the task of a new council to remedy the abuses, but he would argue that for the society to be made aware of the abuses, the debates must continue. Cavendish thought that such an argument from Horsley would carry weight, but there was an effective answer to it. For example, the society could inform itself of any abuses by holding special meetings for the purpose. Then if Horsley persisted with his interruptions, the society would be within its rights to censure him. Blagden gave Banks his opinion after this conference with Cavendish: the resolution Cavendish proposed was probably the best of any proposed so far, and if to it was added another resolution to the effect that any motion had to be announced at the meeting before it was to be debated, the whole affair might be brought to a speedy and favorable conclusion.[17]

But Cavendish's resolution omitted all mention of support for the incumbent president, Banks, which was something less than Blagden and Banks had hoped from him. Cavendish did not even want to talk to Banks about past councils because he would find it awkward. With the help of Blagden's prompting, however, Cavendish recalled past presidents he had served under. Banks's predecessor, the physician John Pringle, Cavendish said, had acted like Banks and had given rise to the same complaint about ineffective councils.[18] Pringle's predecessor, the antiquary James West, was "King Log," but West's predecessor, the astronomer and mathematician Lord Morton, handled the affairs of the society in an unexceptionable way. Cavendish allowed that Banks's method of choosing the candidates for council was fair; but he blamed Banks for not doing as Morton did, which was to "put in people who would have an opinion of their own, without agreeing implicitly with the President in every thing." Cavendish believed that if his resolution carried, it would mean that on election day there would be a contest. He wanted Blagden to reassure Banks that he would support the "House list" on election day unless it was "very exceptionable." He also wanted Blagden to tell Banks that he did not want to be consulted on the list beforehand, as Banks hoped he would (for it would have tended to forestall further criticism of Banks concerning the scientific respectability of the council).[19]

The day after he talked to Cavendish, Blagden went to see Heberden. Heberden had not only talked with Cavendish but also with Banks and with one of Banks's opponents, no doubt Horsley, and his opinion was settled. His opinion was the same as Cavendish's: the proper method of correcting abuses was to choose the proper council, and Banks was fit for his office. To Blagden's proposal of a vote of approbation of Banks, Heberden said he would vote for such a resolution and that if it should pass almost unanimously, the disturbances would be hushed, but he objected to it on the grounds that it would prompt a debate about Banks's conduct and inflame the passions it was intended to quiet. No "method," he believed, would prevent Horsley from bringing motions from time to time. So from Cavendish and Heberden, two highly respected, senior members of the society, Banks received the same advice: let the society affirm that power was invested in the elected council and not in the society acting as a body at any time it should choose, nor, it went without saying, in the person of the president, whoever he happened to be.[20]

Blagden wrote to Banks twice the next day, 24 December. In the morning he wrote to say that Cavendish was probably at his country house at Hampstead. He did not want to go there, since it would appear "too solicitous," and instead he intended to go to Cavendish's townhouse.[21] Later in the day Blagden wrote again, this time to say that he had left a note for Cavendish telling of his meeting with Heberden and conveying Banks's wish that Cavendish come to his house the next day. Cavendish, finding the note, had then called on Blagden to tell him that he could not go to Banks's house. To this, Blagden wrote to Banks that it was "possible" that Cavendish had set aside the following day for doing experiments, but most likely he wanted to avoid an "embarrassing conversation" with Banks. Banks was to be reassured that Cavendish was not "hostile" toward him but wanted to remain on good terms with him. It was only necessary that Banks allow Cavendish to differ with him in opinion at any time "without an open quarrel," which was to repeat what Cavendish wanted of Banks in his dealings with the council.[22]

Blagden then turned their conversation to the principal disrupter of the meetings of the society, Banks's nemesis, Horsley. Blagden put their conversation in quotation marks so that Banks would have Cavendish's exact meaning. (These quotations being the only recorded spoken words by the reserved Henry Cavendish, they hold an interest of their own.)

CAVENDISH: I did not expect any success from the Drs negotiations [Dr. Heberden and, no doubt, Dr. Horsley's]. But whatever violence *they* may express, that is no reason against proceeding with all moderation, as by such conduct the sense of the Society will be ensured against them.

BLAGDEN: I wish you would see Dr. H[orsley] & learn from himself the implacable temper expressed; as I think you would then change the opinion to which you seemed inclined when we conversed last, that those gentlemen might have nothing in view but the good of the Society.
CAVENDISH: I did not say they had nothing else in view, but only that no proof yet appeared of other motives.

At the end of their conversation, Cavendish came around to Blagden's position: he, like Heberden, would approve a vote of confidence in Banks, but only if its wording gave no offense. With this, Blagden declared himself highly satisfied with the results of his mediation.[23] What remained to be done was to bring the right members together to determine a course of action.

The next day was a Thursday, ordinarily a day on which the society met, but this Thursday was Christmas. Blagden did not take a holiday from his politicking but made plans that day to see Banks.[24] On Friday, Banks wrote to Blagden that since his meeting with Heberden, where he learned that Heberden would not support any motion that would suppress debate in the slightest fashion, he was forced to change his "plan" somewhat. Lest his supporters think him cold-blooded and abandon the cause, Banks intended to come to town on Monday with a modified plan. Blagden was to summon certain persons to meet with him. He would "strike while the iron is hot."[25]

In anticipation of a crucial vote to come, some members of the society were busy canvasing against Banks. On Saturday Blagden, who was canvasing for Banks, reported to Banks his findings to date. He named several persons who would definitely support Banks, but some of them would oppose any motion that would limit debate, which meant they would oppose Cavendish's resolution. Their compromise proposal would grant the society both its usual hour for the reading of scientific papers and conducting other normal business and also time for unlimited free debate. Every member would have the right to make a motion and the president would have to remain in his chair for as long after the hour as the debate went on. Blagden thought that the great majority of the society wanted Banks to remain president, but on the question of free debate he did not know how the society would come down.[26]

In his Saturday letter and in another letter on Sunday, Blagden alerted Banks to the serious trouble he was in. "Great opposition is making against you," Blagden said, and he named some members who were "decidedly against [Banks] even on the subject of the Presidency." So far as he could learn, Blagden said, they intended to put Lord Mahon in Banks's place. The alleged injustice done to Hutton as foreign secretary was only the occasion of the dissentions; their real cause was a "grudge of very long standing," backed by many grievances.[27] For example, Bank's opponents charged him with excluding deserving men from the society because they were not of sufficient social rank. The able mathematician Henry Clark, they said, was kept out because he was merely a schoolmaster. And the membership of the last council they held in derision. The battle line, as they drew it, was between Bank's fancy gentlemen, or "Maccaronis," and the "men of Science."[28]

When Banks came to town on Monday, he held a meeting at his house. Cavendish, who already had stayed away from one earlier meeting at Banks's, may have stayed away from this one, too. But whether or not he was there, he entered centrally into the planning done there. To a letter to Banks, Blagden attached a postscript dated Monday, 29 December, which read:

Resolved, That this Society approve of Sir Jos: Banks as their President, and mean to support him in that office.

"Such, my dear friend," Blagden wrote to Banks, "is the resolution Mr C. has just approved at my house." In Bladgen's view, the vote on this resolution would sort out Banks's friends from his foes. Cavendish, he added, still thought that the resolution he first proposed would prove necessary, since the society would not agree that under the present statutes they are forbidden to debate except at the day of elections.[29]

The next day Blagden wrote to Banks that Horsley was busy telling his friends that Banks was going to try to expel him at the next meeting, in that way insuring an ample turnout of Horsley's friends.[30] To ensure that his own friends turned out, Banks sent a card to all members of the society requesting their attendance at the next meeting. When the meeting took place, on 8 January 1784, some 170 members came, fewer than half of whom attended regularly. From the president's chair, facing the massed assembly, Banks watched as "each side took their station and looked as important as if matters of the utmost consequence to the State were the subject of their deliberation."[31] As planned, the accountant general of the society T. Anguish rose to make the motion. The previous two meetings of the society, he reminded his audience, had been disrupted by debates, and at the second of these, Horsley had threatened to keep the society debating the rest of the winter, the obvious intent of which was to unseat Banks. The motion Anguish put to the members was the resolution approving of Banks, which Cavendish had earlier approved. Cavendish now seconded the resolution before the society. Cavendish said nothing in support of it, and there is no evidence that he said anything else during this long night of angry speeches.[32]

The first speech was made by E. Poore, a barrister at law in Lincoln's Inn, who called the motion a dishonorable attempt to evade scrutiny of Banks's conduct by praising it. The attempt would not succeed, he said; it would not stop debate (and did not, as Cavendish and Heberden had predicted). Francis Maseres, cursitor baron of the Exchequer, said that for the society to exercise its power of election of president and council, the society had first to discuss the question of Banks's "abuse of power." Horsley said that the "abuses are enormous," and he went on about them at such length that Banks's supporters clamored for the question, almost drowning him out with their cries and with a clattering of sticks. As a last resort, Horsley said, the "scientific part of the Society" would secede, which would leave Banks leading his "feeble *amateurs*," his mace standing for the "ghost of that Society in which philosophy once reigned and Newton presided as her minister." Maskelyne said that if it proved necessary to secede, the "*best* Society would be the *Royal* Society in fact, though not in name." The mathematician James Glenie was interrupted before he could finish what he had to say, which was that the present council was incapable of understanding mathematics, mechanics, astronomy, optics, and chemistry, and that the society as led by Banks, a natural historian, was degenerating into a "cabinet of trifling curiosities," a "virtuoso's closet decorated with plants and shells." When late in the evening the motion was finally put to a vote, it carried 119 to 42. By a three to one margin, the society wished Banks to continue as their president.[33] This, then, was the outcome of the meetings, letters, maneuverings, and canvasing. The safest course had been taken by Banks's supporters. The resolution contained no detail; it said nothing about limiting debates, nothing about abuses, nothing about reforms, nothing, that is, that might divide the majority.

The opponents of Banks as well as his supporters claimed that they longed for a return of "tranquility, order, harmony, and accord" and the "instructive business of these weekly meetings, *the reading of the learned papers presented to the Society.*"[34] But it was a fact that for three consecutive meetings the debates had prevented the reading of all new scientific papers. Only John Michell's great paper on the distance and other measures of the fixed stars, which Cavendish had communicated to the Royal Society, continued to be read at two of these meetings, on 11 and 18 December, while at the third meeting, on 8 January, no papers at all were read.[35]

The main new paper read together with Michell's at the next, the January 15, meeting was no run-of-the-mill paper. It was a paper by Cavendish, destined to be his most famous, "Experiments on Air," containing his discovery of the production of water from the explosion of airs. Coming after three meetings in which the members had listened to speeches contrasting the present, feeble state of the Royal Society with what it had been in Newton's day, and coming one week after Cavendish had seconded the successful motion approving of Banks's presidency, the reading of Cavendish's work at the first opportunity was clearly a power move, and if by any chance it was not calculated, the effect was the same.[36]

This business-as-usual, the quiet reading of the papers, was not to last. The new statute requiring all motions to be announced in advance did not produce the desired calm. Duly announced was a motion to reinstate Hutton in his office, and it and motions to restrain Banks's interference with elections led predictably to renewed debates in late January and February.[37] At a meeting in March, Maty gave a speech and then went on to read papers, as was his duty. Horsley was at that meeting but few of his supporters came. Banks took hope, writing to Blagden that there was now peace at the society and that it was likely to remain.[38] This was not to be.

The printing of the *Philosophical Transactions* had been held up because of the dissentions, and in general the affairs of the society remained in turmoil.[39] Maty, who had "distinguished himself by his violence against Sir Jos. Banks," in Blagden's words, resigned as secretary of the society.[40] Banks sent another card to all members of the society on 29 March, this one to tell them of the vacancy left by Maty and that, "at his desire," Blagden had declared himself a candidate for the office and that Blagden would make an admirable secretary. Banks's opponents took fresh offense and referred to Banks's card as the "President's Congé d'Elire."[41]

The row over the election of Maty's replacement alarmed Cavendish. New contingency plans were laid, with Cavendish again taking part and for the same reason. On Monday, 5 April, Blagden told Banks that Cavendish and his friend the eminent cartographer Alexander Dalrymple had accompanied him home that evening to determine the "proper measures for preventing a few turbulent individuals from continuing to interrupt the peace of the R. S." Cavendish was willing to join a committee or to call a meeting to form a plan of action and draft appropriate resolutions. The general idea was that the committee would present the resolutions to a much larger meeting of members, the composition of which was to be decided by the committee. If the resolutions were acceptable to these members, they would be expected to vote for them at such times when the dissentions again interrupted the scientific work of the society. From a list of members, Cavendish selected seven who would draft the resolutions. Heberden was one of them, and when Blagden said that Heberden probably would not join them, Cavendish offered to go to Heberden the next morning to try to persuade him. Cavendish had nothing against taking the lead except for his general "unfitness for active exertions."[42] That evening Cavendish wrote to Blagden: "It is determined that Mr Aubert & I shall go to Dr H[eberden] & see what we can do. If it is to no purpose a larger meeting will be called & very likely some resolution similar to what you mentioned proposed to them."[43]

Despite his general disclaimer, Cavendish took an "active part," Blagden wrote to tell Banks the next day, to "render the R. S. more peaceable." Cavendish had called not only on Heberden but also on Francis Wollaston and Alexander Aubert, and he was going to write to William Watson, all of whom were on Cavendish's list of seven, and he had even called for the meeting to take place in his house and had settled on a time for it.[44]

That is the last we hear of Cavendish's efforts to restore peace to the Royal Society. One month later the society voted for the secretary to replace Maty. Hutton, the deposed foreign secretary and still the primary rallying cause for Banks's opponents, ran against Banks's man, Blagden. The vote was again not close, 139 to 39, in favor of Blagden. Banks in effect had made the election of the secretary a vote of confidence in him, since he had endorsed Blagden and Blagden had served throughout the stormy times as Banks's proxy.[45] Banks's victory was conclusive. Blagden wrote to a foreign correspondent that the disaffected members of the society had not only failed in their plan to unseat Banks but in the end had planted him in his seat more firmly than ever.[46] After the event, the dissentions seemed hardly more than a tempest in a teapot to Blagden, who was surprised that foreigners took such interest in that "foolish & trifling affair, as it really was with us."[47] The most important evidence for this was that science had not stopped: to a friend, Blagden wrote that "notwithstanding the interruption given to our business in the Royal Society by some turbulent members ... several valuable papers have been read, and some discoveries of the first magnitude announced," adding that "of these, the most remarkable was made by Mr Cavendish."[48]

During the dissentions, Cavendish was not on the council of the Royal Society, so he had no direct part in the Hutton affair, which had brought them on. (If he had been on the council, the case against Banks would have been substantially weakened. Banks would not be exposed this way again. Before Banks became President of the Royal Society in 1778, Cavendish had frequently sat on the council, but in the years following, 1778–84, he was on it only *once*. In 1785, the year after the dissentions, Cavendish was elected to the council, as he was *every* year after that through 1809, just before his death.) As an ordinary member without office, Cavendish attended all of the meetings of the society at which the great debates took place. Insofar as we have record, he made no public speeches at any of them. He seconded, undoubtedly by prearrangement, the motion approving Banks's presidency, and that was all. And that was all that was needed, for Cavendish was not just another member of the society. First of all, he was a *Cavendish,* a name which carried an authority of its own. He owed nothing to, and needed nothing from, Banks, and for him to act out of personal gain or personal loyalty or disloyalty would have been seen as acting out of character. He was universally respected for his scientific attainments. His field was physical science, not natural history, and he was also known to be a mathematician of a high order of ability. If Cavendish had sided with Horsley and his friends, mathematicians who styled themselves as the genuine scientific element of the society, Banks's credibility would have been shaken and the voting conceivably could have gone differently. Blagden fully understood this, which is why Cavendish was the key to his stratagems to save Banks's presidency, as his letters to Banks reveal. Cavendish's endorsement of Banks by seconding the crucial motion was a *scientific* answer to Horsley's characterization of Banks's men as feeble amateurs.

Blagden, in a letter of 2 April 1784 in which he referred to the dissentions at the Royal Society, also spoke of "our internal operations in politics, & the consequent general election, [which] have set the whole kingdom in a ferment; it is a very interesting scene, which the wisest & steadiest among us contemplate not without emotion."[49] Scientific and general politics were constantly being compared in the course of the dissentions. The one side spoke of the "ruins of liberty," the other side of Englishmen "apt to be mad with ideas of liberty, ill understood."[50] Again, the one side spoke of the "levelling spirit and impatience of all government which infects the present age," the other side of the Royal Society as a "republic," according to which all laws decided by the council are to be debated by the entire membership whenever a mover and a seconder wish it.[51] Or again, the one side urged a democratic solution to the abuses of the society, while the other warned of an illegal "democratic infringement on the principles of the constitution," which was "very much like what was passing in another place."[52] The analogy between scientific debates and those "passing in another place," Parliament, was made explicit. When speakers against Banks were shouted down and the question was demanded, Maskelyne said that he had been at other meetings that modeled their debates after the example of Parliament, and there the question was not put until everyone had had a chance to speak.[53] The favorite analogy was between Banks as president of the Royal Society and the king or some official in government. Horsley described Banks's call upon the members to elect Blagden as their secretary as a "nomination by the president, *as their sovereign,* of the person he would have them chuse; which is exactly similar to the proceeding of the king in the nomination of a new bishop."[54] Horsley's colleague Maty said that his view of the presidency of the Royal Society is of a "presidency of bare order, like that of the Speaker of the House of Commons, and in Council the President ought not to lead more than any other person."[55] Banks's opponents talked of his despotism, of his dictatorial ways, of his wish for dominion. The age of absolute monarchs was over, and Banks seemed not to have noticed, they said. But the supporters of Banks did not wish for an absolute monarch any more than his detractors did, and none was more definite on this subject than Henry Cavendish.

In explaining Cavendish's behavior to Banks, Blagden drew the appropriate parallel between Cavendish's position in science and that of his relatives in politics. "The sum is," Blagden wrote to Banks, "that like his namesakes elsewhere, he is so far loyal as to prefer you to any other King, but chooses to load the crown with such shackles, that it shall scarcely be worth a gentleman's wearing."[56] With regard to Cavendish's "grievance" against Banks, Blagden wrote again to Banks, "It is exactly the old story of an absolute Monarchy, whereas he [Cavendish] thinks the Sovereign cannot be too much limited."[57] In a more reassuring voice, Blagden wrote to Banks after a meeting with Cavendish, "The utmost consequence will be, some diminution of power, but none of dignity."[58]

In drawing comparisons between Henry Cavendish's political views and those of his namesakes, Blagden knew his subject well. He was a frequent caller at Devonshire House, where the Cavendishes came together with Fox and like-minded Whigs. Most important, he was an intimate of Henry Cavendish, whose views on politics were a private matter. The chemist Richard Kirwan wrote to Banks that he was "surprized to hear that Mr Cavendish talks

Politics; for even during Ld North's Rump Parliamant, in w^h his family were so much engaged, he was silent on the subject w^h then agitated the whole nation."[59] In his diary, Blagden wrote that Cavendish was speaking "freer now": "men and measures," Cavendish said in Blagden's presence, had to be changed, and they should "follow Fox."[60] In politics Cavendish was a Whig of the same persuasion as the rest of his family.

Although the arena in which Henry Cavendish acted out his political views was the Royal Society, in his manner of acting he resembled the Cavendishes in Parliament. An appropriate Cavendish to bring up in this connection is William Cavendish, the fourth duke and older first cousin to Henry. The fourth duke held high positions in government including, briefly, the position of prime minister in 1756–57, which was when Henry first entered the circles of the Royal Society.[61] In the political diary he kept, the fourth duke revealed, his editors write, his "complete self-assurance as to his place in the order of the world. He sits in [Privy] Council as naturally as at his dining-room table. Devonshire's assumption was that Great Britain should be governed by an aristocracy, with himself a principal. . . . [His] main concern was always to preserve harmony amongst His Majesty's servants." The fourth duke had no intimate friends in political life. "This detachment was natural to him and inevitably confirmed his exalted station. Here however lay the key to Devonshire's usefulness, recognized by everyone. He was the supremely objective man, never led away by passion, completely reliable and so the ideal receiver of confidences." Devoted to work and duty, everything the fourth duke did he did well.[62] These characteristics of the fourth duke—self-assured, conscientious, cautious, withdrawn, competent, and supremely objective— were those, by and large, of the Cavendish family and, in particular, of that member who distanced himself farthest from the active poilitical life of the nation, Henry Cavendish.[63]

Like the fourth duke and like other politicians of his family, Henry Cavendish preferred to work in committees, to exercise power behind the scenes rather than to come forward into the limelight as a leader. That behavior agreed with his understanding that power should be exercised by councils of serious men of independent judgment. He was not himself ambitious for power. He did not want to be president of the society, nor did he want to make or depose presidents. But he was always ready to advise presidents and others, as a call of duty, and always in the interest of stability and harmony.

Like his namesakes in government, whatever Henry Cavendish did, he did well. Whatever he did not do well—which included delivering speeches, inspiring men to follow him into political battle, his special "unfitness"—he did not do at all. He acted constantly in society, only his was not the given society of high fashion and politics, his birthright, but that of his own choosing, the society of scientific men. He acted from his strengths, which were his intelligence, his sense of fairness, his objectivity, and his ability to work with groups of equals to arrive at decisions for common action. His strengths also included, as his participation in the events of 1783–84 show, an understanding of political behavior; he was a close observer of men just as he was of natural phenomena.

Often in biographical writings about Cavendish and, in the bluntest form, in Francis Bickley's *The Cavendish Family,* he is presented as a pitiable figure.[64] John Pearson, the recent chronicler of the Cavendish family, in *The Serpent and the Stag* places Henry Cavendish within his family and offers a different estimate of him. Henry Cavendish was uninterested in those things that most interested his family, especially national politics, and he was exceedingly shy and at a loss for words in ordinary social settings, yet, "he was more important than all his Cavendish contemporaries put together. . . . [He] was the most original, wide-ranging British man of science since Isaac Newton."

> The more one learns about him, the more enviable and admirable he seems; and the more to be pitied those around him. For Henry Cavendish really was the most fortunate of men, possessed of a great mind, unceasing curiosity, and mental powers that lasted till his death. He was fortunate in being able to stay free from the cares and passions that enslave most lesser men. He had material freedom, and was lucky to have lived in a period when a dedicated aristocratic amateur like him could make discoveries on such a scale and over such a range of subjects.[65]

Robert Schofield, placing Henry Cavendish within his science, writes of the misunderstanding, "embellished by anecdotes about his personal idiosyncrasies," of "this most creative scientist of the eighteenth century."[66]

Notes

1. Charles Richard Weld, *A History of the Royal Society,* 2 vols. (London: J. W. Parker, 1848), 2: 151.

2. Ibid., 2: 152–54.

3. Ibid., 2: 153, 170. Henry Lyons, *The Royal Society, 1660–1940: A History of Its Administration under Its Charters* (New York: Greenwood, 1968), 198–99.

4. Weld, *History,* 2: 162. Lyons, *Royal Society,* 213.

5. The Blagden letters—originals, copies, and drafts—together with the Blagden diaries, both of which I use, are located in the following places: Beinecke Rare Book and Manuscript Library, Yale University, New Haven, Conn.; Fitzwilliam Museum Library, Cambridge; Manuscript Department, British Museum and British Museum (Natural History), London; Royal Society Library, London. I am grateful to these archives for permission to use their materials.

6. The plural "families" was used by Burke because there was more than one politically influential Cavendish family. In the sentence quoted, Burke referred to several political leaders in addition to the Cavendishes. Richard Pares, *King George III and the Politicians* (Oxford: Clarendon, 1953), 59.

7. Ibid., 58–59.

8. There were also politicians on the side of Henry Cavendish's mother, Lady Anne Grey, daughter of the duke of Kent. They were less influential than the Cavendishes, and for the purposes of this article, they may be left out of the discussion.

9. The dukes of Devonshire are discussed individually in Francis Bickley, *The Cavendish Family* (London: Constable, 1911), and in John Pearson, *The Serpent and the Stag* (New York: Holt, Rinehart and Winston, 1983).

10. The subject of the Whigs occupies a large part of Hugh Stokes, *The Devonshire House Circle* (London: Herbert Jenkins, 1917).

11. Pearson, *Serpent,* 128–29.

12. Pares, *George III,* 119–25, 134–35.

13. John Cannon, *The Fox-North Coalition: Crisis of the Constitution, 1782–4* (Cambridge: Cambridge University Press, 1969), x–xi.

14. Lady Sarah Spencer quoted in Stokes, *Devonshire House Circle,* 315.

15. Weld, *History,* 2: 154–60.

16. On Monday evenings, Cavendish and Blagden generally dined together at a club, and I assume that this is what brought them together on Monday, 22 December 1783.

17. Blagden to Banks, 22 December 1783; original letter in Fitzwilliam Museum Library; copy in Department of Manuscripts, British Museum (Natural History), DTC 3. 171–72.

18. Yet Banks's opponents professed to admire Pringle, at least in certain respects, and wished Banks were more like him. Personality and the political temper, not consistency of argument, gave the dissentions of 1783–84 their impetus. Weld, *History,* 2: 160–61.

19. Blagden to Banks, 22 December 1783. Blagden to Banks, Wednesday morning, 24 December 1783; original in Fitzwilliam Museum Library; copy in Department of Manuscripts, British Museum (Natural History), DTC 3. 176.

20. Blagden to Banks, 23 December 1783, Fitzwilliam Museum Library, Perceval H 199.

21. Blagden to Banks, Wednesday morning, 24 December 1783.

22. Blagden to Banks, 24 December 1783; original in Fitzwilliam Museum Library; copy in Department of Manuscripts, British Museum (Natural History), DTC 3. 177–79.

23. Throughout the dissentions, Banks's supporters usually advised, as Cavendish does here, to proceed with moderation. Banks's opponents would either become moderate or by their violence turn the society away from them and into Banks's camp. They were, that is, to be offered the rope to hang themselves, which some of them accepted.

24. Blagden to Banks, 24 December 1783.

25. Banks to Blagden, 26 December 1783, Blagden Letters, Royal Society Library, B. 25.

26. Blagden to Banks, 27 December 1783; original in Fitzwilliam Museum Library; copy in Department of Manuscripts, British Museum (Natural History), DTC 3. 180–81.

27. Blagden to Banks, 23 and 27 December 1783. *Supplement* to Friend to Dr. Hutton, *An Appeal to the Fellows of the Royal Society, Concerning the Measures Taken by Sir Joseph Banks, Their President, to Compel Dr. Hutton to Resign the Office of Secretary to the Society for Their Correspondence* (London, 1784), 11, 15. Lord Charles Mahon, the gifted electrician and inventor, at the close of the meeting of the Royal Society on 8 January discussed below, moved that in the future no motion should be made in the ordinary course of business without giving notice two weeks in advance. This motion, which was supposed to discourage spontaneous agitation at the meetings, was seconded and passed unanimously. Lord Mahon, who was also active in Parliament at the time of the dissentions, would go on to be one of the founders of the Revolution Society in 1788. For a time he was in harmony with the Whig opposition led by Fox. Later he became increasingly isolated, and reviled, because of his persistent championing of the ideals of the French Revolution. F. M. Beatty, "The Scientific Work of the Third Earl Stanhope," *Notes and Records of the Royal Society* 11 (1955): 217–19.

28. Blagden to Banks, 27 December 1783. Blagden to Banks, 28 December 1783, Fitzwilliam Museum Library, Perceval H 202.

29. Postscript dated 29 December 1783, Blagden to Banks, 28 December 1783.

30. Blagden to Banks, 30 December 1783, Fitzwilliam Museum Library, Perceval H 203.

31. Notes of the meeting taken by Banks, quoted in Hector Charles Cameron. *Sir Joseph Banks, K. B., P. R. S.: The Autocrat of the Philosophers* (London: Batchworth, 1952), 134.

32. [Paul Maty], *An Authentic Narrative of the Dissentions and Debates in the Royal Society. Containing the Speeches at Large of Dr. Horsley, Dr. Maskelyne, Mr. Maseres, Mr. Poore, Mr. Glenie, Mr. Watson, and Mr. Maty* (London, 1784), 24–25. *Supplement,* 9.

33. Ibid., 26–77. *Supplement,* 9. "Journal Book of the Royal Society" 31 (1782–85), 270–71, Royal Society Library. Despite charges to the contrary, in the Royal Society at this time, the physical sciences looked to be flourishing and appreciated. At the St. Andrew's Day meeting for

elections on 1 December 1783, Banks gave a discourse on two Copley Medals, one awarded to John Goodricke for his paper on the variation of the star Algol, the other to Thomas Hutchins for his experiments, which Cavendish took part in, on freezing mercury. Entry for 1 December 1783, "Journal Book of the Royal Society."

34. *Narrative,* 30, 70.

35. Blagden to Claude Louis Berthollet, 13 January 1784, draft, Blagden Letterbook, 1783–87, Beinecke Rare Book and Manuscript Library, Yale University. "Journal Book of the Royal Society" 31 (1782–85), 265, 268–71. On 27 November 1783 was begun the reading of the paper by John Michell, "On the Means of Discovering the Distance, Magnitude, etc. of the Fixed Stars, in Consequence of the Diminution of the Velocity of Their Light, in Case Such a Diminution Should Be Found to Take Place in Any of Them, and Such Other Data Should Be Procured from Observations, as Would Be Farther Necessary for that Purpose," *Philosophical Transactions* 74 (1784): 35–57.

36. Henry Cavendish, "Experiments on Air," *Philosophical Transactions* 74 (1784): 119: reprinted in *The Scientific Papers of the Honourable Henry Cavendish, F. R. S.,* vol. 2: *Chemical and Dynamical,* E. Thorpe, ed. (Cambridge: Cambridge University Press, 1921), 161–81. The juxtaposition is reflected in a letter Banks received from abroad at the time. Its author begins by saying that the Royal Society's dissentions "have made a good deal of noise on the Continent," that the opposition to Banks seems to have acted with "extraordinary animosity," and that Banks's report that the troubles are "nearly quelled" is welcome news. The author's next observation is that Cavendish's discovery of the production of water from air is "one of the greatest steps that have been made" towards understanding the elements. T. A. Mann to Banks, 4 June 1784, published in Henry Ellis, ed., *Original Letters of Eminent Literary Men of the Sixteenth, Seventeenth, and Eighteenth Centuries* (London: Printed for the Camden Society, 1843), 426–29, on 426–27.

37. Weld, *History,* 2: 162–64. *Narrative,* 79–134.

38. Banks to Blagden, 6 March 1784, Blagden Letters, Royal Society Library, B. 26.

39. Blagden to le comte de C., 2 April 1784, draft, Blagden Letterbook, 1783–87, Beinecke Rare Book and Manuscript Library, Yale University.

40. Blagden to le comte de C., 14 May 1784, draft, Blagden Letterbook, 1783–87, Beinecke Rare Book and Manuscript Library, Yale University.

41. Weld, *History,* 2: 165. *Supplement,* 12.

42. Blagden to Banks, 5 April 1784, Department of Manuscripts, British Museum (Natural History), DTC 3. 20–21.

43. Cavendish to Blagden, Monday evening [5 April 1784], Blagden Papers, Royal Society Library, c 26.

44. Blagden to Banks, 5 April 1784. Blagden to Banks, 6 April 1784, Department of Manuscripts, British Museum (Natural History), DTC 3. 22–23.

45. Weld, *History,* 2: 165–66.

46. Blagden to le comte de C., 2 April 1784.

47. Blagden to Banks, 9 August 1788. British Museum, Add MSS 33272, pp. 50–51. Blagden believed that the affair was behind them. While it is true that the dissentions did not flare up again, smoldering resentments continued to the end of Banks's long presidency, in 1820. David Philip Miller, "Sir Joseph Banks: An Historiographical Perspective," *History of Science* 19 (1981): 289. Some dozen years after his dismissal as foreign secretary, Charles Hutton gave an embittered description of the Royal Society in his *Mathematical and Philosophical Dictionary.* The entry "Royal Society of London" begins: "This once illustrious body . . ." The meeting hour of the Society had been adjusted to the convenience of "gentlemen of fashion." The *Philosophical Transactions* of the Society "were, till lately, very respectable. . . . Indeed this once very respectable society, now consisting of a great proportion of honorary members, who do not usually communicate papers; and many scientific members being discouraged from making their usual communications, by what is deemed the arbitrary government of the society, the *Philosophical Transactions* have badly deteriorated." Charles Hutton, *A Mathematical and Philosophical Dictionary,* 2 vols. (London: J. Johnson, 1795–96), 2: 399–400.

Cameron, *Banks,* 142, attributes the main reason for the original dissentions to Banks's having removed the selection of the council members from the possessive hands of the secretaries; in correcting an abuse of long standing, Banks incurred the wrath of the secretaries. Miller, above, 288–89—correctly, I think—sees the dissentions as having a more complex origin, involving both scientific and social factors, which were responsible for the long memory of the defeated party of the 1780s.

48. Blagden to Charles Grey, 3 June 1784, draft, Blagden Letterbook, 1783–87, Beinecke Rare Book and Manuscript Library, Yale University.

49. Blagden to le comte de C., 2 April 1784. Writing to Banks three days later, on 5 April, about the dissentions, Blagden added a postscript concerning the elections in London.

50. Glenie's speech on 8 January, quoted in *Narrative*, 70. Blagden to Berthollet, 13 January 1784.

51. Blagden to Banks, 28 December 1783. Letter written by Michael Lort to Lord Percy, 14 February 1784, at the height of the dissentions, quoted in Weld, *History*, 2: 169. Lort amplifies his view of the connection between the politics of the Royal Society and that of the country: at the Royal Society "every fortnight a set of orators get up and fatigue themselves, and much the greater part of the Society, with virulent and illiberal charges against the President. Horsley, Maskelyne, Maty, Maseres, and Poore are the leaders of this band, who are joined by all those turbulent spirits that are impatient of all government and subordination, which is indeed the great evil and disease of the times. I believe I have prolonged and increased my complaints by going out twice to vote against these innovators, who kept the society talking and disputing and balloting till near eleven and twelve o'clock, though they have been baffled in almost every question by near three to one. I will say nothing of our politics; our newspapers contain scarce anything else." Michael Lort to Lord Percy, 24 February 1784, published in *Literary Anecdotes of the Eighteenth Century,* 9 vols., John Nichols, ed. (London: Printed for the author, by Nichols, son, and Bentley, 1812–16), 7: 461.

52. The accountant general Anguish's speech on 12 February, quoted in *Narrative,* 112.

53. Maskelyne's speech on 8 January, quoted in *Narrative,* 62. The Royal Society and Parliament occasionally came together in the same person. Lord Mulgrave, for example, was active both in the debates of the House of Commons and in those of the Royal Society. When Blagden came to see him on the subject of the dissentions, Lord Mulgrave talked to him as much as "his present political agitation would allow." Lord Mulgrave strongly urged Banks and his supporters against temporizing, since discontented men were "never made quiet by coaxing." Blagden, who used the analogy himself, thought that Lord Mulgrave carried the analogy of "H[ouse] of C[ommons] ideas to our Society" farther than was justified. Blagden to Banks, 23 December 1783.

54. Horsley's speech on 1 April, quoted in *Supplement,* 12.

55. Maty's speech on 12 February, quoted in *Narrative,* 99.

56. Blagden to Banks, 22 December 1783.

57. Blagden to Banks, morning, 24 December 1783.

58. Blagden to Banks, 24 December 1783.

59. Richard Kirwan to Banks, 10 January 1789, Department of Manuscripts, British Museum (Natural History), DTC 3. 122–24.

60. Entry for 16 March 1795, Blagden Diary, vol. 3, Royal Society Library.

61. In 1757 Henry Cavendish was proposed by Lord Macclesfield, then president of the Royal Society, as a member of the dining club associated with the society. This meant that he had to wait for a vacancy, which occurred in 1760. Archibald Geikie, *Annals of the Royal Society Club* (London: Macmillan, 1917), 63.

62. P. D. Brown and K. W. Schweizer, eds., *The Devonshire Diary: William Cavendish, Fourth Duke of Devonshire, Memoranda on State of Affairs, 1759–1762,* Camden Fourth Series (London: Royal Historical Society, 1982), 27: 19–21.

63. Caution has been singled out by other writers on Henry Cavendish as a characteristic common to him and to the Cavendishes in general. The family motto *Cavendo tutus,* a play on words meaning "Safe by being cautious," was Cavendish's guide throughout his life according to his main biographer George Wilson, *The Life of the Honble Henry Cavendish* (London, 1851), 190.

64. Bickley, *Cavendish Family,* 207. Pearson, *Serpent,* 121, calls attention to Bickley's estimate of Henry Cavendish.

65. Pearson, *Serpent,* 116, 121.

66. Robert E. Schofield, *Mechanism and Materialism: British Natural Philosophy in an Age of Reason* (Princeton: Princeton University Press, 1970), 254.

Duhem on Gibbs

MARTIN J. KLEIN

During the winter of 1906–7 Pierre Duhem, the professor of Theoretical Physics at the University of Bordeaux, was asked to review a new book for the *Bulletin des Sciences mathématiques*. This was one invitation to review that he would not pass by. The work was the *Scientific Papers* of Josiah Willard Gibbs,[1] collected in two substantial volumes and published less than four years after the untimely death of the great American theorist in April 1903.[2] Duhem had never met Gibbs; the only time during which the two were on the same side of the Atlantic Ocean was the period of Gibbs's postdoctoral study in Europe when Duhem was a small boy just starting primary school.[3] But Duhem did know Gibbs's writings, and could bring twenty years of familiarity with them to bear in his review. He had grown up as a scientist with Gibbs's ideas, had expounded them, criticized them, developed them, and made them the basis for much of his own work. He had even staked his career on the correctness of the Gibbsian approach long before that approach was widely understood. Duhem must have been delighted at this opportunity to write about Gibbs's work as a whole, particularly since he knew that the *Bulletin* would let him write at some length.[4] The review appeared in the August 1907 issue, but Duhem evidently thought so highly of it that he published it twice more the following year—in a more widely read journal and as a separate booklet.[5] His opinion was justified: Duhem's essay on Gibbs tells us a great deal about both men.

Some may be surprised to see Duhem's name linked with that of Gibbs. We historians of science usually think of him as one of our most distinguished colleagues of an earlier generation, as the man who made "the first serious large-scale exploration of the history of medieval science" and established this subject as "an autonomous field of scholarly endeavor."[6] Despite the corrections, modifications, and refinements of Duhem's work over the decades and the changed attitudes toward the field that they led to, Clifford Truesdell's appraisal is still valid: "Only to one can it be given to discover a whole period of history, and for mechanics in the Middle Ages this one was Duhem."[7] Yet notwithstanding this Columbian achievement and the many books and articles he devoted to his historical studies,[8] Duhem had no doubt that his true calling was not that of historian. "I am a physicist," he said to a friend when there was talk of a possible appointment for Duhem as historian of science at the Collège de France. "It is as a physicist that they must take me if I am to return there.''[9] Although that triumphal return to the metropolis never took place, all Duhem's academic appointments were in physics, and it was as a physicist that he was eventually elected to the Académie des Sciences in 1913 as one of its first six members who did not reside in Paris. Duhem continued to teach physics and to do research in physics until the last months of his life. Theoretical physics was always "assigned to first place," and his historical studies had to "mark time" when physics made its demands.[10]

Although his particular investigations into scientific problems and into questions about both the history and the philosophy of science had no obvious relationships to each other, Duhem did not think of his work as neatly compartmentalized. He saw a continuity in his concerns, a vision of the scientific enterprise that tied these seemingly distinct studies together. That unifying vision of Duhem's also came in part from what he found in the writings of Gibbs.

Duhem's first contact with Gibbsian thermodynamics came remarkably early in his life at the Catholic lycée, Collège Stanislas, where he spent ten years, the last as an assistant teacher. His many-sided brilliance was recognized early, and he had the good fortune to study physical science there with a gifted teacher who was also an active scientist. This was Jules Moutier, who worked on his research in the laboratory of Henri Sainte-Claire Deville at the École Normale Supérieure.[11] Moutier was one of the first to apply thermodynamics to chemical problems, and it was to thermodynamics that he introduced his extraordinary student, encouraging him to begin his own research even while he was still at the lycée. As a result Duhem was looked up to by the younger students as a legendary figure who had already "made discoveries."[12] But his most important discovery at that period came in the reading Moutier suggested, which included Hermann von Helmholtz's new paper on chemical thermodynamics[13] and a recently published book by Georges Lemoine on the problems of chemical equilibrium.[14] Lemoine concluded his book with brief accounts of several current theories of dissociation, the phenomenon investigated experimentally by Deville and his collaborators. Since Lemoine's discussion of Gibbs's work on dissociation immediately preceded what he had to say about Moutier's own work, it was sure to catch the eye of Moutier's eager young student. These studies made a deep impression on Duhem. On a significant occasion over thirty years later he referred to this early exposure to the ideas of Helmholtz and Gibbs as having "shown us the path which we then began to follow and from which we have never since departed." On that same occasion Duhem paid tribute to Moutier as the one "who brought us to love the theories of physics."[15]

From the Collège Stanislas Duhem went on to the École Normale Supérieure, entering in 1882 at the head of the new class that was selected by severe competitive examination. As a Normalien he continued to develop his own ideas on thermodynamics, publishing his first paper in 1884. This was a brief note, "On the Thermodynamic Potential and the Theory of the Voltaic Pile," communicated to the Académie des Sciences at its meeting of 22 December by the mathematician, Charles Hermite.[16] In this paper Duhem proposed a general criterion for stable thermodynamic equilibrium, which he saw as a generalization of the principle of virtual velocities in statics, introduced the function that would serve as a potential function for thermodynamics, and modestly noted that his basic theorem had already been "pointed out in a less general form by Gibbs and Helmholtz." Duhem also listed a lengthy set of applications of his newly announced method of basing thermodynamics on the condition that the potential be a minimum at equilibrium, saying that this method had already led to "results too numerous to be summarized even succinctly, in this Note." He limited himself to the single example of the voltaic pile.

It was quite a remarkable debut for one who was still a third year student at the École Normale. Even more remarkable, however, was the fact that Duhem had indeed worked out all this material and had submitted it two days earlier to the Faculty of the Sorbonne as a thesis for the doctorate. The work was referred to a committee of examiners whose chairman was Gabriel Lippmann, and six months later it was flatly rejected as "not worthy of being defended as a thesis before the Faculty of Sciences of Paris."[17]

Following the lead of Gibbs and Helmholtz, Duhem had made the second law of thermodynamics the basis for the study of chemical and electrochemical equilibrium. In so doing he had explicitly denied the general validity of the principle of maximum work, which claimed that energy considerations based on the first law of thermodynamics were sufficient for fixing the equilibrium state. Unfortunately for Duhem this principle of maximum work had been formulated by Marcellin Berthelot, and Berthelot was one of the most powerful scientists in France. Despite criticism of his principle from various quarters, Berthelot would go on defending its universal validity for another twenty years.[18] Duhem was convinced that Berthelot was behind the rejection of his thesis by Lippmann's committee of examiners, that Berthelot had decided to suppress the implied and perhaps also explicit criticism of his principle of maximum work found in Duhem's submitted thesis.[19]

Whether or not Duhem was right in attributing his rejection to Berthelot, as well as to Lippmann, is not our concern here. The rejection certainly did not cause Duhem to give up his views. On the contrary, his reaction was to publish a book on the subject in 1886, a book that presumably was a revised and perhaps expanded version of the rejected thesis.[20] With this defiant gesture Duhem took the chance of stunting a potentially brilliant career in order to reaffirm his trust in the thermodynamics of Gibbs and Helmholtz. He did receive his doctorate two years later on the basis of another thesis, welcomed by the mathematics faculty even though it dealt with the theory of magnetism. But the position that he received on his graduation from the École Normale was at the University of Lille and not in Paris. Duhem was certain that this was the result of Berthelot's decision: "This young man shall never teach in Paris."[21]

(Although I do not intend to discuss Duhem's book, *The Thermodynamic Potential and Its Applications,* I should mention that one important result, derived by Duhem in his chapter on the vapor pressure of solutions,[22] has helped to keep his name alive among chemists and physicists. It had already been obtained in a much more general form by Gibbs[23] a decade earlier, and it has come to be known in the literature as the Gibbs-Duhem equation.[24])

During the years when Duhem was working on his rejected thesis and the book he based on it, his knowledge of the American physicist's works was largely second hand. One would never know this from Duhem's writings where Gibbs's papers are referred to directly and often. But on 10 January 1887 Duhem wrote to Gibbs from Paris, requesting copies of his papers on themodynamics.[25] "Unfortunately," wrote Duhem, "the journals in which these memoirs were published are not very widely distributed in Paris; I have not been able to obtain them for myself up to now, and I know your important works only from the rather short excerpts published in Wiedemann's *Beiblätter.* And so, in trying to understand your ideas from these excerpts and to reconstruct your theories completely in a series of works on the same subject that I have published in the past two years, I must have often failed to recognize your claims and involuntarily committed injustices with respect to you." Like others, Duhem had been frustrated in his attempts to find the *Transactions* of the Connecticut Academy of Arts and Sciences, the journal in which Gibbs's major works had been published.

In his reply Gibbs pointed out to Duhem that the Connecticut Academy did exchange its publications regularly with a number of institutions and learned societies in Paris, and that copies of its *Transactions* ought to be available to him in at least four and possibly in as many as eight different libraries. Gibbs wrote that he was sending Duhem copies of the papers he had on hand. These did not include his major work, "On the Equilibrium of Heterogeneous Substances," reprints of which were long since exhausted. The volume of the *Transactions* in which it had appeared was, however, on its way to the Library of the École Normale Supérieure with an invitation from the Connecticut Academy's librarian to exchange publications on a regular basis. Gibbs also expressed his "grateful appreciation of the very flattering manner" in which his work had been discussed in Duhem's "valuable treatise," which he had seen at the Yale University Library.[26]

Two weeks later on 16 February 1887 Duhem was answering Gibbs, having already received his letter and reprints and having had time to read enough of Gibbs's papers to recognize that Gibbs had anticipated much of what Helmholtz had written on the thermodynamics of the electrochemical cell.[27] Duhem wrote that he was sending Gibbs two packets containing almost everything that he had published, and hoped that Gibbs would be willing to continue receiving his works. By July Duhem had published a lengthy two-part study of Gibbsian thermodynamics analyzing its structure, discussing its historical development from the earlier work of Clausius, and showing how it could be applied to the phenomenon of dissociation.[28]

Duhem continued to send Gibbs many of his papers during the following years and Gibbs reciprocated,[29] but there is no record of any correspondence between them for well over a decade. It is hard to know what Gibbs thought of Duhem's work. Their styles were so very different. Duhem wrote at great length and with great facility: "He composed in his mind, and then when he sat down at his desk he covered pages with his beautiful handwriting, regular and legible, without crossing out a word, without stopping except for the time to refer to a book in order to transcribe a citation."[30] In the thirty-four years of his career he wrote well over three hundred works, many of them books or lengthy articles. Gibbs's writing is best described as spare, laconic. I have seen a page of proof from his "On the Equilibrium of Heterogeneous Substances" which read originally, "It is indeed evident that. . . ." The only change Gibbs made was to delete the word *indeed*. Some years after his death, Gibbs's niece, Miss Theodora Van Name, gave his reprints to the Yale University Library, and I can report that a number of the papers Duhem had sent still had their pages uncut. I have often wondered whether Duhem was the unnamed author mentioned by Gibbs's student E. B. Wilson. Gibbs gave Wilson permission to cut the pages of some books he had never opened, saying he could do so "if you think it worthwhile," and then adding: "The author kindly sends me all he writes; there is a great deal of it; I sometimes feel that a person who writes so much must spread his message rather thin."[31] Of course, it may have been someone else entirely.

By the time Duhem was beginning his scientific career with his work on thermodynamics, Gibbs had turned away from that subject and was thinking about other aspects of theoretical physics. He seems to have felt that his book-length memoir "On the Equilibrium of Heterogeneous Substances" contained all that he wanted to say about thermodynamics, perhaps even all that needed to be said about it. During the 1880s Gibbs developed his vector analysis and worked extensively on problems of optics and electromagnetic theory.[32] Duhem, however, continued to make thermodynamics the central theme in his research, trying to clarify its basic concepts and logical structure and also extending the range of its applications throughout physics and chemistry. By 1894 he had formulated his vision of a general thermodynamics or energetics that would encompass all changes of state of bodies, changes of position as well as changes of physical properties, on the basis of one set of principles.[33] Dynamics would then become a particular case of this general thermodynamics. This was a very different approach from that adopted by physicists for several centuries as they tried to reduce thermodynamics and all other branches of physics to a presumably more fundamental mechanics. Duhem saw this as one of the greatest advantages of his proposed approach: "If the science of motion is no longer the first of the physical sciences, logically speaking, and becomes only a particular case of a more general science whose equations encompass all changes of bodies, we think there will be less temptation to reduce the study of all physical phenomena to the study of motion; ... then we shall more readily avoid what has up to now been the most dangerous stumbling-block for theoretical physics—the search for a mechanical explanation of the universe."[34]

Duhem's research prospered, but his career did not.[35] After six years at Lille he was transferred, but to Rennes, in Brittany. His hopes for a position in Paris were dashed again a year later in 1894 when he was moved to Bordeaux. This time he was given to understand by Louis Liard, director of higher education at the Ministry of Public Instruction that he should accept the change, "that he must understand that Bordeaux is the road to Paris."[36] Duhem took this reassurance so seriously that he did not unpack all his family's belongings after moving to Bordeaux, waiting for the call to Paris that he expected. That call never came, and the professorship of theoretical physics created for him at Bordeaux in 1895 proved to be his final academic appointment.

In the spring of 1900 the Académie des Sciences elected two new corresponding members to its Section of Mechanics. The first of these was Gibbs, elected on 21 May;[37] a week later Ludwig Boltzmann's name was added to the list. Duhem was quick to express his pleasure over the choice of Gibbs. "Allow me to congratulate you," he wrote; "allow me above all to congratulate the Académie for having associated itself with the one whose work 'On the Equilibrium of Heterogeneous Substances' crowns the XIXth century as Lagrange's *Mécanique analytique* crowns the XVIIIth century. Please forgive one of the first apostles of your ideas for wanting to express his joy at this election."[38] Duhem's joy must have been enormously enhanced two months later when he too was elected a corresponding member of the Académie.[39] His satisfaction with this recognition from Paris at long last was coupled with the pleasure of seeing his name listed with Gibbs in the Académie's Section of Mechanics.

Duhem was particularly conscious of his link to Gibbs at just this time. For almost two years he had been guiding the research of an American mathematician of French descent, Paul Saurel, who had come to Bordeaux to work with Duhem.[40] On 28 June, 1900 Saurel defended a thesis on Gibbsian thermodynamics for the new degree, Docteur de l'Université. (This was a degree that had been created to attract foreign students to France, since it did not demand that candidates first acquire a French baccalaureate.[41]) Duhem described the background of Saurel's work in an article published in Bordeaux in September of the same year.[42] He referred to Gibbs's memoir whose publication was "one of the most significant events in the history of science during the last three centuries," comparing it once again to Lagrange's famous treatise. Just as Lagrange's methods allowed one to formulate any problem of mechanics in mathematical form, so did Gibbs's work permit the mathematical formulation of all the problems of "chemical mechanics." But Gibbs's work needed extended commentary in the form of further researches, both theoretical and experimental, that would develop the points that Gibbs had left implicit in his highly condensed exposition. "Nowhere, perhaps, has this commentary been pushed further than at the Faculty of Sciences of Bordeaux," wrote Duhem, referring of course to his own efforts and those of his students. Duhem was satisifed that Saurel's work clarified and extended what Gibbs had left behind, and Saurel's papers do indeed contain elegant treatments of a number of nice questions in thermodynamics.[43]

By 1900 Duhem must have felt quite confident that the general tendency of his own work harmonized with that of Gibbs. His plans for the creation of a generalized thermodynamics or energetics would not be fully realized until his treatise on that subject appeared in two large volumes in 1911,[44] but these plans were surely seen by their author as Gibbsian in spirit. Duhem could not have anticipated the surprise that Gibbs had in store for him.

In 1902, as part of a series of volumes in celebration of Yale University's bicentennial, Gibbs's book on statistical mechanics was published.[45] Very few of his colleagues outside of New Haven were even aware of his interest in the subject, though he had been thinking about it, giving courses on it, and working towards a publication for almost twnety years.[46] For those whose admiration of Gibbs's thermodynamics was heightened by his apparent avoidance of references to the atomic structure of matter or the possible mechanical basis for thermodynamic principles, this book must have been a shock. Duhem was surely one of that group, but when he received his presentation copy of *Elementary Principles in Statistical Mechanics,* his letter of thanks showed no trace of the tremors.[47] "Permit me to thank you most deeply for the honor you do me in sending me your new work. A book by you cannot fail to be a scientific event. I am happy to be one of the first to know this event, thanks to you." What Duhem's letter does not contain is any reference at all to the subject of Gibbs's book. Scientific events are, after all, of many sorts, and this one would take some getting used to.

Gibbs died the following year, and three years later his *Scientific Papers* were published. When Duhem received this work to review, he must have been particularly interested in the biographical sketch that served as an introduction.[48] Never having met Gibbs he now at least had an opportunity to learn something about the life of the man whose work he had admired and whose ideas he had been building on for a quarter of a century. The biographical sketch was written by Gibbs's former student and Yale colleague, Henry Andrews Bumstead, a man whose "keenness of perception was thoroughly appreciated by Gibbs."[49] Bumstead began with a brief description of his subject's family background and education and ended with a few personal remarks about the man he admired so highly, but most of his dozen or so pages were devoted to an accurate and illuminating account of Gibbs's work and its place in the structure of physics.

What Duhem found in Bumstead's sketch about Gibbs the man seemed to correspond remarkably well to the intellectual characteristics of Gibbs the scientist as Duhem had come to know him from his writings. Duhem made this close relationship between the man and his work the theme of his essay-review, using the biographical material to try to account for some of the particular features of Gibbs's writings. "That the most abstract and most algebraic scientific work can nevertheless reflect as in a true mirror its author's character is an idea that continually came to us as we were perusing the two volumes in which someone had the very happy idea of gathering the memoirs of Josiah Willard Gibbs."[50]

Duhem commented first on Gibbs's success in achieving an "absolute calm" in the circumstances of his quiet, regular life, this being "only the image of the calm he had realized within himself." Free of ambition, "the only passion capable of robbing a thinker of the full possession of his genius," Gibbs could pursue a life of untroubled scientific thought. This struck Duhem as especially noteworthy in the midst of a nineteenth-century America "that appears to us as burning with feverish activity, devoured by the thirst for gold." (Remember, Duhem had never been in New Haven.) Seeking a comparison, Duhem reached back to the thirteenth century—he had been spending much of his own time in the medieval world for the last few years—and evoked the image of a monk rigorously developing a chain of syllogisms to prove some abstract metaphysical thesis, while the world outside his cell resounded with the clangour of armor, the shouts of battle, and the cries of the massacred.[51]

One phrase of Bumstead's kept returning to Duhem as he thought about Gibbs. "Of a retiring disposition," Bumstead wrote, "he went little into general society and was known to few outside the university."[52] Duhem came to realize that the phrase "of a retiring disposition" characterized much more about Gibbs than his way of dealing with his fellow citizens in New Haven. It meant, for example, that he did not share the feeling, so widespread among scientists, that they must tell others about their newest ideas, no matter how preliminary and uncertain they might be. Gibbs did not want or need a sounding board for his ideas. He was more than content to be his own and only critic. Preferring to think his problems through in solitude, Gibbs would discuss only his fully matured, finished work. Rather than radiating outward Gibbs condensed inward. As Duhem put it: *"Concentration,* it seems, is the essential characteristic of his intellectual and moral physiognomy."[53]

Bumstead had noted that even the graduate students who worked with Gibbs did not hear about his work in progress. "He did not take them into his confidence with regard to his current work," and so they "were deprived of the advantage of seeing his great structures in process of building, of helping him in the details, and of being in such ways encouraged to make for themselves attempts similar in character, however small their scale."[54] Such were the hazards of working with a man "of retiring disposition."

The spare style in which Gibbs wrote was another manifestation of that disposition, that tendency to concentration rather than diffusion. What a later editor described as "austerities of style and extreme economy (one might almost say parsimony) in the use of words" characterize all his writing.[55] But, as Duhem pointed out, no language offers greater possibilities for conciseness than the language of algebra, and Gibbs expressed himself algebraically when the situation allowed it. Duhem quoted Gibbs's own description of algebra as "this most refined and most beautiful of [labor-saving] machines."[56] The vector analysis Gibbs developed, taught for years, and wrote up for the convenience of his students was another device that made it possible "to condense a difficult calculation in a few wonderfully concise symbols." Duhem described Gibbs as cultivating his vector analysis "with real love," a love whose strength was demonstrated by Gibbs's readiness to engage in an extended polemic on behalf of vectors.[57] He had been willing to sacrifice his jealously guarded calm and quiet to repulse the attacks on vectors made by partisans of quaternions led by that belligerent Scot, Peter Guthrie Tait. Duhem himself disliked both vectors and quaternions, and wrote elsewhere that "the taste for symbolic algebras is an index of that ampleness of mind" characteristic of English physicists, not a term of praise in his vocabulary.[58] He preferred the "classical Cartesian algebra," as a glance at any of his lengthy calculations will immediately confirm, but he certainly thought of himself as algebraically rather than geometrically minded. So much so that his old comrade from the École Normale, the distinguished mathematician Jacques Hadamard, commented on it: "It is a striking thing in one whose whole life was devoted to studying the physical world that Duhem's mathematics appears ... to be purely algebraic;" it was far removed from geometrical interpretations or concrete images of any sort.[59] This trait in Duhem may well have led him to underestimate the degree to which Gibbs's work was genuinely geometrical in spirit.[60]

Gibbs's "retiring disposition," interpreted by Duhem as his tendency to condense and concentrate his thought, could help to explain several character-istic but puzzling features of his memoir on heterogeneous equilibrium. Since much of that work dealt with problems of direct concern to chemists, one might have expected that the exposition would be addressed to these potential readers. Gibbs's theoretical analysis of such subjects as coexistent phases and equilibrium in chemical reactions could have been made more accessible to chemists if he had provided examples and discussed particular chemical systems. "If one wants the chemist to recognize these properties, if one wants him to grasp a thermodynamic equation, if one wants him to be able to apply it to the reactions he studies, it is necessary to make these quantities take on again, for a moment, the concrete and particular state from which they were abstracted and generalized."[61] It was not as though Gibbs were incapable of dealing with experimental results and specific systems. He did not suffer from "the awkwardness of the pure mathematician, ignorant of observational science, unskilled in finding among the multitude of experimental laws those that could illustrate his equations."[62] He did work out the detailed consequences of his theory for one class of chemical reactions in gases, and meticulously compared his results with the available experimental data from several sources.[63] Normally, however, "once he had condensed the truth into a concise, very general formula, he was loath to develop the infinite series of particular cases encompassed by this universal general proposition."[64]

Because he would not come out to meet his readers, Gibbs failed to call attention to some of his most noteworthy results. Duhem described how J. D. van der Waals and H. W. Bakhuis Roozeboom had brought the phase rule to light ten years after its publication, showing its power to bring order to a mass of experimental material. "M. van der Waals demonstrated a remarkable perspicacity in perceiving the phase rule among the algebraic formulas where Gibbs had to some extent hidden it."[65] How many more such seeds of potentially major experimental discoveries, Duhem wondered, "had remained sterile because no physicist or chemist had noticed them under the algebraic shell that concealed them?"[66] The sad truth was that Gibbs's work had much less impact on the development of chemical thermodynamics than it should have had. Others, who independently rediscovered results published years before by Gibbs, exerted more influence by their partial results than Gibbs had had with his general theory. This never seemed to bother Gibbs, who made no effort to claim the priority he evidently deserved for so many results, though priorities are defended vigorously, if not bitterly, by most scientists. On the one occasion he did point out something he had done a decade earlier, his tone was not that of "an inventor laying claim to something that others had unjustly forgotten to attribute to him," but rather "a kindly notification from one who has already found the truth and wants to save unnecessary labor by those who are still looking."[67]

No one before Gibbs had looked at thermodynamics as a general theory of the equilibrium states of material systems and of the stability of those equilibrium states. Once Gibbs came to this understanding of the subject, he chose his postulates to make his treatment of thermodynamics as logically direct as possible. He hoped he had found "the point of view from which the subject appears in its greatest simplicity."[68] The simplicity of his exposition, however, was a logical simplicity and not the simplicity that allows the student to grasp the subject most readily. Gibbs's postulates, stated at the outset of his memoir, "On the Equilibrium of Heterogeneous Substances," were actually new to the literature. They represented a distillation of the results of decades of work by many investigators beginning with Carnot. Yet, as Duhem pointed out, Gibbs stated them "without commentary, without a single example, without any historical introduction" that might have given his readers some insight into what these postulates meant and where they had come from. Someone who approached Gibbs's memoir without any prior knowledge of the subject might well have thought that these postulates were "accepted by all physicists," that they were "in common use." This was not the case at all. The concept of entropy and that of reversible process and "Clausius's audacious assertion" about the behavior of entropy in irreversible processes were all far from general acceptance when Gibbs wrote. "But who would have suspected the existence of all these reasons for hesitation and doubt when contemplating the serene assurance with which the professor from New Haven wrote the two equations that contain his entire thermodynamic work?"[69] Duhem had tried to provide the kind of introduction to Gibbs's memoir that he thought necessary in the pair of papers he wrote right after his first direct encounter with that work.[70] He was conscious of the essential novelty of Gibbs's whole approach to thermodynamics, and not just of his results.[71] By masking this essential novelty behind his very general postulates, Gibbs had given a drastic example of just how retiring his "retiring disposition" could be.

Duhem had been struck by "the almost unbelievable care" Gibbs had taken to formulate every physical hypothesis in its most general form and to refrain from developing its detailed consequences. Since he thought that Gibbs's book on statistical mechanics offered a particularly curious example of this practice, Duhem took the opportunity to comment on this work,[72] which had not been included in the *Scientific Papers*. Judging by these comments, Duhem never really mastered the contents of Gibbs's last work. He apparently did not even grasp the real point of the statistical approach. The crucial distinction between the individual physical *system,* which could be anything from a sample of gas in a container to a block of metal subjected to external stresses, and the *ensemble* or population, which is a theoretical construct for thinking about the statistics of the incompletely specified physical system, seems to have eluded Duhem. He writes about canonical *systems* and takes Gibbs to task for not demonstrating that such systems exist and what in their internal structure makes them canonical.[73] Duhem seems to have missed Gibbs's discussion of the appropriateness of the canonical *ensemble* for representing essentially any physical system in thermodynamic equilibrium at a definite temperature, or even (to an excellent approximation) any system whose total energy is fixed.

Knowing Duhem's negative attitude to the program of mechanical explanation, we can hardly be surprised that he did not make the effort to penetrate Gibbs's difficult book setting forth a new approach to this subject. He certainly did read and ponder Gibbs's preface to his book, the preface in which Gibbs raised some very serious problems that blocked attempts "to explain the mysteries of nature" by applying statistical mechanics to the molecules that presumably constituted all matter. Because of these difficulties

Gibbs had limited himself to pursuing "statistical inquiries as a branch of rational mechanics," assuming nothing about the constitution of matter. In this way his only errors could be "the want of agreement between the premises and the conclusions, and this, with care, one may hope, in the main, to avoid," as Gibbs put it with his usual precision.[74]

What intrigued Duhem was what Gibbs really thought about the future of mechanical explanation, statistical or otherwise. He knew that most physicists were reluctant to write about those aspects of their thought that did not form part of the deductive systems set forth in their papers and books. But he also knew that just those ideas, "perceived only more or less clearly, formulated only more or less explicitly," were often what lay behind the published work and determined its direction. Even though they remained "vague and indistinct" and could not be expounded systematically, they might still be the real basis for a physicist's conviction that he was following the right track. This philosophical background for the particular physical investigations, carefully guarded from view by most scientists, had to be known, if one were ever to understand the origins of those works that constituted the public record of science.[75]

But could one expect Gibbs to have shared his innermost thoughts with his readers? Duhem knew very well that the answer was: "Certainly not." Gibbs, who would publish his scientific work only when it had fully matured, only when he was fully satisfied that his formulation was as clear, precise, and general as possible; this Gibbs could not even consider publishing whatever "vague and indistinct" ideas lay behind his investigations. "We must resign ourselves to ignorance about the philosophical ideas that no doubt presided over the birth of physical theories in Gibbs's mind."

Duhem had to conclude regretfully that we would never know what Gibbs thought about the future of mechanical explanation. The very existence of his *Statistical Mechanics* demonstrated the depth of Gibbs's interest in the possibility of such explanations, despite the contrary impression made on many readers (including Duhem) by his earlier, purely thermodynamic papers. But the cautionary remarks and doubts expressed by Gibbs in the preface to his book still left open the possibility that he would eventually have rejected all further attempts at mechanical explanation. One had no way of getting behind the words of this physicist "of a retiring disposition."

Duhem's essay succeeded in showing the ways in which Gibbs's work did "reflect as in a true mirror its author's character." Could one try to show the same sort of connection between the man Pierre Duhem and his work? To what extent do his writings reflect a true image of that "uneasy genius"?[76] It is not a question that can easily be answered. Duhem wrote some thirty books and hundreds of articles dealing with physics, chemistry, and both the history and philosphy of science. He was out of step with most of the major tendencies of the intellectual life of his period: the list of ideas he vigorously opposed runs from Maxwellian electrodynamics and atomism through the secular democracy of the Third Republic, and even includes the neo-Thomist scholastic revival that dominated the thought of the Roman Catholic Church of which he was such a devout communicant.[77] Duhem's writings clearly portray a forceful personality,[78] but not one that can be readily characterized. He was certainly not of a retiring disposition.

Notes

1. Henry Andrews Bumstead and Ralph Gibbs Van Name, eds., *The Scientific Papers of J. Willard Gibbs*, 2 vols. (New York: Longmans Green and Co., 1906).

2. For information on Gibbs's life, see Lynde Phelps Wheeler, *Josiah Willard Gibbs. The History of a Great Mind* 2nd ed. (New Haven: Yale University Press, 1952).

3. Duhem's life and work are discussed in detail in Stanley L. Jaki, *Uneasy Genius: The Life and Work of Pierre Duhem* (The Hague, Boston, Lancaster Pa.: Kluwer Academic Publishers, 1984).

4. Duhem had published lengthy reviews in the *Bulletin* before. For references, see Jaki, *Uneasy Genius*, 444, 446, 448.

5. P. Duhem, "Étude sur le caractère de l'oeuvre de Gibbs (à propos de *The Scientific Papers of Willard Gibbs*)," *Bulletin des Sciences mathématiques* 31 (1907): 181–211; P. Duhem, "Josiah-Willard Gibbs à propos de la publication de ses Mémoires scientifiques," *Revue des Questions scientifiques* 13 (1908): 5–43; and P. Duhem, *Josiah-Willard Gibbs à propos de la publication de ses Mémoires scientifiques* (Paris: A. Herman 1908). Page references will be to this version, hereafter cited as Duhem, *Gibbs*.

6. David C. Lindberg, preface in *Science in the Middle Ages*, Edited by D. C. Lindberg (Chicago: University of Chicago Press, 1978), vii–viii.

7. Clifford Truesdell, *An Idiot's Fugitive Essays on Science. Methods, Criticism, Training, Circumstances.* (New York, Berlin, Heidelberg, Tokyo: Springer-Verlag, 1984), 174.

8. See the chronological bibliography of Duhem's writings in Jaki, *Uneasy Genius,* 437–55. A bibliography arranged by journal of publication can be found in the Duhem memorial volume: *Mémoires de la Societé des sciences physiques et naturelles de Bordeaux,* 7th ser., *1* (Cahier 1, 1917), (Cahier 2, 1927), 41–70. This volume will be referred to as *Bordeaux.*

9. Édouard Jordan, "Biographie de Pierre Duhem," *Bordeaux,* 9–39, 16.

10. Hélène Pierre-Duhem, *Un Savant Français, Pierre Duhem* (Paris: Plon, 1936), 194.

11. See, for example, Jules Moutier, *La Thermodynamique et ses principales applications* (Paris: Gauthier-Villars, 1885). Jaki discusses Moutier and his career in some detail; see, Jaki, *Uneasy Genius,* 164–65, and 260–65.

12. Jordan, "Biographie," 11.

13. Hermann von Helmholtz, "Zur Thermodynamik chemischer Vorgänge," *Sitzungsberichte der Preussischen Akademie der Wissenschaften, Physik. Mathematische Klasse,* Berlin (1882): 22.

14. Georges Lemoine, *Études sur les équilibres chimiques* (Paris: A. Hermann, 1881), 304–18.

15. Duhem, "Notice sur les Titres et Travaux scientifiques de Pierre Duhem," *Bordeaux,* 72.

16. P. Duhem, "Sur le potentiel thermodynamique et la théorie de la pile voltaïque," *Comptes Rendus Hebdomadaires des Séances de l'Académie des Sciences, Paris, 99* (1884): 1113–15, hereafter, *Comptes Rendus.*

17. See Jaki, *Uneasy Genius, 50–53.*

18. For a recent and balanced discussion of this issue, see R. G. A. Dolby, "Thermochemistry versus Thermodynamics: The Nineteeth Century Controversy," *History of Science* 22 (1984): 375–400.

19. Jordan, "Biographie," 15–16; and, Duhem, *Duhem, 50–53.* See also Donald G. Miller, "Pierre–Maurice–Marie Duhem," in *Dictionary of Scientific Biography,* 16 vols., Charles C. Gillispie, ed., 4 (New York: Scribners, 1971), 225–33; and his article, "Ignored Intellect: Pierre Duhem," *Physics Today* 19, no. 12 (1966): 47–53.

20. P. Duhem, *Le potentiel thermodynamique et ses applications à la mécanique chimique et à l'étude des phénomènes électriques* (Paris: A. Hermann, 1886).

21. Duhem, *Duhem,* 53, 146. Hélène Duhem was presumably quoting her father, but where he would have heard it is not apparent. However, the remark is attributed to both Lippmann and

Berthelot on p. 53, but only to Berthelot on p. 146. It is quoted uncritically in almost all the works on Duhem.

22. P. Duhem, *Le potentiel thermodynamique,* 33.

23. Gibbs, *Scientific Papers* 1, 88, 97.

24. Jaki, *Uneasy Genius,* discusses the history of the name, Gibbs–Duhem equation, on p. 308, footnotes 215, 216. I know of no earlier use of the name than the 1929 reference he gives.

25. Duhem to Gibbs, 10 January 1887. The letters from Duhem to Gibbs and a draft of Gibbs's letter to Duhem are in the Beinecke Library of Yale University. They are quoted by permission of Yale University.

26. Gibbs to Duhem, 2 February 1887 (draft).

27. P. Duhem to Gibbs, 16 February 1887.

28. P. Duhem, "Étude sur les travaux thermodynamiques de M. J. Willard Gibbs," *Bulletin des Sciences mathématiques* 11 (1887): 122–48, 159–76.

29. Wheeler, *Gibbs,* contains Gibbs's mailing lists for reprints in Appendix IV, 235–48. Works sent to Duhem are listed on pp. 238 and 248.

30. Jordan, "Biographie," 20.

31. Edwin Bidwell Wilson, "Reminiscences of Gibbs by a Student and Colleague," *Scientific Monthly* 32 (1931): 217.

32. See Wheeler, *Gibbs,* 106–33. Also see Ole Knudsen, "An Eclectic Outsider: J. Willard Gibbs on the Electromagnetic Theory of Light," in *The Michelson Era in American Science, 1870–1930,* Stanley Goldberg and Roger H. Stuewer, eds., (New York: American Institute of Physics, 1988), 224–34.

33. P. Duhem, "Commentaire aux principes de la Thermodynamique," *Journal de Mathématiques pures et appliquées* 8 (1892): 269–330; 9 (1893): 293–359; 10 (1894): 207–85.

34. See Duhem, "Commentaire," *J. de Maths. pures et appliqués,* 10 (1894): 231. Duhem quoted this material in his "Notice," *Bordeaux,* 75–76.

35. Duhem's life in Lille, Rennes, and Bordeaux is discussed by his daughter, in H. Duhem, *Duhem,* and in more detail by Jaki, *Uneasy Genius.*

36. Quoted in Duhem, *Duhem,* 98.

37. *Comptes rendus* 130 (1900), 1371.

38. P. Duhem to Gibbs, 29 May 1900.

39. *Comptes rendus* 131 (1900), 325. see Jaki, *Uneasy Genius,* 144–45.

40. See Jaki, *Uneasy Genius,* 137–41.

41. Mary Jo Nye, *Science in the Provinces: Scientific Communities and Provincial Leadership in France, 1860–1939* (Berkeley, Los Angeles, London: University of California Press, 1986), 27–28.

42. P. Duhem, "Un doctorat de l'Université de Bordeaux," *Revue Philomathique de Bordeaux et du Sud–Ouest* 3 (1900): 396–97.

43. See, for example, Paul Saurel, "On the Critical State of a One–Component System," *Journal of Physical Chemistry* 6 (1902): 474–91.

44. P. Duhem, *Traité d'Énergetique ou de Thermodynamique générale,* 2 vols. (Paris: Gauthiers-Villars, 1911).

45. J. W. Gibbs, *Elementary Principles in Statistical Mechanics, Developed with Especial Reference to the Rational Foundation of Thermodynamics* (New York: C. Scribners and Sons, 1902).

46. Martin J. Klein, "Some Historical Remarks on the Statistical Mechanics of Josiah Willard Gibbs," in *From Ancient Omens to Statistical Mechanics: Essays on the Exact Sciences Presented to Asger Aaboe,* edited by J. L. Berggren and B. R. Goldstein, *Acta Historica Scientiarum Naturalium et Medicinalium* 39 (Copenhagen, 1987), 281–89.

47. Duhem to Gibbs, 9 April 1902.

48. Henry Andrews Bumstead, "Josiah Willard Gibbs," in *The Scientific Papers of J. Willard Gibbs,* 1, pp. xi–xxvi. The shorter original version of Bumstead's paper appeared in *American Journal of Science* 16 (1903): 187–202. I will quote from the longer version in the *Scientific Papers.*

49. Leigh Page, "Biographical Memoir of Henry Andrews Bumstead, 1870–1920," *Biographical Memoirs of the National Academy of Sciences* 13 (1929): 107.

50. Duhem, *Gibbs,* 6.

51. Ibid., 9–10.

52. Bumstead, *Gibbs,* xxiii.

53. Duhem, *Gibbs,* 10.

54. Bumstead, *Gibbs,* xxiv.
55. L. P. Wheeler. E. O. Waters, and S. W. Dudley, eds., *The Early Work of Willard Gibbs in Applied Mechanics,* (New York: H. Schuman, 1947), 43. The remark is by E. O. Waters and refers in particular to Gibbs's thesis, printed here for the first time.
56. J. W. Gibbs, "On Multiple Algebra," in *Scientific Papers* 2, 91.
57. Duhem, *Gibbs,* 16–20.
58. P. Duhem, *La théorie physique: son objet et sa structure* (Paris: Chevalier et Rivière, 1906). I quote from the English translation, *The Aim and Structure of Physical Theory,* translated by Philip P. Wiener (Princeton: Princeton University Press, 1954), 77.
59. Jacques Hadamard, "L'oeuvre de Duhem dans son aspect mathématique," *Bordeaux,* 638.
60. Duhem evidently ignored Gibbs's first two papers (*Scientific Papers* 1 3–54), which are geometrical in title and in spirit. (Note especially Gibbs's remarks on p.32.) For further discussion of these papers and their geometrical character, see M. J. Klein, "The Scientific Style of Josiah Willard Gibbs," in *Springs of Scientific Creativity,* edited by Rutherford Aris, H. Ted Davis, Roger H. Stuewer (Minneapolis: University of Minnesota Press, 1983), 142–62.
61. Duhem, *Gibbs,* 21.
62. Ibid., 26.
63. Gibbs, *Scientific Papers* 1, 172–84; 372–403.
64. Duhem, *Gibbs,* 26.
65. Ibid., 23–24.
66. Ibid., 25.
67. Ibid., 15–16.
68. Wheeler, *Gibbs,* 89.
69. Duhem, *Gibbs,* 32–33.
70. See note 28.
71. This remark applies just as well to Gibbs's first two papers on thermodynamics. See the references in note 60.
72. Duhem, *Gibbs,* 35.
73. Ibid., 35–38.
74. Gibbs, *Statistical Mechanics, x.*
75. Duhem, *Gibbs,* 38–43.
76. Jaki used this description of Duhem as the title of his biography. It comes from the reports on Duhem by Gustave Demartres, dean of the Faculty of Sciences at the University of Lille during Duhem's years at that institution. See Jaki, *Uneasy Genius,* 102.
77. Duhem's relationships to the complex Catholic thought of his time are discussed by Niall Martin. See R. N. D. Martin, "The Genesis of a Mediaeval Historian: Pierre Duhem and the Origins of Statics," *Annals of Science* 33 (1976): 119–29; "Darwin and Duhem," *History of Science* 20 (1982): 64–74; "Saving Duhem and Galileo: Duhemian Methodology and the Saving of the Phenomena," *History of Science* 25 (1987): 301–19.
78. See the remarks by E. J. Dijksterhuis in the Preface to his book, *Val en Worp: Een bijdrage tot de Geschiedenis der Mechanica van Aristoteles tot Newton* (Groningen: P. Noordhoff, 1924), vi.

Life Span Developmental Psychology and the History of Science

MICHAEL M. SOKAL

Many late twentieth-century historians sometimes imply that biographical inquiry has nothing to say to their concerns, and explicit doubts as to the genre's value often pepper their conversations and, in some cases, their writings. Certainly, the current vogue of social history, particularly as it has developed within the community of American historians with its stress on those who left no self-conscious record, reinforces this attitude. Many historians believe that this concentration on the mass is overdue even though they do not see it as overriding other interests. But they do feel put upon when social historians in their desire to promote a particular approach to the past sometimes suggest that any individual who historians can identify does not deserve their attention and is definitely not worth writing about.[1]

But those of us who have been fortunate enough to have worked with Robert E. Schofield know better. His scholarly career to date provides several excellent illustrations of the ways in which biography addresses significant historical questions. His first published paper, for example, explains astutely, within a broad overview of his scientific interests "Wesley's persistent objections to Newtonianism," and thus illustrates effectively the extent of Newton's influence. Schofield's brilliant first book, *The Lunar Society of Birmingham,* contains insightful chapters on "Lunar Personnel." These biographical sketches discuss quite appropriately just what each member brought to the society and asked of it, and, throughout his institutional analysis Schofield demonstrates effectively just how these personal characteristics and concerns shaped the Lunar Society and its role in eighteenth-century English science. His biography of Stephen Hales, written with David G. C. Allan, does much to illuminate how one significant individual worked with Newtonian ideas in the same time and place. And his current major project, a full biography of Joseph Priestley, promises to demonstrate effectively just how an even more important individual developed Newtonian thought in a way that has long influenced the evolution and practice of the physical sciences.[2]

Through all of his work, even so, Schofield remains proudly and clearly a historian of science, as he focuses his attention on scientific ideas and institutions and examines his subjects' lives primarily for the light his studies shed on their science and its influence. Some of his students, however—myself included—have begun to take their biographical concerns a bit further. While we have not abandoned our interest in the development of science, in at least some of our work we aim primarily to illuminate the experiences and interests of particular individuals who happened to have devoted most of their working lives to science.[3]

In this paper, however, I would like to show that these two approaches to the past do not necessarily represent two completely independent concerns. Instead, I believe strongly that they complement each other effectively and that concentration upon the history of scientific ideas and institutions and the development of an individual life mutually reinforce one another. In particular, I want to make an audacious claim: that an understanding of the most personal details of an individual's private life can help historians of science understand the development of their subject's scientific interests, including both the most trivial and most significant details of his scientific career. Furthermore, I want to claim that a sensitivity to an individual's psycho-social development through his life—based in my case at least on a study of one recent conception of adult life-span developmental psychology—enables historians of science to see things that they otherwise might miss.

I want to approach my thesis by outlining (and, through overemphasis, making explicit) several of the approaches I am taking in my current major historical project: a full analytic biography of James McKeen Cattell (1860–1944). Best known to many interested in past science as perhaps the first experimental psychologist who argued that he and his colleagues should apply their science to practical problems in the world in which they lived, in 1890 Cattell invented the first self-identified program of mental tests, and in using these tests as a Columbia University professor achieved a great reputation. To be sure, by 1901, others had shown that Cattell's tests yielded inconsistent results that revealed nothing of interest to anyone about the tested individual. But his continually reiterated stress on an applied, or rather applicable, psychology meant much to his younger colleagues throughout the field. In many ways he set the course that American psychology followed through the twentieth century. Better known to others as the long–time (1894–1944) owner and editor of *Science,* Cattell by the 1910s had converted this failing journal into the leading American general scientific periodical that many scientists found required reading. In 1900 *Science* became the official journal of the largest scientific society in the country, the American Association for the Advancement of Science, even while privately owned. Through this journal Cattell played a series of major organizational roles in the association. He largely wrote the revised constitution the association adopted around 1920 and for the next twenty years chaired its executive committee. Many observers saw Cattell as the dictator of the AAAS, and he did little to counter the perception.

In addition, Cattell became well known as the editor of several other important scientific journals, ranging from the highly technical, such as *The Psychological Review* (1894–1903) and *The American Naturalist* (1907–39), to the ostensibly popular, such *Popular Science Monthly* (1900–1915), its successor *The Scientific Monthly* (1915–39), and *School and Society* (1915–38). His work as the compiler of *American Men of Science,* first issued in 1906 and continued today in a seventeenth edition as *American Men and Women of Science,* (1989) attracted much attention, especially as he used the data he collected for his extensive statistical studies of the American scientific community. In 1910 he also issued the first formal ratings of American universities. Long widely known as a crusader for the professorial governance of these universities, he attracted much national notoriety in 1917 when Columbia University president Nicholas Murray Butler dismissed Cattell from the professorship he had held since 1891. Indeed, he is probably best known to the general scholarly community as a wartime martyr to academic freedom. Historians of science concerned with a wide vartiety of "internal" and "external" topics have thus long been interested in his career, and I began my own studies of his life as a student at Case Western Reserve, inspired by Robert E. Schofield, and under the direction of Robert C. Davis.

I have written extensively on Cattell's life and career,[4] but I strongly believe that the biography I am now preparing will enrich and extend my past work considerably. In particular, my familiarity with some recent well-received work in adult life-span developmental psychology, especially that of Daniel J. Levinson, has enabled me to see an important unity throughout Cattell's intellectual, institutional, and even personal activities that has much to say to those interested in his scientific work.[5] Further I believe that Levinson's work leads directly to a richer understanding of the origins of Cattell's specific scientific ideas about mental testing. Through the biography, I use these psychological ideas to help shape (but not determine) the biographical and more general historical material I deal with, much as a skeleton gives form to a vertebrate's flesh and muscle. Despite its important function, the skeleton remains hidden through the organism's normal life and, though its existence can be inferred, it cannot be detected without the intrusion of scalpel or X rays. In the same way, my use of Levinson's ideas in the biography, though fully and explicitly documented, remains (for the most part) unseen and a reader uninterested in bones need not consult the X rays that my citations provide. This approach differentiates my work from that of most psychohistorians, who often investigate the past solely to "demonstrate" the "validity" of a particular psychological theory. The present paper, however, written in tribute to an individual whose seminar on the historiography of science has proved a major factor in shaping the historical careers of dozens of his students, serves appropriately to dissect the figure I am creating and thus illustrates its anatomy. To mix metaphors, it presents (and probably overemphasizes) the removable scaffolding I am using to erect the biography. Throughout this essay, then, I assume the role of anatomist or civil engineer.

But butchers and builders both know that skeletons, whether of bone or steel, have weaknesses, and I realize the limitations of the psychological ideas on which I rely. Levinson himself has recently argued that he can find evidence supporting his general conception of the adult life course in many past and current cultures.[6] With several critics, however, I believe that their applicability is much more limited.[7] Though influenced by both Freudian and Jungian theory, Levinson based his ideas primarily on extensive research on a group of mid-twentieth-century American men, and his concepts seem most clearly useful for studies of career-oriented males living and working in an achievement-oriented culture like that of the modern (though not perhaps the post-modern) United States. As his book's title *The Seasons of a Man's Life* indicates, Levinson himself realizes (or at least realized at one time) the gender limitations of his work and is in fact now completing a study of modern American women that traces developmental patterns that differ, partially, from those he found in his studies of men. That being said, Levinson's concepts can help a historian develop an understanding of (if not explain) the evolving ideas and institutional activities of many American scientists of the late nineteenth and early twentieth centuries. After all, the robber barons and their followers of the Gilded Age oriented themselves as clearly toward secular achievement as any late-twentieth-century individual.

In his studies, Levinson found that when he could identify an individual's "Dream"—that is, a man's goal, his purpose for life, and his "vague sense of self-in-adult-world"—he could then better understand the person's concerns, life-direction, personality, and career path. In doing so, Levinson carefully avoided assigning too much precision to the concept, and continually stressed that a Dream is usually "more formed than a pure fantasy, yet less articulated than a fully thought-out plan." Though the assortment of men he studied included many who worked directly with ideas—in particular, a bevy of biologists and a nest of novelists—Levinson for his purposes avoided examining the minute details of his subjects' work.[8] However, in my studies of Cattell's scientific ideas, I found that when I identified Cattell's Dreams (for I believe he had more than one), I gained new insight into many details of his intellectual development.

Cattell had both personal and professional Dreams, which were tightly interwoven throughout his life. His professional Dream emerged through the 1870s, as he studied at Lafayette College, in Easton, Pennsylvania, a small school where his father, William C. Cattell, served as president. During the elder Cattell's twenty-year term, the college's endowment grew a hundred-fold. It began to offer well respected programs in engineering and the applied sciences and even attracted an international reputation for its instruction in several areas. William Cattell's great administrative achievement, based primarily on his personal contacts with trustees, other supporters, and the members of his faculty, strongly influenced his son's approach to those around him. This aspect of the younger Cattell's temperament did much to shape the journals he later edited and institutions he later controlled.[9] The major overt influence on his intellectual development was, however, the college's best known and most distinguished scholar, Francis Andrew March. A noted philologist and the prime American contributor to the *Oxford English Dictionary,* March also preached the understanding of Baconian empiricism shared by many mid-nineteenth century Americans—emphasizing systematic collection, depreciating hypothesis, and stressing a concern for usefulness— and used this method in his detailed reconstruction of Anglo-Saxon grammar. He also stressed this approach to knowledge through his teaching at Lafayette. Cattell, following March in a vague plan to devote his life to scholarship, later adopted it as a component of his own scientific ideology.[10] Cattell's professional Dream of a scholarly life also led him in 1880 to leave America for eight years of study at German universities. Its later articulation in Europe, under the influence of his personal Dream, led him to aim for a career in the emerging American university system as an experimental psychologist.

The personal Dream that helped shape Cattell's life emphasized an ideal vision of a warm and mother-centered family life that, most importantly, also influenced his scientific ideas directly. This Dream had many sources. Cattell grew up in a family whose members shared an obsessive devotion to each other. His father William Cattell, a nervous man who exhibited a genteel fretfulness all his adult life, in particular doted on his wife. The father never allowed the younger Cattell to attend school and instead tutored him at home. As a result, those who knew Cattell as a Lafayette freshman thought him spoiled, and throughout his life he always saw the family as the center of his existence.[11] In the twentieth century, he viewed the family as a genteel haven into which a highly competitive professional could retreat to escape the pressures of his career. As a college student, Cattell explored these ideals directly in his philosophical studies and devoted a great deal of attention to "The Ethics of Positivism," the topic of his honorary philosophical oration at the 1880 Lafayette commencement. While Auguste Comte's Positivism continuously stressed the authority of science, he had also sought to develop a complete philosophical system. In doing so, he based his ethical views on the virtue of altruism, and had taken the sacrifice of the mother in childbirth as his model. With its similarity to the veneration of the Virgin Mary, this notion has led some to characterize Comte's philosophy as Catholicism without Christianity. Significantly, Cattell's first major purchase as a graduate student in Europe was an expensive copy of Raphael's Sistine Madonna, which hung and still hangs in Dresden.[12] More importantly, by leading him to the study of Comtean ethics, Cattell's personal Dream helped steer his scholarly interests toward science, determined the shape of his scientific ideology, and led eventually to his program of mental tests.

Initially, Cattell focused his studies in Europe on philosophical topics, and his reading of Positivism led him to seek out those philosophers who approached their subject most scientifically. At Göttingen in 1880, the lectures of Herman Lotze convinced him that the subject's future lay in the new experimental approach to philosophical questions then emerging as physiological psychology. Meanwhile, Comte's influence also led Cattell to the scientific ideology that determined the way in which he did science throughout his life. Cattell's studies of Positivism introduced him to Comte's law of three stages and his hierarchy of the sciences, both of which emphasize the authority of mathematics. Even as a senior at Lafayette, he began fusing these concepts, and their stress on quantification, with the Baconian focus on empiricism and usefulness he got from March to form his own scientific ideology. The Baconian aspects of this ideology demanded that Cattell avoid hypothesis and collect, much as March's work involved the gathering and collation of word forms, while its Comtean facets led Cattell to emphasize quantitative bits of information. This approach to science had an immediate effect on his earliest experimental work in the 1880s, both as a graduate student working at Johns Hopkins and, much more importantly, with Wilhelm Wundt at Leipzig after Lotze's death. In Germany, despite his earlier philosophical interests, Cattell explicitly ignored his teacher's broader psychological theories, and his laboratory notebooks demonstrate his sharp disdain for Wundt's concern with an experimenter's self-observation of his mind. Instead, the American explicitly differentiated (for apparently the first time) the experimental subject from the experimenter himself, and his laboratory procedures emphasized the conclusions that an observer could draw from measuring his subject's reactions during the experiments.[13] Long before others, Cattell introduced a quantitative natural history approach that established the behavioral perspective that most American psychology would take in the twentieth century.

In 1886, after earning his German Ph.D., Cattell established himself at St. John's College, Cambridge.[14] While in England his continual contacts with Francis Galton led Cattell to develop further his scientific ideology within the context of the articulation of both of his Dreams. In this case, another of Levinson's concepts, that of the role of a mentor in a man's life, leads far beyond past accounts of Cattell's dealings with Galton and illuminates a much more richly textured relationship. Cattell first met Galton in 1885, after years of reading his work and exchanging letters with him, and found in him another scientist obsessed with measurement. On moving to Cambridge, Cattell began looking to Galton as a personal guide to the scientific world of late nineteenth-century England. His view that professional success depended heavily on personal contacts, derived from his father's experience and reinforced throughout his years of study in Europe, thus found further confirmation in his dealings with Galton. In all, Cattell saw Galton as "the greatest man" he ever met, and in many ways Galton's influence on Cattell went far beyond the professional.[15]

On one level, Cattell adopted and used much of the late Victorian and Edwardian vocabulary of Galton and his contemporaries. Like R. T. A. Crawford, the second of *The Masters* on whom C. P. Snow's novel focuses, Cattell always spoke of himself and his colleagues as men of science, rather than scientists. This usage, suitably modified for the late twentieth century, continues today in the directories edited by his successors. More significantly, like many other Americans of his generation, Cattell found himself attracted by Galton's program of eugenics as a means to improve the human race. Cattell, however, ignored those aspects of eugenics that many of his American colleagues found congenial: the elimination and segregation of the "unfit," primarily though sterilization and immigration restriction. He concentrated instead, under the influence of his personal Dream, with its stress on motherhood and family life, on the propagation of the "fit" within a stable family structure. He and his wife thus had seven children, and throughout his scientific and editorial career he worked to promote large and stable families. He regularly argued from census data that the college educated were not producing enough children to replace themselves. Similarly, he once used his *American Men of Science* questionnaires to investigate the birth-control practices of scientists. And, though he typically defended almost any professor involved in almost any academic freedom case, he joined several times in condemning those charged with moral turpitude, for their actions struck, he believed, at the very basis of family life. He also attacked the common early-twentieth-century practice of prohibiting married women from teaching in the public schools. He did so for two reasons: such practices prevented the best-educated women who wanted to keep their jobs from contributing to the mental quality of the next generation; and they set before girls just the model be believed they ought to avoid, the unmarried and hence childless woman.[16]

Galton's main intellectual influence on Cattell, however, took a different form. He introduced Cattell to the Darwinian concepts of function and variation. These ideas did little to redirect Cattell's commitment to an empirical approach to science. They did, however, prove important by helping him focus on the collection of quantitative data demanded by Cattell's scientific ideology. Functional concerns for life in the world meshed well with Cattell's Baconian thrust on utility, and thus led him to stress the applicability of his psychology. Yet Galton's work never emphasized function and other Darwinian interests proved more important for Cattell's psychology. Galton's studies always stressed the differences between individuals, the variations that made natural selection possible and on which his eugenics focused. Through (for example) his Anthropometric Laboratory, Galton led Cattell to realize that the psychological techniques he had used in Germany to study the mind could be used to study specific psychological differences between people.[17]

Therefore, in the late 1880s, Cattell set out to create a series of psychological tests, designed in part to provide data on human variability. More important, Cattell wanted his tests to be useful, and he stressed that they would delineate an individual's mental traits and help him adapt to the world. But the kind of tests that Cattell developed reflected his scientific ideology, which itself had been shaped by his personal Dream. Instead of concentrating on complex functions and what were known as the "higher mental processes," as Galton himself had hoped to in the early 1880s and Alfred Binet did fifteen years later, Cattell focused his attention on those simple psychological phenomena he could measure easily, such as an individual's reaction time under different conditions, or the smallest stimulation that his different senses could detect. Expanded through the 1890s and applied to hundreds of subjects at Columbia University, Cattell's program of tests thus yielded thousands of bits of quantitative data whose precision Cattell bragged about regularly. To him, such exactitude made his results worthwhile. While stressing the ultimate applicability of testing as such, he ignored the fact that the traits his tests measured had limited functional meaning. By 1901 others had shown that Cattell's tests yielded inconsistent results that revealed nothing of interest about the individuals tested. Rarely have historians been faced with such a clear case of scientific failure, and historians of psychology have long attempted to explain just why Cattell, despite his Darwinian stress on variation and his functional belief in the applicability of his science, structured his tests as he did. The answer lies in his scientific ideology, shaped by his personal Dream.[18]

Cattell's editorship of *Science* and his other journals, his control of the AAAS, and some of his other institutional efforts might also be termed failures, though none of these cases is as clear cut. Certainly the journals played important roles in the American scientific scene throughout the first third of the twentieth century, and *Science* today thrives as the most important general American scientific periodical. But as I have shown elsewhere,[19] by the 1930s Cattell had alienated many members of the American scientific community, and through that decade many leading physicists actually avoided sending the results of their research to *Science*. These men and women found *Nature* and *The Physical Review* attractive outlets for their work, but they also actively distrusted Cattell and his editorship. Similarly, in the 1920s and 1930s, when Cattell openly controlled it, the AAAS actually lost much of the importance it had been able to claim earlier in the century, and many influential members tended to blame the decline on Cattell. During these two decades Cattell frequently feuded with five successive AAAS permanent secretaries, the society's executive officers, and succeeded in replacing four of them. By 1940 his colleagues felt forced to free the association from his control, and they made the eighty-year-old Cattell yield the chair of the AAAS executive committee that he had held for twenty years: the length of his father's term as Lafayette's president. Soon afterward they even tried unsuccessfully to wrest from him *Science*'s editorship, though they knew he owned the journal. Finally, as Carol Singer Gruber has shown convincingly, Cattell's dismissal from Columbia University in 1917, long seen as Butler's tyrannical attack on the principles of academic freedom, resulted primarily from Cattell's constant and bitter personal attacks on his Columbia colleagues. Cattell was a difficult man to deal with, and his contemporaries regularly complained to each other about him.[20]

As these examples show, Cattell's temperament, particularly the way in which he dealt with those around him, did much to shape his scientific career and the institutions in which he worked. They also demonstrate that a sensitivity to their subjects' personal traits, focused by ideas like Levinson's, can help historians illuminate the course of the past. As suggested earlier, however, many previous attempts to inform historical analyses of an individual's personality development with psychological ideas have failed, as scholars lack relevant sources and inappropriately apply ideas out of context. In contrast, Levinson's work focuses less on an individual's psycho-sexual development before the age of three and more on the formation, in adolescence, of his approach to others, including his peers and those with authority over him.[21] Fortunately, Cattell's teenage years are well document-ed and historians can trace effectively the origins of his style of personal interaction that played such an important role throughout his scientific career.

As sketched earlier, Cattell had been brought up at the center of one of the wealthiest families in Easton, and as the first–born of two brothers he had always been his parent's favorite child. He never attended school but studied instead in his father's library, often with his doting parents as his tutors. As such, he never had to compete with others for his teacher's ear and thus grew to expect excessive attention. An older friend wrote of him as a "Mama's boy" who became angry when others would not bow to his desires. Anson Hunter, the protagonist of F. Scott Fitzgerald's 1922 story, "The Rich Boy," found that his friends' parents "were vaguely excited when their own children were asked to [his] house." Even as a child, he noticed "the half-grudging American deference that was paid to him." Cattell experienced much the same attention, and like most children raised within any given setting, he saw nothing unusual in it. Instead, he gradually grew to expect deference as simply his due. As a student at Lafayette, he became further used to the deference of the college faculty. To be sure, his professors expected him to do as much and to work as hard as his classmates, and they always held his behavior up to certain standards. But Cattell was bright and hardworking and fulfilled their expectations readily and he was, after all, their president's son and the grandson (and namesake) of the college's major benefactor. Their attitude thus reinforced his expectation of deference from all he met, including those with authority over him. Like Anson Hunter, Cattell egocentrically "accepted this as the natural state of affairs" and thus developed "a sort of impatience with all groups of which he was not the center ... which remained with him for the rest of his life."[22]

In addition, Cattell always believed that through personal contact individuals could shape institutions and confirmed this point of view through his fathers's and Galton's experiences. Cattell always acted as if his desires should determine the course of events, and even in his early twenties this attribute caused him trouble. In 1883, when a graduate student at Johns Hopkins, he complained that the university's president, Daniel Coit Gilman, did not pay him the attention he believed was his due, and he publicly insulted Gilman when his fellowship was not renewed.[23] More significantly, with his need for deference and his reliance on personal contact, he had trouble dealing later even with the philanthropic foundations and federal bureaucracies that emerged during the early twentieth century. At many crucial instances his approach to others almost isolated him personally and the institutions within which he worked.

In coming to terms with this and other evidence, one of Levinson's best known ideas does much to clarify just why Cattell allowed his approach to others to dominate his temperament. Although Levinson's concept of the Dream has attracted some attention and many have begun to search openly for a mentor in their biographical research, it is the idea of the "mid-life crisis" that probably represents the best known term derived from life-span development psychology.[24] Levinson himself avoids the phrase and especially the meanings some have attached to it and prefers the more general, "mid-life transition." For most of the men he studied, this transition involved adjusting themselves, at about age forty, to the real world and, in particular, to modifying their Dreams in the face of changing circumstance. During this transition, individuals typically come to terms with reality and realize the many illusory qualities of their Dreams; after all, only a few can win Nobel or Pulitzer Prizes or become company presidents or play for the Boston Celtics, and nobody has a perfect marriage or family.[25]

Scientists in particular find that they can not do all that they had planned and that not every experiment or calculation supports their ideas. They learn further that no laboratory or institute can provide the unlimited support they crave and that other scientists do not share their total commitment to their personal research program. They discover that they are not the center of their field and that their Dreams of revolutionizing their sciences single-handedly must fail. In coming to terms with all of these realizations an individual has to modify his Dream and accept his true place in the world. In doing so, Levinson found that many men seem to want to redefine themselves and thus seek what he calls, following José Ortega y Gasset, a new "life structure." Levinson's studies revealed various "sequences through the mid-life transition" toward a new "life structure," including several that comprise what he calls "advancement." Others, however, clearly involve decline,[26] and in many ways these seem more useful to a historian interested in understanding Cattell's career.

The period of Cattell's life that began just before his fortieth birthday (in 1900) and continued for the next several years was clearly a time of transition and, perhaps, even of serious crisis. In general he found that all he wanted to do—his teaching and research, his editing, his active role as a father of a large family, and even his long commute from his cherished mountain-top home, fifty miles north of New York—simply required more time than he had, so he began to devote less and less time to his work at Columbia. Professionally, by his forty-first birthday, the majority of American psychologists became convinced that his program of mental tests, to which he had devoted most of his scientific attention through the previous decade, had failed. His goal of an applicable science of psychology, an important part of his professional Dream, thus seemed beyond his reach. At the same time, Cattell's relations with his colleagues grew worse as his approach to those around him brought him into conflict with other strong-willed men. Butler became Columbia's president in 1902 and the two men began feuding with each other within weeks of the inauguration. The following year Cattell's continual quarrels with Princeton psychologist James Mark Baldwin, who shared with him the ownership and editorship of *The Psychological Review,* came to a head and attracted much notoriety in the American psychological community. Cattell abandoned the journal that had been his first editorial venture.[27] With the birth of his seventh child about the same time, he came to realize that with the expensive tastes he had developed as a spoiled young man, he could not raise his large family in its spacious home without greatly supplementing his Columbia salary.

All of these factors gradually came together and led Cattell step-by-step out of his laboratory and testing room and bit-by-bit into his editorial office. The transition actually began earlier because in 1894, when he took control of *Science,* he looked on the journal at least partially as an investment. In 1900 he even borrowed money from his mother to buy *Popular Science Monthly,* which he saw primarily as an eventual source of income. By 1903, when he began the first edition of *American Men of Science,* Cattell was clearly more an editor than a working scientist, although the change occurred gradually and cannot be dated with precision.

From 1900, then, Cattell began to devote more and more time to editorial and organizational problems. Before 1905, he actively involved himself in several serious debates about the policies of the Carnegie Institution of Washington and the Smithsonian Institution. He approached both controversies as he did most issues, by personally setting his views before those involved and expecting them to defer to his opinions. These actions and attitudes only caused him trouble, and in carrying them out he even alienated several colleagues who had supported him and had been his friends for decades. Most notably, his campaign to get the Carnegie Institution to abandon its planned approach to science led to bitter arguments with John Shaw Billings and Robert S. Woodward. Both had been members of *Science*'s editorial board from 1894, and Billings, who had nominated Cattell for membership in the National Academy of Sciences, had first expressed interest in Cattell's work in the early 1880s. In 1903, however, despite Cattell's pleas, Billings resigned from *Science,* and the two men never again shared a friendly conversation.[28]

In 1906, at the age of forty-six, Cattell sought appointment as Secretary of the Smithsonian Institution. In many ways this candidacy can best be seen as his attempt to complete his mid-life transition by finally resolving the tensions he felt between his old and new life structures. He actively invested much time, effort, and emotion in his campaign, which involved a direct personal appeal to the institution's regents. He asked hundreds of scientists, most of whom he knew only through letters as contributors to *Science*, to write directly to the regents they knew best and urge his appointment. His efforts failed. The majority of the regents never saw him as a viable candidate, and the deluge of letters alienated others. At age forty-six, then, Cattell failed to complete the transition he had entered into half a decade earlier, and he never achieved the new life structure he sought. Resulting in part from his expectation of deference and impatience with those around him, this failure, especially when seen in conjunction with the more obvious failure of his program of mental testing, merely reinforced this attitude and set the course for the rest of his career. From about 1906 onward, Cattell approached the world more egocentrically than ever, with a sense of bitterness about his role in psychology and science in general. No wonder his colleagues found him hard to deal with and sought eventually to displace him.[29]

Cattell continued to lose friends through the 1910s, as his colleagues in the scientific community and at Columbia began complaining to each other about him: "it is too bad that Cattell is opposed to so many things." His 1917 dismissal from Columbia, however, only served to increase his bitterness, and his experiences in the 1920s reinforced these feelings. Early in the decade, he sought again to apply psychology through his Psychological Corporation, but this venture failed as he relied entirely on his scientific ideology, which had been out of date forty years earlier. In 1929, even as American psychologists honored his seniority by electing him president of that year's Ninth International Congress of Psychology, the first ever held in America, he could not respond gracefully. He went out of his way to attack publicly the work of William MacDougall, an English-born psychologist then at Duke University. AAAS officials began consoling themselves with the fact that Cattell had turned sixty in 1920 and thus, they assumed, would probably not be active much longer. But he continued editing *Science* through 1944, and the association and its members continued to suffer.[30]

I know that many historians of science will find the ideas set forth here uncongenial. Nevertheless, most will agree that a sensitive feel for the significance of appropriate biographical detail, like that exhibited by Robert E. Schofield throughout his career, adds much to and enhances the value of even the most "internal" of studies in our field. Not all have this feel however, and, as I hope my discussion has demonstrated, such a scholarly sensitivity can sometimes be subtly shaped (and sharpened) through the study of some recent socio-psychological scholarship. At the very least, such notions deserve exploration and, as I think I have shown here, they might even profit from historians' attention. And I know that Robert E. Schofield would heartily approve any such attempt to extend our discipline's influence.

Notes

This essay has evolved through three successive versions: as part of a proposal to the National Endowment for the Humanities program in Humanities, Science, and Technology, which led to the major research grant that has supported my work (including this chapter) since January 1985, and under which I am now preparing a full analytic biography of James McKeen Cattell; at the Seventeenth International Congress of History of Science, University of California, Berkeley, 9 August 1985; and at the Boston Colloquium for the Philosophy of Science, Boston University, 15 October 1985. Its adaptation has been shaped largely by the warm and very helpful comments of many kind and thoughtful friends and colleagues. Its conclusions rest primarily upon the riches found in the two hundred manuscript boxes of the James McKeen Cattell papers, deposited by his children in the Manuscript Division of the Library of Congress, Washington, D.C.

1. This sketch may overstate the case a bit, but contains at least "symbolic truth." For a more temperate view of social history's relations with other genres, see the essays in Stanley I. Kutler and Stanley N. Katz, eds., *The Promise of American History: Progress and Prospects,* the tenth-anniversary number of *Reviews in American History* (Baltimore: Johns Hopkins University Press, 1982).

2. R. E. Schofield, "John Wesley and Science in 18th Century England," *Isis* 44 (1953): 331–40; Schofield, *The Lunar Society of Birmingham: A Social History of Provincial Science and Industry in Eighteenth-Century England* (Oxford: Clarendon Press, 1963); Schofield and David G. C. Allan, *Stephen Hales: Scientist and Philanthropist* (London: Scolar Press, 1980).

3. See, for example, Alice Kimball Smith and Charles Weiner, eds., *Robert Oppenheimer: Letters and Recollections* (Cambridge: Harvard University Press, 1980); Michael M. Sokal, "The Unpublished Autobiography of James McKeen Cattell," *American Psychologist* 26 (1971): 626–35.

4. See primarily Michael M. Sokal, ed., *An Education in Psychology: James McKeen Cattell's Journal and Letters from Germany and England, 1880–1888* (Cambridge: MIT Press, 1981), and the other articles and chapters cited here.

5. Daniel J. Levinson (with Charlotte N. Darrow, Edward B. Klein, Maria H. Levinson, and Braxton McKee), *The Seasons of a Man's Life* (New York: Alfred A. Knopf, 1978).

6. Daniel J. Levinson, "A Conception of Adult Development," *American Psychologist* 41 (1986): 3–13.

7. See, for example, Robert R. Sears, "Mid-Life Development," *Contemporary Psychology* 24 (1979): 97–98.

8. Levinson, *Seasons,* 91–93.

9. On William C. Cattell and Lafayette College, see *Memoir of William C. Cattell* (Philadelphia: Lippincott, 1899); Frank B. Copp, *Biographical Sketches of Some of Easton's Prominent Citizens* (Easton, Pa.: Hilburn & West, 1879), 193–205; David B. Skillman, *Biography of a College: Being a History of the First Century of Life of Lafayette College,* 2 vols. (Easton, Pa: Lafayette College, 1932).

10. On Francis A. March, see James W. Bright, "Address in Commemoration of Francis Andrew March, 1925–1911," *Publications of the Modern Language Association* 29 (1914): 1–24; Phyllis Franklin, "English Studies: The World of Scholarship in 1883," *Publications of the Modern Language Association* 99 (1984): 356–70.

11. See family correspondence, William C. Cattell papers, Presbyterian Historical Society, Philadelphia; "A Boy's Trip to Europe, 1869–1870," typescript copy of a diary kept by Israel Platt Pardee, D. B. Skillman Library, Lafayette College, Easton, Pa.

12. Seldon J. Coffin, *The Men of Lafayette, 1826–1893: Lafayette College. Its History. Its Men. Their Record* (Easton, Pa: George W. West, 1891), 9; Giacomo Barzellotti, *The Ethics of Positivism: A Critical Study* (New York: Charles B. Somerby, 1878). See also Sokal, *An Education in Psychology,* 16–18, and the highly positive opinion of Brooke E. Westcott, "Aspects of Positivism in Relation to Christianity," *The American Presbyterian and Theological Review,* new series, 6 (1868): 589–609.

13. See Sokal, *An Education in Psychology,* 96–105, 132–35. See also Sokal, "Graduate Study With Wundt: Two Eyewitness Accounts," in Wolfgang G. Bringmann and Ryan D. Tweney eds., *Wundt Studies: A Centennial Collection* (Toronto: C. J. Hogrefe, 1980), 210–25.

14. See Sokal, "Psychology at Victorian Cambridge—The Unofficial Laboratory of 1887–1888," *Proceedings of the American Philosophical Society* 116 (1972): 145–47.

15. Levinson, *Seasons,* 97–101; Sokal, *An Education in Psychology,* 218–23.

16. C. P. Snow, *The Masters* (New York: Charles Scribner's Sons, 1951), 185; Sokal, "The Unpublished Autobiography of James McKeen Cattell," 633–34; Cattell, "The Causes of the Declining Birth Rate," in *Proceedings of the First National Congress on Race Betterment* (Battle Creek, Mich.: The Race Betterment Foundation, 1914), 67–72; Cattell, "The Diminishing Family," *The Independent* 83 (1915): 422–24; Cattell, "The School and the Family," *Popular Science Monthly* 74 (1909): 84–95.

17. On Galton's interest in human variation and his anthropometric laboratory, see Karl Pearson, *The Life, Letters, and Labours of Francis Galton* (Cambridge: Cambridge University Press, 1914–1930), 2:357–86; Galton, "On the Anthropometric Laboratory at the Late International Health Exhibition," *Journal of the Anthropometric Institute* 14 (1885): 205–21.

18. See Sokal, "James McKeen Cattell and the Failure of Anthropometric Mental Testing, 1890–1901," in William R. Woodward and Mitchell G. Ash eds., *The Problematic Science: Psychology in Nineteenth-Century Thought* (New York: Praeger, 1982), 322–45.

19. See Sokal, "*Science* and James McKeen Cattell, 1894 to 1945," *Science* 209 (4 July 1980): 43–52.

20. Carol Singer Gruber, "Academic Freedom at Columbia University, 1917–1918: The Case of James McKeen Cattell," *AAUP Bulletin* 58 (1972): 297–305.

21. See Levinson, *Seasons,* 90–91.

22. F. Scott Fitzgerald, "The Rich Boy," reprinted in *All the Sad Young Men* (New York: Charles Scribner's Sons, 1926).

23. See James McKeen Cattell Journal, 13 April 1883: "Pres. Gilman has not taken as much interest in me, as he might have done," as quoted in Sokal, *An Education in Psychology,* 74.

24. For a popularized and greatly oversimplified version of Levinson's ideas, see Gail Sheehy, *Passages: Predictable Crises of Adult Life* (New York: Dutton, 1976). For the ways in *Passages* distorts and depends upon Levinson's work and other serious studies of adult life-span developmental psychology—such as Roger L. Gould, *Transformations: Growth and Change in Adult Life* (New York: Simon & Schuster, 1978); and George E. Vaillant, *Adaptation to Life* (Boston: Little, Brown, 1977);—see Diane White, "The Upbringing of Adult Males," *Boston Globe,* 21 May 1978, A9, A14.

25. Levinson, *Seasons,* 194–244.

26. Ibid., 200–208.

27. On Cattell's relations with Butler, see Gruber, "Academic Freedom at Columbia University;" on his relations with Baldwin, see James Mark Baldwin, *Between Two Wars, 1861–1921: Being Memories, Opinions, and Letters Received,* (Boston: Stratford, 1926), 1: 64–65.

28. See Sokal, "*Science* and James McKeen Cattell," 45–47.

29. Ibid., 48–50.

30. Sokal, "The Origins of the Psychological Corporation," *Journal of the History of the Behavioral Sciences* 17 (1981): 54–67; Sokal, "James McKeen Cattell and American Psychology in the 1920s," in Josef Brozek ed., *Explorations in the History of Psychology in the United States* (Lewisburg, Pa: Bucknell University Press, 1984), 273–323.

Collective Lives of American Scientists:
An Introductory Essay and a Bibliography
CLARK A. ELLIOTT

Part 1:
Introduction

Biography in the History of Science (General)

An approach to the history of science is possible from several directions, each of which is valid and none of which (ideally) is undertaken without incorporating aspects of the others. For example, a student or scholar can trace the development of a particular idea over a period of time, or examine the collective concerns of an established or incipient discipline within varying time frames. It is fruitful to examine the mutual relations of science with technology, religion, literature, and other areas of concern. The history of institutions of various types—journals, professional and learned societies, universities, industrial corporations, government—is an important aspect of scientific history. Other studies approach the subject from an analytical or sociological point of view; they examine the intellectual and social relationships among or within groups or categories of scientists, and in the context of the larger social, economic, and political sphere.

The study of the lives of scientists is one among these other approaches and one of the longest and best established means of investigating scientific history. Thomas Hankins has written of an apparent decline in the place of biography in the history of science.[1] Judging by recent recipients of the Pfizer Award from the History of Science Society,[2] however, and by the completion of the *Dictionary of Scientific Biography*,[3] biographical studies are holding their own. Rather than a decline, it is more likely that better biographies will follow, but increasingly this will be considered as only one historiographic approach. This paper is concerned with science prior to 1941. Nonetheless, it is a legitimate question for consideration by all historians whether the study of post-World War II science will require that the place of biography be reexamined. The reason is not because of changing fashions in historiography but because of changes in the organizational aspects of science itself. It is a question of whether "Big Science," and team research, leave a legitimate place for the study of individual lives.[4]

Granting the value of biography, at least for pre-World II science, and recognizing it as one of a number of approaches to historical study, one still is left with the notion that only leaders or contributors to scientific thought—and especially to its theoretical aspects—are legitimate or worthwhile subjects. Hankins, for example, states that "the philosopher-scientist" is an especially good subject for the biographical approach, while the "'little man'" is unlikely to be the subject of such a study.[5] In a somewhat similar vein, while recognizing the value of an "'ethnographic biography'" in the history of anthropology (and the history of science more generally), Jacob Gruber sees the ultimate arena for biography as that of intellectual history.[6]

Lives in American Science

An approach to biography that stresses leaders and theoretical work poses particular problems for the history of American science. With some notable exceptions, Americans made few significant contributions to theoretical science before this century. George Sarton wrote: "There is no American science, but there are American scientists.... The best way to explain American achievements is to focus the reader's attention upon a few of the leading scientists."[7] His point was not to denigrate American science in general but to argue that America must be seen in the context of international science. By asking, in effect, for emphasis on the leaders of science, however, Sarton overlaid his own conception of the history of science on the American frame, and in doing so clouded peculiar and in many ways more interesting features.

It is true that certain notable scientists emerged during every period of American history, but for each notable there were a dozen—a hundred—others who collected, arranged, and named insects or rocks, analyzed or manufactured chemical compounds, wrote textbooks, taught in the schools or colleges, recorded observations of the planets or the weather, surveyed the coast or the interior terrain, presided over a local natural history society, or directed a museum or government agency. These persons, as more typical, disclose a truer picture of the place of science in American society. Their individual and collective aspirations toward a scientific career and a self-constituted intellectual community touch American history more deeply and with a clearer ring of historical reality. The achievements for American science, during much of its history, were not in contributing to or participating in the development of an internationally constructed body of theoretical knowledge. For some time, the chief international involvement of Americans was to contribute data and material (e.g., biological specimens) that were interpreted by Europeans.

Biographical studies intersect the historiography of American science on three levels. No one would question the value of full and definitive biographies of persons such as Benjamin Franklin or Josiah Willard Gibbs. There are, of course, aspects of lives even on this plane that reveal much about the social and structural context of science in America, but here the emphasis is on matters of the mind. It approximates a purer form of intellectual history, that at its best is connected to the general ideational and societal climate of its time. It resembles Hankins's view of biography as pertinent particularly to the philosopher-scientist. Important as it is to investigate and to understand these lives, the historically significant accomplishment for American science prior to the twentieth century was at the next level, where achievement was marked through the establishment of necessary institutions and the legitimatization of career avenues that made possible the practice of science in America. The lives of a number of American scientists (perhaps of secondary rank on the international stage although leaders in America) were tied to the task of building those necessary institutions for the support of science. Sally Gregory Kohlstedt has observed, "Because many of the early leaders [of American science] established and then sustained organizations from which their lives and work seemed almost indistinguishable, the line between biography and institutional history has not been precisely drawn in the history of science."[8] The lives of such persons can take on a double value, focusing on their contributions to a body of knowledge, and to the development of organizations for the support and furtherance of scientific study and research. The lives of persons such as Joseph Henry, Louis Agassiz, and George Ellery Hale are links between the intellectual and social (organizational) history of science in America.

The third level of concern for biographical studies forms the base of the pyramid, which through its anchorage in the broader historical population of America may be conceived as having given support to the upper levels. If at any point in early American history there were relatively few leaders of scientific thought, or builders of scientific institutions of national importance, there were many Americans who wanted to understand the physical environment. Over time, there were increasing numbers who wanted to make scientific concerns the focal point of their lives. The story of science in America cannot be understood without reference to this base population whose historical significance is appreciated from several points of view. These perspectives from biography include (among others): the contributions of individuals to the scientific literature (even when they are inconsequential contributions to knowledge); the establishment and working of local, regional, and national institutions (educational, governmental, and scientific); the experience of individuals as they encountered the reality of America in their efforts to formulate for themselves the elements of a scientific career; and the revelation in individual lives of the relations of science to the institutions and the content of general learned culture, religion, medicine, and technology. Charles Rosenberg, in considering the place of biography in the history of American science, observed that the study of lives is an approach that provides "an inclusive way of framing questions and ordering data; it is an approach that necessarily includes social, institutional, and intellectual factors in a structured context."[9] The historiographic contributions of biography suggested above are applicable to lives at all levels of import, and the study of relative failure can reveal as much as unalloyed success. As an illustration of these points, the biographical factors in the recent study of black American scientist E. E. Just[10] have been delineated in terms of institutional history, personal aspects, social history, national differences in scientific style, and the subject's scientific work; the same review suggests (but does not pursue) the issue of the relations between Just's scientific abilities and the racial prejudice that he encountered.[11]

The general characteristics of the scientific population in America prior to World War II have been examined statistically, and I have published a bibliographic guide to the chief quantitative studies available at that time.[12] The same paper includes a statistical profile of the scientist during five successive historical periods, viewed in terms of birthplace, parental occupation, education, employment, and scientific fields. This was accompanied by brief sketches of five scientists, each representative of the five historical periods.[13] In that study based on lives of scientists is evident a cycle undoubtedly to be found in other aspects of American science and, while perhaps on a somewhat different time schedule, in other national contexts as well. It is the development of science from a period of dependency and integration with generalized learned culture during the colonial period (youth), to a time of development of relative independence within the insular walls of academia in the later nineteenth century (adolescence). This is followed by a period in this century when science is a functioning and contributing element within the economic, social, and political structure of the nation, but presumably subject also to degrees of imposed or of self denial that reflect the acknowledged need for science to serve a larger communal purpose (maturity).[14]

Definitions of a Scientist and Identification of the Population

An understanding of the evolving definition of the scientific population has been one of the central concerns among historians of American science, which contrasts with the concerns of historians in Europe. In its plainest view, the discussion has been in terms of amateurs and professionals, but Nathan Reingold has demonstrated the limitations of that view. For the crucial period of the mid-nineteenth century, he has constructed a three-part structure of researchers (generally but not necessarily in scientific employment), practitioners (ordinarily employed in science but not major publishers or contributors to scientific knowledge), and the cultivators of science.[15] The historical trend since the Civil War has been to make the social definition and therefore the identification of the scientist more certain. The central elements of that definition, characteristic of the scientific population after 1900, in Reingold's view are education, employment that draws upon the prescribed educational background, and membership in a professional association.[16]

Paralleling Reingold's straightforward operational definition of the scientist in the twentieth century is the availability of the serial directory, *American Men of Science,*[17] in which the scientific population of this century is well and probably largely represented. The situation for the colonial period and to the end of the nineteenth century is less certain because unequivocal definitions of the scientific population are difficult to construct, and because systematically compiled directories or lists are not available.[18]

Researchers who have attempted systematically to compile lists of American scientists for particular purposes have used several approaches or criteria. In the most obvious and straightforward approach, the compiler scans existing general biographical works for the names of scientists or persons involved in science. Two other primary strategies have been used, either as alternatives to or as supplements to this approach. They involve consulting membership lists for scientific and other learned societies and bibliographies of scientific works. The use of these two latter approaches constitutes an effective *ad hoc* definition of the American scientist population, insofar as the focus of interest is a self-selected group based on participant's action rather than an after-the-fact evaluative selection as represented by biographical works.

Society membership and publication as definitions of the scientist at various points in American history have been used by several researchers, to good effect. Margaret Young used publication as the chief criterion for identifying or designating colonial and early American scientists.[19] During the nineteenth century when the journal article came to characterize scientific communication, the Royal Society of London's *Catalogue of Scientific Papers,* which covers the entire nineteenth century, is a good means of developing a list of persons actively interested in science.[20] That source was used to compile a directory of pre–Civil War scientists included in my doctoral thesis.[21] In addition, I have in manuscript an author index to all articles published in American journals that are indexed in the Royal Society's *Catalogue,* covering the entire nineteenth century.[22]

Sally Gregory Kohlstedt examined the membership of the scientific community in the mid-nineteenth century as coterminous with membership or activity within the American Association for the Advancement of Science. Significantly, her criteria for leadership within the AAAS involved either publication *or* membership on committees.[23] James McKeen Cattell, in initially compiling names for *American Men of Science,*[24] found most convenient the use of membership lists of professional societies, which by the early twentieth century were more highly developed and specialized than they had been earlier.

The historical relations between society membership and publication, as characteristics of scientists, are intriguing, both on the individual and the collective levels. During the early colonial era there were no native scientific or learned societies, and later during that period the number of such societies was too restrictive to encompass the varied population of persons contributing to or seriously interested in science. However, it is probable for all periods of American history since the early nineteenth century, that society membership at one level or another is the more inclusive net, while publication (above a certain number) is the best qualitative gauge. The two approaches are largely interdependent,[25] making allowance for society officers who never published and published scientists who never were society members. The relations between the two activities warrant further study, but are beyond the immediate concerns of this paper. The central point in the present context is that lists of society members, and bibliographies of scientific writings (especially journal articles), at least for the period before 1900, should be used to supplement indexes to, or lists of names in, the available biographical reference works. This fact is especially germane for the study of science and scientists at the local or regional level, or within limited chronological periods. It is in these settings that interest most fruitfully could dwell on the seemingly peripheral characters whose names are not found in lists or directories of scientists that focus on the national or international level.

Biographical-Bibliographical Sources: An Historical Overview

The collective biographical literature for American science, as reflected in the bibliography to which this essay is a preface, is of two types. First, there are bibliographies and indexes to biographical sources. Secondly, there are works that in themselves contain biographical information. In both categories there are general sources in which scientists are included along with other groups of persons and sources that relate only to scientists (or to closely collateral groups). Each of the categories is of potential value for researchers seeking information on known persons. To a certain degree they also can be used in compiling lists of names of scientists not already known to the researcher To encompass the widest view of American science, however, (incorporating seemingly peripheral characters as discussed in the previous section), additional sources such as society membership lists, bibliographies of scientific writings, and the like would be necessary, and those sources are not listed directly in the bibliography.

A number of major and some second-rank American scientists have been the subjects of book-length, professionally authored biographies.[26] Many others are warranted. In the vast majority of cases, however, access to obituaries, memoirs, brief biographical data in local histories and elsewhere, or to sketches in biographical directories or dictionaries will suffice. Much of this literature is characteristic not only of science. It is part of the general output of variant biographical writings that flow through American history.

What follows is a brief historical survey of the American scientific biographical literature introductory to the bibliography in Part 2.

Widely dispersed obituaries, memoirs, and other biographical and auto-biographical items form the bedrock for the study of the lives of American scientists. These appear in many places—newspapers, professional and popular journals, proceedings of scientific and other societies, pamphlets, etc. The major scientific serial of the nineteenth century, *The American Journal of Science* (*Silliman's Journal*), included obituaries from the beginning. Many such miscellaneous biographical sources for the recognized scientists have found their way into bibliographies appended to the sketches in standard works such as the *Dictionary of American Biography*. While always to be approached with a degree of skepticism, obituaries, memoirs, and the like should not be overlooked as a direct source useful also (among other things) as a gauge of the scientist's reputation among contemporaries (allowing for the requirements of proper respect for the dead). Even the mere fact of authorship of an obituary, or its prominent (or obscure) placement in publication is a useful piece of evidence. Some of the most interesting (as well as many pedestrian) biographical sources are memoirs of minor scientific characters that appear in the publications of the local scientific societies. These sources often give a revealing grassroots view of American science.

Over the years, various bibliographies, indexes, and the like have appeared that permit access to the scattered obituary and other biographical literature of American science. With volume 7 (covering publications appearing 1864–73), the Royal Society's *Catalogue of Scientific Papers* began to index biographical items, listing them under the name of the subject. Max Meisel's masterful work on natural history lists a number of biographies,[27] and other specialized indexes or bibliographies are available. These supplement more recent general developments, such as the *Biography and Genealogy Master Index*[28] to biographical dictionaries. The latter represents a significant development in biographical research, in which commercial publishers are developing large on-going databases of biographical and bibliographical references.

The collective biography or biographical dictionary has a long history in Western culture, dating more or less in its modern form from the sixteenth and seventeenth centuries.[29] The writing of biography in America began in the mid-seventeenth century. Biographies in the colonial period usually were fairly brief (the longer and more definitive biography was the product of the nineteenth and twentieth centuries), and Cotton Mather's *Magnalia Christi Americana* was one of the first extensive works of collective biography, with sketches of varying length.[30] William Allen's 1809 volume was an early instance of an American biographical reference work.[31] There are sketches there of scientists such as John Bartram, Thomas Brattle, Mark Catesby, the two scientific John Winthrops, and others. The author, who cites sources for his information, states, in defense of the deficiencies of his work, that the modern compilers of similar works in Europe have little else to do but to combine or abridge the labors of their predecessors, and employ the materials previously collected to their hands. But in the compilation of this work a new and untrodden field was to be explored.[32]

This preface to Allen's work is a revealing statement of the demands on the collective biographer in the early stages of that genre in the United States, a development that culminated in this century with the *Dictionary of American Biography* (1928–).

Aspects of the history of science were addressed at an early date in the publication of collective biographies of physicians. These began in 1828 with the work of James Thacher,[33] produced at a time when science and medicine were more closely allied in American intellectual life. The preface to Thacher's work shows how firmly the moral lesson and pious example still attached itself to the function of biography. Yet Thacher, in compiling the work, recognized the primacy of fact over eulogy.[34]

Scientists in ninteenth-century America never formed a fraternity of profession to the degree that physicians did. In consequence, separate works of scientific biographical reference were slow to emerge. The appearance of collective works on American science, however, is an interesting historical footnote to the development of science in America. In imitation of European counterparts, the National Academy of Sciences began publication of its serial *Biographical Memoirs* in 1877, and it continues today as an essential bio-bibliographical source for the leaders of American science. One of the first retrospective works of collective American scientific biography appeared in 1896 under the editorship of William Jay Youmans.[35] The volume included fifty sketches, arranged chronologically by year of birth; it ended at 1810, with expressions of hope for a second volume never published. Ten years later James McKeen Cattell made the compilation and publication of biographical data on living scientists an adjunct to his work on psychological studies and a contribution to the organizational structure of science, with the appearance of the first edition of *American Men of Science*.[36] My biographical diction-ary, published seventy-three years thereafter, was presented as a retrospective companion to *American Men of Science*.[37]

Through the years, and through the *ad hoc* efforts of numerous persons and organizations, an impressive array of collective biographical works on American scientists have appeared. Although they vary greatly in type, content, and quality, taken together they constitute an effective source for the study of lives in American science. Among the more useful works are those listed in the bibliography that follows.

Notes

1. Thomas L. Hankins, "In Defence of Biography: The Use of Biography in the History of Science," *History of Science* 17 (1979): 2–3.

2. Since 1980, three of the annual awards have been for biographical studies: Sigmund Freud (1980), Isaac Newton (1983), and Ernest Everett Just (1984).

3. Charles Coulston Gillispie, ed., *Dictionary of Scientific Biography,* vols. 1–16 (New York: Scribner, 1970–1980).

4. This eventuality is considered in the context of documentation strategy for postwar science, in Clark A. Elliott, ed., *Understanding Progress as Process: Documentation of the History of Post-war Science and Technology in the United States: Final Report of the Joint Committee on Archives of Science and Technology* (HSS-SHOT-SAA-ARMA) (Chicago, Ill.: Distributed by the Society of American Archivists, 1983), especially 28–29.

5. Hankins, "In Defence of Biography," 8, 11.

6. Jacob Gruber, "In Search of Experience: Biography as an Instrument for the History of Anthropology," in *Pioneers of American Anthropology: The Uses of Biography,* edited by June Helm (Seattle and London: University of Washington Press, 1966), especially 22–23.

7. George Sarton, foreword to Bernard Jaffe, *Men of Science in America: The Role of Science in the Growth of Our Country* (New York: Simon and Schuster, 1944), xiii.

8. Sally Gregory Kohlstedt, "Institutional History," in "Historical Writing on American Science," edited by Sally Gregory Kohlstedt and Margaret W. Rossiter, *Osiris* 2d series vol. 1 (1985): 17–18.

9. Charles Rosenberg, "Science in American Society: A Generation of Historical Debate," *Isis* 74 (September 1983): 366.

10. Kenneth B. Manning, *Black Apollo of Science: The Life of Ernest Everett Just* (New York: Oxford University Press, 1983).

11. Jane Maienschein, "History of Biology," in "Historical Writing on American Science," edited by Kohlstedt and Rossiter, p. 153.

12. Clark A. Elliott, "Models of the American Scientist: A Look at Collective Biography," *Isis* 73 (March 1982): 77–93.

13. Colonial Period, Thomas Brattle (1658–1713); Early Republican Period, Benjamin Waterhouse (1754–1846); Antebellum period, Jeffreys Wyman (1814–74); Pre–World War I, Charles S. Hastings (1848–1932); Between the World Wars, Homer B. Adkins (1892–1949).

14. Elliott, "Models," 93.

15. Nathan Reingold, "Definitions and Speculations: The Professionalization of Science in America in the Nineteenth century," in *The Pursuit of Knowledge in the Early American Republic: American Scientific and Learned Societies from Colonial Times to the Civil War,* edited by Alexandra Oleson and Sanborn B. Brown (Baltimore: Johns Hopkins University Press, 1976), 33 (comparison to European historiography) and 38 (outline of the three groups of Americans in science).

16. Ibid., 37.

17. James McKeen Cattell, ed., *American Men of Science: A Biographical Directory* (New York: The Science Press, 1906), and subsequent issues.

18. I have in preparation a volume under the title *Biographical Index to American Science: The Seventeenth Century to 1920,* to be published by Greenwood Press. This work will include about 3,000 scientists who died before 1921 and give references to some standard biographical sources for each. Although it will, in fact, name a large number of the scientists for the period covered, it is by no means a complete census based on systematic definition. The primary source for names was the biographical works themselves, supplemented to a certain extent by lists of authors of journal articles in the nineteenth century.

19. Margaret Ann Young, "Early American Scientists, 1607–1818" (M. A. thesis, University of Oklahoma, 1957), 21.

20. Clark A. Elliott, "The *Royal Society Catalogue* as an Index to Nineteenth Century American Science," *Journal of the American Society for Information Science* 21 (November/December 1970): 396–401. This study is based on the first series of the *Catalogue,* covering the period 1800–1863.

21. Clark A. Elliott, "The American Scientist, 1800–1863: His Origins, Career and Interests" (Ph.D. thesis, Case Western Reserve University, 1970), 290–365.

22. My preliminary studies of these lists indicate that a very large percentage of the names do not appear in the standard biographical sources. Many of the authors individually would be of only the most marginal concern for historians of science in America, although the overall list could have interest from a quantitative point of view. More refined lists of scientists could be created by choosing those authors who published above a set number of articles.

23. Sally Gregory Kohlstedt, *The Formation of the American Scientific Community: The American Association for the Advancement of Science 1848–60* (Urbana and Chicago: University of Illinois Press, 1976).

24. See the preface to the first edition of *American Men of Science* (1906).

25. Kohlstedt, *Formation of the American Scientific Community*, 192, concluded, "Correlation of officers and publishing authors in the AAAS is . . . high."

26. See the index to Marc Rothenberg, *The History of Science and Technology in the United States: A Critical and Selective Bibliography,* Bibliographies of the History of Science and Technology, vol. 2 (New York and London: Garland, 1982).

27. Max Meisel, *A Bibliography of American Natural History: The Pioneer Century, 1769–1865* (Brooklyn, N.Y.: Premier Publishing Co., 1924–29).

28. Miranda C. Herbert and Barbara McNeil, eds., *Biography and Genealogy Master Index: A Consolidated Index to More Than 3,200,000 Biographical Sketches* (Gale Biographical Series, no. 1), 8 vols. (Detroit: Gale, 1980), with supplements.

29. Robert B. Slocum, *Biographical Dictionaries and Related Works* (Detroit: Gale, 1986), 14.

30. Dana Kinsman Merrill, *American Biography: Its Theory and Practice* (Portland, Maine: Bowker Press, 1957), 99–108.

31. William Allen, *An American Biographical and Historical Dictionary . . .* (Cambridge: Hilliard & Metcalf, Printers, 1809).

32. Ibid., iv.

33. James Thacher, *American Medical Biography,* 2 vols. (Boston: Richardson & Lord and Cottons & Barnard, 1828).

34. Ibid., 1: v–vi.

35. William Jay Youmans, ed., *Pioneers of Science in America: Sketches of Their Lives and Scientific Work,* reprinted with additions from The Popular Science Monthly (New York: D. Appleton and Company, 1896).

36. *American Men of Science* (1906).

37. Clark A. Elliott, *Biographical Dictionary of American Science: The Seventeenth through the Nineteenth Centuries* (Westport, Conn.: Greenwood Press, 1979).

Part 2:
Bibliography of Bibliographies, Indexes and Collective Lives of American Scientists Deceased Before 1941

What follows is an enumerative bibliography of works that aid in access to the biographical literature, or which are themselves works of concise biographical information. The bibliography is intended to relate especially to the study of American scientists deceased before 1941, although certain items will incorporate more recent scientists (and scientists from other countries). The bibliography is divided into several numbered sections which reflect the different types of collected biographical sources. These consist of: (1) general guides; (2) general and (3) science-related bibliographies and indexes; (4) general and (5) science-related biographical directories, dictionaries, and compilations; (6) collective biography; and (7) histories of science with a biographical emphasis.

1. General Guides to Biographical Sources

Slocum, Robert B. *Biographical Dictionaries and Related Works.* 2d ed. 2 vols. Detroit: Gale, 1986.

Terner, Janet R. *Biographical Sources in the Sciences*. Washington D.C.: Library of Congress, Science and Technology Division, Reference Section, 1978. 25 pp.

2. Bibliographies and Indexes—General

The American Genealogical-biographical Index to American Genealogical, Biographical and Local History Materials. Edited by Fremont Rider [through volume 53]. Middletown, Conn.: Godfrey Memorial Library, 1952–, vol. 1–. In progress; vol. 1–151, A–Sabin appeared as of 1988. Earlier edition published, complete in 48 volumes, under title *The American Genealogical Index* (1952).

Arnim, Max. *Internationale Personalbibliographie 1800–1943*. 2d. ed. 2 vols. Leipzig and Stuttgart: Karl W. Hiersemann, 1944–1952. Volume 3, 1944–1959 (1961–63). Volume 3–[5], 1944–1975, A–[T] (1981–), in progress.

Beers, Henry P., comp. *Bibliographies in American History, 1942–1978: Guide to Materials for Research*. 2 vols. Woodbridge, Conn.: Research Publications, Inc., 1982. See chapter 11: "Biography and Genealogy," 425[426]–511.

Biographical Books 1876–1949: Vocation Index, Name/Subject Index, Author Index, Title Index. New York and London: R. R. Bowker, 1983. 1768 pp.

Biographical Books 1950–1980: Vocation Index, Name/Subject Index, Author Index, Title Index, Biographical Books in Print Index. New York: R. R. Bowker, 1980. 1557 pp.

Biography Index: A Cumulative Index to Biographical Material in Books and Magazines. New York: Wilson, 1947–, vol. 1–.

Carrell, William B. "Biographical List of American College Professors to 1800." *History of Education Quarterly* 8 (1968): 358–74.

Dargan, Marion. *Guide to American Biography: Part 1—1607–1815; Part 2—1815–1933*. 2 vols. Albuquerque: University of New Mexico Press, 1949, 1952. Reprinted: Westport, Conn.: Greenwood Press, 1973. 510 pp.

Herbert, Miranda C., and Barbara McNeil, eds. *Biography and Genealogy Master Index: A Consolidated Index to More Than 3,200,000 Biographical Sketches*. 8 vols. Gale Research Series, no. 1. Detroit: Gale, 1980. Also supplementary volumes. Issued also in a cumulative microfiche form under the title *Bio-base* and on-line by way of DIALOG Information Services, Inc. as *Biography Master Index (BMI)*.

Herman, Kali. *Women in Particular: An Index to American Women*. Phoenix, Ariz.: Oryx Press, 1984. 740 pp.

O'Neill, Edward Hayes. *Biography by Americans, 1658–1936: A Subject Bibliography*. Philadelphia: University of Pennsylvania Press; London: Milford, 1939. 465 pp.

Riches, Phyllis M. *Analytical Bibliography of Universal Collected Biography,*

Comprising Books Published in the English Tongue in Great Britain and Ireland, America and the British Dominions. . . . London: Library Association, 1934. 709 pp. Reprinted: Detroit: Gale, 1980.

Stetler, Susan L., ed. *Biography Almanac: A Comprehensive Reference Guide to More than 24,000 Famous and Infamous Newsmakers from Biblical Times to the Present as Found in Over 550 Readily Available Biographical Sources.* 3d. ed. 3 vols. Detroit: Gale, 1987.

3. Bibliographies and Indexes—Science and Related Subjects

Barr, Earnest Scott. *An Index to Biographical Fragments in Unspecialized Scientific Journals.* University: University of Alabama Press, 1973. 294 pp.

Bearman, David, and John T. Edsall, eds. *Archival Sources for the History of Biochemistry and Molecular Biology: A Reference Guide and Report.* Boston: American Academy of Arts and Sciences; Philadelphia: American Philosophical Society, 1980. 388 pp. and 6 microfiche. See "Introduction to the Bibliographic file," 37–38; "Name Index to the Bibliographic File," 309–22; and 2 related microfiche.

Bell, Whitfield J., Jr. *Early American Science: Needs and Opportunities for Study.* Williamsburg, Va.: Institute of Early American History and Culture, 1955. 85 pp. See Part 5: "Fifty Early American Scientists: Selected Bibliographies," 45–80.

Boivin, B. "A Basic Bibliography of Botany Biography, and a Proposal for a More Elaborate Bibliography." *Taxon* 26 (1977): 75–105, 603–11.

Bolton, Henry C. "Index to Scientific Portraits in the 'Popular Science Monthly,' Vols. I–XXXV, 1872–1889." *Library Journal* 14 (November 1889): 435–36.

———. *Select Bibliography of Chemistry, 1492–1904.* 4 vols. Smithsonian Miscellaneous Collections, vols. 36, 39, 41(3), 44(5) Washington, D.C.: Smithsonian Institution, 1893–1904. Reprinted, Volume for 1492–1892: New York: Kraus Reprint Corp., 1966.

Brush, Stephen G., and Lanfranco Belloni. *The History of Modern Physics: An International Bibliography.* Bibliographies of the History of Science and Technology, vol. 4. New York: Garland, 1983. 334 pp.

Carnegie Library. *Men of Science and Industry: A Guide to the Biographies of Scientists, Engineers, Inventors, and Physicians in the Carnegie Library of Pittsburgh.* Pittsburgh: Carnegie Library, 1915. 189 pp.

Carpenter, Mathilde M. "Bibliography of Biographies of Entomologists." *The American Midland Naturalist* 33 (1945): 1–116; 50 (1953): 257–348.

Corgan, James X., and Sharon R. Tanner. "Biographical Data in Serial Publications of the Tennessee Academy of Science." *Journal of Tennessee Academy of Science* 54, no. 1 (1979): 5–10.

DeVorkin, David H. *The History of Modern Astronomy and Astrophysics: A*

Selected, Annotated Bibliography. Bibliographies of the History of Science and Technology, vol. 1, New York: Garland, 1982. 434 pp.

Edwards, Everett E. *A Bibliography of the History of Agriculture in the United States.* U.S. Department of Agriculture Miscellaneous Publication no. 84. Washington, D.C.: Government Printing Office, 1930. Reprinted: New York: Franklin, 1970. 307 pp. See "Agricultural Leaders," 218–233.

Elliott, Clark A. *Biographical Index to American Science: The Seventeenth Century to 1920.* Westport, Conn.: Greenwood Press. In progress.

Fahl, Ronald J. *North American Forest and Conservation History: A Bibliography.* Santa Barbara, Calif.: Clio Press for the Forest History Society, 1977. 408 pp. For biographies, see individual names in the index.

Fruton, Joseph S. *A Bio-bibliography for the History of the Biochemical Sciences Since 1800.* Philadelphia: American Philosophical Society, 1982. 886 pp. Also, *A supplement,* 1985.

Gascoigne, Robert Mortimer. *A Historical Catalogue of Scientists and Scientific Books: From the Earliest Times to the Close of the Nineteenth Century.* New York and London: Garland, 1984. 1177 pp.

Gilbert, Pamela. *A Compendium of the Biographical Literature on Deceased Entomologists.* Publications-British Museum (Natural History) no. 786. London: British Museum (Natural History), 1977. 455 pp.

Grainger, Thomas H., Jr. *A Guide to the History of Bacteriology.* Chronica Botanica no. 18. New York: Ronald Press, 1958. 210 pp. See biographical listings, 135–80.

Heilbron, J. L., and Bruce R. Wheaton. *Literature on the History of Physics in the 20th Century.* Berkeley Papers in History of Science. Berkeley: Office for History of Science and Technology, University of California, 1981. 485 pp.

Higgins, Thomas James. "A Biographical Bibliography of Electrical Engineers and Electrophysicists." *Technology and Culture* 2 (1961): 28–32, 146–65.

————. "Biographies and Collected Works of Mathematicians." *American Mathematical Monthly* 51 (October 1944): 433–45; 56 (May 1949): 310–12.

————. "Book-length Biographies of Chemists." *School Science and Mathematics* 44 (October 1944): 650–65; 48 (June 1948): 438–40; 65 (February 1965): 139–42.

————. "Book-length Biographies of Engineers, Metallurgists and Industrialists." *Bulletin of Bibliography* 18 (1946): 207–10, 235–39; 19 (1947): 10–12.

————. "Book-length Biographies of Physicists and Astronomers." *American Journal of Physics* 12 (1944): 31–39, 234–36.

Holloway, Lisabeth M. *Medical Obituaries: American Physicians' Biographical Notices in Selected Medical Journals before 1907.* Garland Reference Library of Social Science, 104. New York: Garland, 1981. 513 pp. Incorporates brief biographical data with bibliographic entries.

Hoy, Suellen M., and Michael C. Robinson. *Public Works History in the United States: A Guide to the Literature.* Nashville, Tenn.: American Association for State and Local History, 1982. 477 pp. Biographies are listed in topical sections by author but are not indexed.

Ireland, Norma. *Index to Scientists of the World from Ancient to Modern Times: Biographies and Portraits.* Useful Reference Series no. 90. Boston: Faxon, 1962. 662 pp.

Isis Cumulative Bibliography: A Bibliography of the History of Science Formed from Isis Critical Bibliographies. . . . London: Mansell, in conjunction with the History of Science Society, 1971–. See *1913–65,* vols. 1–2; *1966–1975,* vol. 1.

John Crerar Library. *Author-title Catalog.* 35 vols. Boston: G. K. Hall, 1967. Includes biographies under name of scientist.

Lambrecht, Kalman, and W. Quenstedt. *Palaeontologi: Catalogus bio-bibliographicus.* Fossilium catalogus. I. Animalia, edited by W. Quenstedt, pars 72. 's-Gravenhage: W. Junk, 1938. 495 pp.

May, Kenneth Ownsworth. *Bibliography and Research Manual on the History of Mathematics.* Toronto: University of Toronto Press, 1973. 818 pp.

Meisel, Max. *A Bibliography of American Natural History: The Pioneer Century. 1769–1865.* 3 vols. Brooklyn, New York: Premier Publishing Co., 1924–29. Reprinted: New York: Hafner Publishing Co., 1967.

Miller, Genevieve. *Bibliography of the History of Medicine of the United States and Canada, 1939–1960.* Baltimore: Johns Hopkins University Press, 1964. 428 pp.

Mitterling, Philip I. *United States Cultural History: A Guide to Information Sources.* Gale Information Guide Library: American Government and History Information Guide Series, vol. 5. Detroit: Gale, 1979. 581 pp.

New York Academy of Medicine. *Catalog of Biographies in the Library of the New York Academy of Medicine.* Boston: G. K. Hall, 1960. 165 pp.

Oehser, Paul H. "A Handlist of American Naturalists Based on the Dictionary of American Biography." *American Naturalist* 72 (1938): 534–46.

Pelletier, Paul A. *Prominent Scientists: An Index to Collective Biographies.* 2d ed. New York: Neal-Schuman, 1985. 356 pp.

Poggendorff, Johann C. *Biographisch-literarisches Handwörterbuch zur Geschichte der exakten Wissenschaften.* Leipzig: Barth, 1863–1904; Leipzig: Verlag Chemie, 1925–1940; Berlin: Akademie-Verlag, 1955–. Title varies. Reprinted, Band 1–6, to 1931: Ann Arbor, Mich.: Edwards, 1945.

Porter, Roy. *The Earth Sciences: An Annotated Bibliography.* Bibliographies of the History of Science and Technology, vol. 3. New York: Garland, 1983. 192 pp.

Pure and Applied Science Books 1876–1982. 6 vols. New York: R. R. Bowker, 1982.

Rothenberg, Marc. *The History of Science and Technology in the United States: A Critical and Selective Bibliography.* Bibliographies of the History of Science and Technology, vol. 2. New York and London: Garland, 1982. 242 pp.

Royal Society of London. *Catalogue of Scientific Papers, 1800–1900.* 19 vols. London: Clay, 1867–1902; Cambridge: University Press, 1914–25. See vol. 7 (1864–73) and subsequent volumes, listing biographies under the name of the scientist.

Sarjeant, William A. S. *Geologists and the History of Geology: An International Bibliography from the Origins to 1978.* 5 vols. New York: Arno, 1980. Also: Supplement, 1987.

Sarton, George. "Historians and Philosophers of Science: Biographies Available in *Isis* (1–45) and *Osiris* (1–11)." *Isis* 46 (December 1955): 360–66.

Smit, Pieter. *History of the Life Sciences: An Annotated Bibliography.* New York: Hafner Press, 1974. 1036 columns and pp. 1037–71.

Stuckey, Ronald. "Index to Biographical Sketches and Obituaries in Publications of the Ohio Academy of Sciences 1900–1970." *Ohio Journal of Science* 70 (1970): 246–55.

Thornton, John L. *A Select Bibliography of Medical Biography. With an Introductory Essay on Medical Biography.* 2d ed. London: Library Association, 1970. 170 pp.

Tullis, Carol. "Black American Scientists." In *Black American Biography, Black American Scientists, Black Americans in Public Affairs,* Robert Swisher, Carol Tullis, and Richard Hicks, 15–37. Indiana University/ Focus: Black American Bibliography Series. Bloomington: Indiana University Libraries and Focus: Black America, Summer 1969. 52 pp.

U. S. Geological Survey. *Bibliography of North American Geology. 1785–1918. 1919–1928. 1929–1939. 1940–1949.* Washington, D.C.: Government Printing Office, 1896–. Issued as bulletins of the U. S. Geological Survey. Title varies. First section: John M. Nickles, *Geologic Literature on North America, 1785–1918.*

U.S. National Library of Medicine. *Index Catalogue of the Surgeon General's Office, U. S. Army (Army Medical Library), Authors and Subjects.* Washington, D.C.: Government Printing Office, 1880–, series 1–. Series 1–4, vols. 1–11 (in 58 vols), 1880–1955; Series 5, vols. 1–3, 1959–1961.

Wade, J. S. "A Bibliography of Biographies of Entomologists with Special Reference to North American Workers." *Annals of Entomological Society of North America* 21 (September 1928): 489–520.

Watson, Robert Irving. *Eminent Contributors to Psychology.* 2 vols. New York: Springer Publishing Co., 1974–76.

Wells, John W., and George W. White. "Biographies of Geologists." *Ohio Journal of Science* 58 (1958): 285–98.

4. Biographical Directories, Dictionaries, and Compilations—General

Adams, Oscar Fay. *A Dictionary of American Authors.* 5th ed., revised and enlarged. Boston and New York: Houghton, Mifflin, 1905.

Allibone, S. Austin. *A Critical Dictionary of English Literature and British and American Authors Living and Deceased from the Earliest Accounts to the Latter Half of the Nineteenth Century.* 3 vols. Philadelphia: J. B. Lippincott & Co., 1858–71. Also: Kirk, John Foster. *A Supplement to Allibone's Critical Dictionary of English Literature and British and American Authors.* 2 vols. Philadelphia: J. B. Lippincott & Co., 1891. Reprinted (all volumes): Detroit: Gale, 1965.

The American Biographical Archive. New York: K. G. Saur, Inc., 1986–[88]. Will consist of a projected 1,200 microfiche and printed index. Covering the period up to the early twentieth century, entries from 367 biographical reference works are arranged in a single alphabetical sequence.

Cullum, George Washington. *Biographical Register of the Officers and Graduates of the U. S. Military Academy at West Point, N. Y., from Its Establishment in 1802, to 1890.* 3 vols. 3rd ed., revised and extended. Boston: Houghton, Mifflin, 1891. Also: Supplement, 1890/1900-, vol. 4–.

Dictionary of American Biography. 20 vols. Published under the auspices of the American Council of Learned Societies. New York: Scribner, 1928–1937. Also: Supplements, 1944–, 1–.

Downs, Robert B., John T. Flanagan, and Harold W. Scott. *Memorable Americans, 1750–1950.* Littleton, Colorado: Libraries Unlimited, 1983. 379 pp.

Hinding, Andrea, ed. *Women's History Sources: A Guide to Archives and Manuscript Collections in the United States.* 2 vols. New York: R. R. Bowker, 1980. In addition to information on archival and manuscript collections, also gives brief data or biographical hints for women who are the subjects of the collections. Especially useful for little known persons.

James, Edward T., ed. *Notable American Women 1607–1950.* 3 vols. Cambridge: Belknap Press, Harvard University Press, 1971.

Levernier, James A., and Douglas R. Wilmes, eds. *American Writers before 1800: A Biographical and Critical Dictionary.* 3 vols. Westport, Conn.: Greenwood Press, 1983.

Nason, Henry B., ed. *Biographical Record of the Officers and Graduates of the Rensselaer Polytechnic Institute, 1824–1886.* Troy, N.Y.: William H. Young, 1887. 614 pp.

National Cyclopaedia of American Biography. New York: J. T. White, 1892–94. vols. 1–N63.

Ohles, John F., ed. *Biographical Dictionary of American Educators.* 3 vols. Westport, Conn.: Greenwood Press, 1978.

Ruffner, James A., ed. *Eponyms Dictionaries Index: A Reference Guide to Persons Both Real and Imaginary, and the Terms Derived from Their Names....* Detroit: Gale, 1977. 730 pp.

U. S. Library of Congress. *National Union Catalog of Manuscript Collections, 1959–*. Ann Arbor, Mich.: J. W. Edwards, Publisher, 1962; Hamden, Conn.: Shoe String Press, Inc., 1964; Washington, D.C.: Library of Congress, 1965–. Frequently gives brief biographical identifications of creators or main subjects of collections. See *Index to Personal Names in the National Union Catalog of Manuscript Collections 1959–1984*. 2 vols. Alexandria, Va.: Chadwyck-Healey, 1988.

Wallace, William S. *A Dictionary of North American Authors Deceased before 1950*. Toronto: Ryerson Press, 1951. 525 pp. Reprinted; Detroit: Gale, 1968.

Wilson, J. G., and John Fiske, eds. *Appleton's Cyclopaedia of American Biography*. 7 vols. Now York: Appleton, 1894–1900. There also are other editions of this work.

5. Biographical Directories, Dictionaries, and Compilations—Science and Related Subjects

Abbott, Robert Tucker, ed. *American Malacologists: A National Register of Professional and Amateur Malacologists and Private Shell Collectors and Biographies of Early American Mollusk Workers Born Between 1618 and 1900*. Falls Church, Va.: American Malacologists, 1973. 494 pp.

American Men of Science: A Biographical Directory. New York, N.Y.; Garrison, N.Y.; Lancaster, Pa.: Science Press, 1906–.

American Men and Women of Science Editions 1–14 Cumulative Index. New York and London: R. R. Bowker, 1983. 847 pp.

American Society of Civil Engineers. *A Biographical Dictionary of Amercian Civil Engineers*. ASCE Historical Publication no. 2. New York: Committee on History and Heritage of American Civil Engineering, American Society of Civil Engineers, 1972. 163 pp.

American Society of Mechanical Engineers. *Mechanical Engineers in America Born Prior to 1860: A Biographical Dictionary*. Sponsored by the History and Heritage Committee. New York: American Society of Mechanical Engineers, 1980. 330 pp.

Asimov, Isaac. *Asimov's Biographical Encyclopedia of Science and Technology: The Lives and Achievements of 1510 Great Scientists from Ancient Times to the Present Chronologically Arranged*. 2d revised ed. Garden City, N.Y.: Doubleday and Co., Inc., 1982. 941 pp.

Atkinson, William B. *Biographical Dictionary of Contemporary American Physicians and Surgeons*. 2d enlarged ed. Philadelphia: D. G. Brinton, 1880. 747 + 21 pp.

The Auk. *Biographies of Members of the American Ornithological Union*. By T. S. Palmer and others. Washington, D.C.: 1954. 630 pp.

Barnhart, John H., comp. *Biographical Notes upon Botanists, Maintained in the New York Botanical Garden Library*. 3 vols. Boston: G. K. Hall, 1965.

Bulloch, William. *The History of Bacteriology*. London: Oxford University Press, 1938. 422 pp. Reprinted: New York: Dover Publications, 1979. See "Biographical Notices of Some of the Early Workers in Bacteriology," 349–406.

Casey, Albert Eugene. *Biographical Encyclopedia of Pathologists: Southern United States of America. Persons Trained in Pathology before 1937 and Resident in the South before the Golden Anniversary Meeting of the Southern Medical Association, Washington, D.C., 12–15 November, 1956; and Including Observations on the Training for Research, Teaching, and Practice in Pathology*. Birminghan, Ala.: Published for Memorial Institute of Pathology by the Amite and Knocknagree Historical Fund, 1963. 920 pp. (Pp. 377–[806] are photo-reproduced on [105] pp.)

Chemical Who's Who. New York: Lewis Historical Co., 1928–[56]. Title and name of publisher vary; 1st ed., 1928, *Who's Who in the Chemical and Drug Industries*.

Cortada, James W. *Historical Dictionary of Data Processing—Biographies*. Westport Conn.: Greenwood Press, 1987. 321 pp.

Daintith, John, Sarah Mitchell, and Elizabeth Tootill. *Chambers Biographical Encyclopedia of Scientists*. Edinburgh: W & R Chambers, 1983. 599 pp. First published by Facts on File, Inc., 1981, to which material has been added.

Daniels, George H. *American Science in the Age of Jackson*. New York and London: Columbia University Press, 1968. 282 pp. See Appendix I: "Biographical and Bibliographical Sketches of Fifty-five Leading American Scientists of the Period 1815–1845," 201–28.

Davis, Henry B. O. *Electrical and Electronic Technologies: A Chronology of Events and Inventors to 1900*. Metuchen, N.J.: Scarecrow, 1981. 213 pp.

Davis, Richard C., ed. *Encyclopedia of American Forest and Conservation History*. 2 vols. Forest History Society. Riverside, N.J.: Macmillan Publishing Co., 1983.

Dunlap, Orrin Elmer. *Radio's 100 Men of Science: Biographical Narratives of Pathfinders in Electronics and Television*. New York and London: Harper and Bros., Publishers, 1944. 294 pp. Reprinted: Freeport, N.Y.: Books for Libraries Press, 1970.

Elliott, Clark A. "The American Scientist, 1800–1863: His Origins, Career and Interests." (Ph.D. diss., Case Western Reserve University, 1970). 383 pp. (Xerox University Microfilm order no. 71–01685). See Appendix VII: "Biographical Directory of Scientists," 290–365.

———. *Biographical Dictionary of American Science: The Seventeenth through the Nineteenth Centuries*. Westport, Conn.: Greenwood Press, 1979. 360 pp.

Ewan, Joseph, and Nesta Dunn Ewan. *Biographical Dictionary of Rocky Mountain Naturalists: A Guide to the Writings and Collections of Botanists, Zoologists, Geologists, Artists and Photographers 1682–1932*.

Regnum Vegetabile 107. Utrecht and Antwerp: Bohn, Scheltema and Holkema; The Hague and Boston: W. Junk; distributed by Kluwer Boston, Inc., 1981. 253 pp. A revision of Joseph Ewan's *Rocky Mountain Naturalists* (1950).

Farber, Eduard, ed. *Great Chemists*. New York and London: Interscience Publishers, 1961. 1642 pp. Coverage is international and from ancient times.

Freeman, T. W., Marguerita Oughton, and Philippe Pinchemel, eds. *Geographers: Biobibliographical Studies*. [London]: Mansell, 1977–.

Geiser, Samuel W. *Men of Science in Texas, 1820–1880*. Dallas: Southern Methodist University Press, 1958. 256 pp.

Gillispie, Charles Coulston, ed. *Dictionary of Scientific Biography*. New York: Scribner, 1970–[80-], vols. 1–[16–].

Gross, Samuel D., ed. *Lives of Eminent American Physicians and Surgeons of the Nineteenth Century*. Philadelphia: Lindsay and Blakeston, 1861. 836 pp.

Guralnick, Stanley M. *Science and the Ante-bellum American College*. APS Memoirs vol. 109. Philadelphia: American Philosophical Society, 1975. See Appendix V: "Biographies of Science Professors," 167–221.

Harshberger, John W. *The Botanists of Philadelphia and Their Work*. Philadelphia: [Press of T. C. Daivs & Son], 1899. 457 pp.

Howard, Arthur V. *Chambers Dictionary of Scientists*. New York: Dutton, 1966. 500 columns. First issued: London: W & R Chambers, 1951.

Humphrey, Harry B. *Makers of North American Botany*. New York: Ronald Press Co., 1961. 265 pp.

Hunt Institute for Botanical Documentation. *Biographical Dictionary of Botanists Represented in the Hunt Institute Portrait Collection*. Boston: G. K. Hall, 1972. 451 pp.

Kaufman, Martin, Stuart Galishoff, and Todd L. Savitt, eds. *Dictionary of American Medical Biography*. 2 vols. Westport, Conn.: Greenwood Press, 1984.

Kelly, Howard A., and Walter L. Burrage. *Dictionary of American Medical Biography*. New York: Appleton, 1928. 1364 pp. Reprinted: Boston: Milford House, 1971.

Killeffer, David Herbert. *Eminent American Chemists: A Collection of Portraits of Eminent Americans in the Field of Chemistry from the Earliest Days of the Republic to the Present Together with Short Sketches of the Work of Each*. New York: D. H. Killeffer, 1924. 33 pp.

Kohlstedt, Sally Gregory. *The Formation of the American Scientific Community: The American Association for the Advancement of Science, 1848–1860*. Urbana; University of Illinois Press, 1976. 246 pp. and unpaginated appendix. See appendix: "Biographical Directory of the AAAS, 1848-60."

Lawry, John D. *Guide to the History of Psychology*. Totowa, N.J.:

Littlefield, Adams, 1981. 114 pp. See Section I: "Innovators," 1–50.

Library-Anthropology Resource Group (LARG), comp. *Biographical Directory of Anthropologists Born before 1920.* Edited by Thomas L. Mann. New York: Garland, 1988. 245 pp. International in coverage.

Miles, Wyndham D., ed. *American Chemists and Chemical Engineers.* Washington, D.C.: American Chemical Society, 1976. 544 pp.

Murchison, Carl, ed. *Psychological Register.* Worcester, Mass.: Clark University Press, 1929–32. vols. 2–3 [retrospective section, vol. 1, never published].

National Academy of Sciences. *Biographical Memoirs.* 1877–, 1–.

The Naturalists' Directory. Salem, Mass.; Boston, Mass.: Cassino Press; [etc.], 1877–. Title varies: *The International Scientists' Directory,* 1881/82–1882/83, 1885, 1888; *The Scientists' International Directory,* 1892, 1894, 1896; *The Naturalists' Universal Directory,* 1904–?

Ogilvie, Marilyn Bailey. *Women in Science: Antiquity through the Nineteenth Century: A Biographical Dictionary with Annotated Bibliography.* Cambridge: MIT Press, 1986. 254 pp.

Osborn, Herbert. *A Brief History of Entomology, Including Time of Demosthenes and Aristotle to Modern Times, With Over 500 Portraits.* Columbus, Ohio: Spahr and Glenn, 1952. 303 pp. See Part III: "Personal Sketches Arranged Alphabetically," 173–236, and portraits, 237–94.

Roysdon, Christine, and Linda A. Kharti. *American Engineers of the 19th Century: A Biographical Index.* New York: Garland, 1978. 247 pp.

Schapsmeier, Edward L., and Frederick H. Schapsmeier. *Encyclopedia of American Agricultural History.* Westport, Conn.: Greenwood Press, 1975. 467 pp.

Siegel, Patricia Joan, and Kay Thomas Finley. *Women in the Scientific Search: An American Bio-bibliography, 1724–1979.* Metuchen, N.J., and London: Scarecrow Press, Inc. 1985. 399 pp.

Smart, Charles E. *The Makers of Surveying Instruments in America since 1700.* 2 vols. Troy, N.Y.: Regal Art Press, 1962–67. Note: Vol. 2 was not seen.

Smithsonian Institution. *Guide to the Smithsonian Archives.* Archives and Special Collections of the Smithsonian Institution 4. Washington, D.C.: Smithsonian Institution Press, 1983. 431 pp. Gives brief biographical information as well as descriptions of manuscript collections.

Stearns, Raymond P. "Colonial Fellows of the Royal Society of London, 1661–1788." *Notes and Records of the Royal Society of London* 8, no. 2 (April 1951): 178–246.

Sterling, Keir B., ed. *Biographical Dictionary of North American Environmentalists.* Westport, Conn.: Greenwood Press. In progress.

Stone, Richard French, ed. *Biography of Eminent American Physicians and Surgeons.* Indianapolis: Carlon & Hollenbeck, 1894. 729 pp. Note: A second edition was issued 1898 but was not seen.

Stroud, Richard H., ed. *National Leaders of American Conservation.*

Sponsored by the Natural Resources Council of America. Washington, D.C.: Smithsonian Institution Press, 1985. 432 pp. Earlier (1st) edition published as: Clepper, Henry Edward, comp. *Leaders of American Conservation* (1971).

Thacher, James. *American Medical Biography; or, Memoirs of Eminent Physicians Who Have Flourished in America, to Which Is Prefixed a Succinct History of Medical Science in the United States from the First Settlement of the Country.* 2 vols. Boston: Richardson & Lord and Cottons & Barnard, 1828. Reprinted: New York: Milford House, 1967.

Turner, Roland, and Steven L. Goulden, eds. *Great Engineers and Pioneers in Technology.* Vol. 1, *From Antiquity through the Industrial Revolution.* New York: St. Martin's Press, 1981. 488 pp. Three volumes are projected.

Who Was Who in American History: Science and Technology. A component of *Who's Who in American History.* Chicago: Marquis's Who's Who, 1977. 688 pp.

Who's Who in Engineering: A Biographical Dictionary of the Engineering Profession. 1st–9th eds. New York: Lewis Historical Publishing Co., 1922/23–1964.

Williams, Stephen West. *American Medical Biography; or, Memoirs of Eminent Physicians; Embracing Principally Those Who Have Died since the Publication of Dr. Thacher's Initial Work in 1828 on the Same Subject.* Greenfield, Mass.: L. Merriam and Co., 1845. 664 pp. Reprinted: New York: Milford House, 1967.

Williams, Trevor I., ed. *A Biographical Dictionary of Scienists.* 3d ed. New York: John Wiley & Sons, 1982. 674 pp.

World Who's Who in Science: A Biographical Dictionary of Notable Scientists from Antiquity to the Present. Chicago: Marquis-Who's Who, 1968. 1855 pp.

Young, Margaret Ann. "Early American Scientists, 1607–1818." (Master's thesis, University of Oklahoma, 1957). 192 pp. Available from University of Oklahoma Library, Norman. See Appendix I: "Brief Sketches of the Scientists," 65–129; and Appendix II: "Checklist of Persons Not Included in the Main List of Scientists," 130–92.

Zusne, Leonard. *Biographical Dictionary of Psychology.* Westport, Conn.: Greenwood Press, 1984. 563 pp. This is a 2d ed. of his *Names in the History of Psychology* (1975).

6. Collective Biography

Alden, Roland H., and John D. Ifft. *Early Naturalists in the Far West.* Occasional Papers of the California Academy of Sciences no. 20. San Francisco: California Academy of Sciences, 1943. 59 pp.

Crowther, James Gerald. *Famous American Men of Science.* New York: W. W. Norton and Co., Inc., 1937. 414 pp.

Dies, Edward Jerome. *Titans of the Soil: Great Builders of Agriculture.*

Chapel Hill: University of North Carolina Press, 1949. 213 pp. Reprinted: Westport, Conn.: Greenwood Press, 1976.

Elman, Robert. *First in the Field: America's Pioneering Naturalists.* New York: Mason/Charter, 1977. 231 pp.

Gee, Wilson. "South Carolina Botanists: Biography and Bibliography." *South Carolina Bulletin* no. 72 (1918). 52 pp.

Geiser, Samuel Wood. *Naturalists of the Frontier.* 2d. ed. revised and enlarged. Dallas, Texas: University Press, Southern Methodist University, 1948. 296 pp.

Goddard, Dwight. *Eminent Engineers: Brief Biographies of 32 of the Inventors and Engineers Who Did Most to Further Mechanical Progress.* New York: Derry-Collard, Co., 1906. 280 pp.

Hanley, Wayne. *Natural History in America: From Mark Catesby to Rachel Carson.* New York: Quadrangle/The New York Times Book Co., 1977. 339 pp.

Harris, Jonathan. *Scientists in the Shaping of America.* Specialized Studies in American History Series. Menlo Park, Calif.: Addison-Wesley Publishing Co., 1971. 142 pp.

Haynes, Williams. *Chemical Pioneers: The Founders of the American Chemical Industry.* New York: D. Van Nostrand Co., 1939. 288 pp.

Hume, Edgar Erskine. *Ornithologists of the United States Army Medical Corps.* Publications of the Institute of the History of Medicine, Johns Hopkins University, 1st series: Monographs, vol. 1. Baltimore: Johns Hopkins University Press, 1942. 583 pp.

Jaffe, Bernard. *Men of Science in America: The Story of American Science Told through the Lives and Achievements of Twenty Outstanding Men From Earliest Times to the Present Day.* Revised ed. New York: Simon and Schuster, 1958. 715 pp.

Jones, Bessie Z., ed. *The Golden Age of Science; Thirty Portraits of the Giants of 19th-century Science, by their Scientific Contemporaries.* New York: Simon and Schuster in cooperation with the Smithsonian Institution, 1966. 659 pp.

Jordan, David Starr, ed. *Leading American Men of Science.* New York: H. Holt and Co., 1910. 471 pp.

Kelly, Howard A. *Some American Medical Botanists Commemorated in Our Botanical Nomenclature.* New York and London: D. Appleton and Co., 1929. 215 pp.

Klein, Aaron E. *The Hidden Contributors: Black Scientists and Inventors in America.* Garden City, N.Y.: Dubleday, 1971. 203 pp.

Mallis, Arnold. *American Entomologists.* New Brunswick, N.J.: Rutgers University Press, 1971. 549 pp.

Miller, Lillian B. *The Lazzaroni: Science and Scientists in Mid-nineteenth Century America.* Washington, D.C.: Published for the National Portrait Gallery, Smithsonian Institution, by the Smithsonian Institution Press, 1972. 121 pp.

Strong, Douglas H. *The Conservationists.* Reading, Mass.: Addison-Wesley, 1971. 196 pp.

Stuart, Charles B. *Lives and Works of Civil and Military Engineers in America.* New York: D. Van Nostrand, 1871. 343 pp.

Tracy, Henry Chester. *American Naturists.* New York: E. P. Dutton and Co., Inc., 1930. 282 pp.

Wilson, Leonard G. *Benjamin Silliman and his Circle: Studies on the Influence of Benjamin Silliman on Science in America.* New York: Neale Watson, 1979. 227 pp.

Yost, Edna. *Modern Americans in Science and Technology.* Revised ed. with additions. New York: Dodd, Mead, 1962. 175 pp. Note: Originally published as *Modern Americans in Science and Invention* (1941).

Youmans, William Jay, ed. *Pioneers of Science in America: Sketches of Their Lives and Scientific Work.* Reprinted with additions from the Popular Science Monthly. New York: D. Appleton and Co., 1896. 508 pp.

7. Histories of Science with Biographical Emphasis (Selective)

Bedini, Silvio. *Early American Scientific Instruments and Their Makers.* Smithsonian Bulletin no. 231. Washington, D.C.: Smithsonian Institution, 1964. 196 pp.

Gies, Joseph, and Frances Gies. *The Ingenious Yankees.* New York: Crowell, 1976. 376 pp.

Hughes, Arthur F. W. *The American Biologist through Four Centuries.* Springfield, Ill.: James C. Thomas Publisher, 1982. 386 pp.

Johnson, Thomas C. *Scientific Interests in the Old South.* University of Virginia Institute for Research in the Social Sciences, Institute Monograph no. 23. New York: D. Appleton Co., for the Institute for Research in the Social Sciences, University of Virginia, 1936. 217 pp. See index; also Chapter 7: "Scattered Scientists," 172–96.

Merrill, George P. *First One Hundred Years of American Geology.* New Haven: Yale University Press, 1924. 773 pp. Reprinted: New York: Hafner Publishing Co., 1964.

Smith, David Eugene, and Jekuthiel Ginsburg. *A History of Mathematics in America before 1900.* Chicago: Mathematical Association of America, 1934. 209 pp.

Smith, Edgar F. *Chemistry in America.* New York: D. Appleton & Co., 1929. 356 pp. Reprinted: New York: Arno Press, 1972 (copyright 1914).

Stearns, Raymond P. *Science in the British Colonies of America.* Urbana: University of Illinois Press, 1970. 760 pp. See index; also Appendix 6: "A Check-list of the Colonial Fellows of the Royal Society of London, 1661–1783," 708–11.

Stevens, Gwendolyn, and Sheldon Gardner. *The Women of Psychology.* Vol. 1, *Pioneers and Innovators.* Cambridge, Mass.: Schenkman, 1982. 241 pp.

Voss, Edward Groesback. *Botanical Beachcombers and Explorers: Pioneers of the 19th Century in the Upper Great Lakes.* Contributions from the University of Michigan Herbarium vol. 13. Ann Arbor: University of Michigan Herbarium, 1978. 100 pp.

Part II
The Disciplines of Science

The Limits of Observation and the Hypotheses of Georges Louis Buffon and Charles Bonnet

VIRGINIA P. DAWSON

About midcentury, the emphasis on observation and experiment characteristic of the natural history of the early eighteenth century gave way to a preoccupation with more speculative explanations of classic problems in the life sciences. Reproduction, nutrition, and the nature of living matter became the elements of philosophical systems and their empirical examination received considerably less emphasis. In the Leiden tradition of the early part of the century, often associated with the influence of Isaac Newton's *Opticks,* naturalists had emphasized facts and experiments carefully observed and repeated. Following Newton's famous dictum, *Hypotheses non fingo,* they had studiously avoided "framing hypotheses," a reaction against the prior system building of René Descartes.

Scholars have called this experimental tradition "conservative," "providential," or "positivistic" to distinguish it from the later, more speculative and radical ideas of French *philosophes* like Denis Diderot, Julien Offray de La Mettrie and Baron d'Holbach. Members of the later group are often referred to as materialists because they granted to matter self-active properties, thereby limiting or dispensing with God's creative and continuing presence in the natural world. The most radical materialists were avowed atheists, but not all eighteenth-century materialists were as extreme. Less radical thinkers such as Georges Louis Buffon and John Turberville Needham, as well as Charles Bonnet and Albrecht von Haller, who were far more conservative, were called materialists by their contemporaries. For each, possibly through the influence of Gottfried Wilhelm Leibniz, matter in a living body of an animal or plant had properties that went beyond the brute and passive stuff of the Cartesian system.[1]

The recognition of this shift in the biological sciences from the empiricism of the early part of the century to more speculative theories of the late eighteenth century challenges more general interpretations of the Enlightenment. Eighteenth-century science cannot be viewed as a cumulative and progressive program to rid itself of metaphysical, ethical, and theological preoccupations as it pushed toward greater objectivity.[2] The denial that nature was providentially ordered, and the progressive stripping of God from an active role in his creation did not encourage greater objectivity based on more precise observations. On the contrary, despite the popularity of the microscope after 1745, the empirical base that Buffon and Needham created to justify their theories was less solid than prior observations of "insects" by René Antoine Ferchault de Réaumur, Abraham Trembley, and Charles Bonnet, or even the great seventeenth-century microscopists, Anton Leeuwenhoek, Jan Swammerdam, and Marcello Malpighi. The microscopes that Buffon and Needham used were no doubt more powerful than those of their predecessors, providing in some cases a magnification as high as 400 times.[3] But, in the eyes of some contemporaries, the observations of Buffon and Needham were flawed. Réaumur, for example, claimed that the observations that supported Buffon's *Histoire naturelle* could not be repeated. He accused Buffon and Needham of letting a system dictate what they saw. More radical materialists like Diderot, La Mettrie, and d'Holbach made no pretense of actually involving themselves in the arduous study of natural objects. In the late eighteenth century, systems could be erected on the most flimsy of observational scaffolding. The French *philosophes* were readers and thinkers, not observers. Although ownership of a microscope was fashionable, as were popular excursions into the countryside to collect insects, no backpocket was big enough for Réaumur's ponderous volumes of the *Mémoires pour servir à l'histoire des insectes* (1734–42). Buffon was among those who had little patience for Réaumur's observational methodology and remarked, with a thinly veiled allusion to the aging academician's monumental study of insects, that a fly "should not hold any greater position in the head of a naturalist than it occupies in Nature."[4]

In 1749 Buffon published his controversial account of the generation and development of animals in volume two of his *Histoire naturelle*. Boldly calling upon Newtonian forces, he proposed that living corpuscles enmeshed in internal molds were responsible for the origin and progressive determination of the external and internal shapes of living beings. Repelled by the implications of Buffon's theory, Bonnet, a semireclusive and pious scion of one of Geneva's patrician families, pitted himself against the powerful *Intendant* of the Paris *Jardin des plantes.* Though they never met, there is little doubt that each paid close attention to the publications of the other. Several years before the publication of the *Histoire naturelle,* Bonnet had already begun to direct his thoughts to metaphysics and had abandoned the empirical orientation of his earlier studies of insects and plant physiology. He responded to Buffon's "system," by proposing an equally problematic solution to the questions of generation. The irony of their long antagonism is that, in hindsight, neither could satisfactorily account for the facts. To adequately explain the generation of living organisms, not only did old ideas have to be rethought, but biologists also needed better instruments than the simple microscopes with their single lenses or the elegant, but optically crude, compound microscopes of the period. Nevertheless, as new scholarship on Buffon has revealed, the rethinking of the prerequisites for a history of nature as opposed to a natural history was an indispensible legacy to the science of the nineteenth century.[5]

The Empirical Challenge to the System of Eggs

There is no question that the shift toward materialistic explanations was assisted by the discovery of the the freshwater polyp, or hydra, by Abraham Trembley in 1740. The regeneration of the polyp from sectioned parts and Trembley's study of the phenomenon of budding challenged the dominant idea of the early part of the century that all life comes from eggs. Only with the approach of winter did Trembley's polyps lay eggs. When spring came, new polyps hatched from these eggs. Normally, throughout spring, summer, and fall, Trembley's polyps formed out of the very substance of the parent, at first resembling raised lumps on the exterior skin of the little animal's body. Little by little, they grew tentacles and finally separated.[6]

Born in Calvinist Geneva, Trembley had studied at the University of Leiden before becoming the tutor of the children of Count Willem Bentinck at Sorgvliet, near The Hague. Undoubtedly he profited from the availability of the skilled instrument makers of the Netherlands and his association with the circle of Leiden Newtonians.[7] For his observations, Trembley used an instrument of his own design, a type of simple microscope, or magnifying glass, mounted on a moveable arm that could be positioned on the outside of the jar in which he kept his supply of polyps. To observe through the magnifying lens, Trembley darkened his chamber and placed a single candle in back of his jar. His book, *Mémoires, pour servir à l'histoire d'un Genre de Polypes d'eau douce à bras en forme de cornes,* is by modern standards a surprisingly accurate record of the physiology of this simple animal, now classified not as an insect, but among the coelenterates.[8]

In his careful descriptions of budding and regeneration, Trembley unwittingly produced a convincing challenge to the preexistence theory, the accepted explanation of animal generation, popularly known as the "system of eggs." I say unwittingly, because nothing in Trembley's Calvinist background suggests that he was predisposed to call the established order into question. Nevertheless, both regeneration and budding were difficult to reconcile with the idea ' that offspring were not created but existed "preformed" within one of the parents before birth. According to the preexistence theory, not only was the newborn believed to be preformed, but also miniatures or germs of all preceding generations were encased within the egg or sperm placed at creation in the progenitor of the species. Though the discovery of spermatic animalcules by Anton Leeuwenhoek provided seeming evidence that the germs were carried by the male, observations by Malpighi, Antonio Vallisnieri, and Swammerdam gave increasing authority to the view that the germ was carried in the egg. Regnier de Graaf's (actually incorrect) identification of the ovum in mammals bolstered the rational symmetry of the ovist position. Thereafter eggs of viviparous animals were seen as analogous to those of oviparous animals.[9]

The preexistence theory had reached its most perfect philosophical expression in the first volume of Nicholas Malebranche's *Recherche de le verité,* published in 1674. Malebranche could not accept the Cartesian idea that chance combinations of particles in male and female seminal fluid could explain generation. The idea of the encasement, or *emboîtement,* of germs provided a way to adjust the mechanical vision of the world to the belief in the mystery of the supreme architect's creative act in forming the earth and his creatures. Malebranche reasoned that a single seed or womb might contain an infinite number of seeds or eggs to supply an infinite number of centuries. Since it was possible to demonstrate mathematically the divisibility of matter to infinity, "that is sufficient to have us believe that there could be smaller and smaller animals to infinity, though our imagination is shocked at this thought."[10]

Whether egg or sperm carried the miniature was less important than the affirmation of preexistence. With the weight of theology and science behind them, Christians were united in the belief that God had created a rationally ordered universe, order that extended from the laws describing the motion of the planets to the most minute germs in the womb of the least insect. The problem with Trembley's careful study of the polyp's structure was that it appeared to contradict the ancient idea that all life comes from eggs: *ex ovo omnia.* In short, Trembley's polyp was an anomaly at a time when the facts of science necessarily complemented ethical and religious views. The polyp unbalanced the rational symmetry of biological explanations and demanded a rethinking of cherished presuppositions.

However, it is unlikely that facts revealed by observation alone could have produced the shift to materialism. Equally important, though difficult to pinpoint with the same certitude as the empirical discoveries, were changes in the philosophical climate in England and France that began in the first decades of the century. In addition to the challenge to the impious determinism of Descartes's explanation of generation, the Cambridge neo-Platonists objected to his rigid distinction between soul and body. Ralph Cudworth, for example, challenged the idea of the Cartesian soulless animal automaton, arguing that animals were not like clocks and watches. "But on the contrary," he wrote, "if it be evident from the Phenomena, that Brutes are not meer *Senseless Machins* [sic] or *Automata* . . . therefore they must have something more than matter in them."[11] Neo-Platonists explained this "something more" by the idea of "plastic forces" added to brute matter through the action of God. This idea took root particularly in Protestant thought because, unlike the Cartesian God who operated on his creation only indirectly through natural laws, forces assured the continuing active providential intervention of God in nature.

Well before the popularization of Newton in Voltaire's *Lettres philoso-*

phiques (1733), continental Protestant thinkers had begun to assimilate the ideas of their British counterparts. Through Jean Le Clerc's translation and publication of excerpts of Cudworth's *The Intellectual System of the Universe* in his *Bibliothèque choisi,* a popular periodical published in Holland in the early years of the eighteenth century, neo-Platonic ideas were disseminated to the French-speaking world.[12] Trembley's fascination with various aspects of the polyp's sensitivity to light, cold and touch may have been influenced indirectly by neo-Platonism. Although Trembley carefully refrained from mixing knowledge founded on observation with abstract theories, he had also studied the writings of John Locke, an author strongly influenced by neo-Platonism.[13]

The materialism that surfaces after 1740 coincided with a revival of interest in Leibnizian philosophy, possibly because Leibnizian metaphysics offered a better solution to the problems of animal generation, vaguely perceived before the discovery of polyps but brought into focus by Trembley's careful and irreproachable experiments.

By far the most influential work that gave the French reading public access to the ideas of Leibniz appeared shortly after his death. This was the famous correspondence with Samuel Clarke, first published in London in 1717. A French edition, published in Amsterdam in 1720, edited by Pierre Des Maizeaux, remarked that "what is certain is that the metaphysics of Locke has found a great number of Partisans; but up to now, that of Leibniz has not had great fortune."[14] In the early years of the century, a time when the popularity of the empirical tradition of Newton of the *Opticks* was at its height, Leibniz repelled readers with the excessively scholastic style of *Theodicy* (1710), the most important statement of his philosophy.[15] The scarcity of Leibniz's published works coincided with his bitter rivalry with Newton over the invention of the calculus. This relative obscurity was to change, stimulated not only by the sustained interest in Leibnizian mathematics by generations of Swiss students who studied with the Bernoulli family in Basle, but also by a renewal of interest in the more metaphysical aspects of his thought.

Abraham Trembley discovered the polyp at a time when the practice of natural history was tied to natural theology. He had little interest in metaphysics, but demonstrated an extraordinary patience for the study of what appeared at first inspection to be a most insignificant example of God's creation. Regeneration pushed biology down new philosophical paths, away from the empirical style that characterized Trembley's work.

Buffon's Theory of Generation

In Buffon's intellectual formation, John Lyon and Philip Sloan date the period of transformation from what they refer to as the "positivistic" Newtonianism of John Keill in England and Willem Jacob 'sGravesande, and Pieter van Musschenbroek in Holland to a more Leibnizian stance between

1739 and 1745. They suggest that Buffon was probably drawn to Leibnizian metaphysics through his association with individuals such as Pierre de Maupertuis, Samuel Koenig, Jean Jallabert, and particularly the scintillating and brilliant Gabrielle-Émilie du Châtelet. Du Châtelet's *Institutions de physique* became a vehicle for the presentation of Newtonian mechanics within a Leibnizian ontology. Lyon and Sloan see the influence of Leibniz on Buffon as primarily methodological. By 1749, as a result of his study of the *Institutions de physique,* in the "Initial Discourse" to the *Histoire naturelle* Buffon had come to disparage all mathematics as a science that dealt only with abstractions. He cast natural history in a more positive light: it "was to be a science dealing with the concrete and specific aspects of the natural world."[16]

Buffon may have had his first taste of Leibnizian mathematics through Gabriel Cramer, professor of mathematics at the Academy of Calvin in Geneva and a former, highly regarded student of the Bernoullis. Their correspondence reveals that Buffon briefly visited Geneva in 1731. In the years that followed, there ensued a warm exchange of letters containing mathematical problems for solution. About 1740 or 1741, in a missing letter, Cramer asked Buffon a question about Leibniz. Although Buffon appears never to have answered Cramer directly, his letter of 4 April 1744 revealed the impact of Cramer's suggestion on the fledgling mathematician. Buffon wrote: "I will tell you that the long letter that you wrote to me in defense of Leibniz occupied me for several days and I wanted to send you a long response nearly three years ago."[17]

At first, Buffon's interest in mathematics, nurtured through his correspondence with Cramer, complemented his early preoccupation with the problems of generation. As early as 1733, Buffon was involved in discussions of the problems of animal reproduction with Louis Bourguet, known as the Pliny of Neuchâtel, Switzerland. Bourguet, a correspondent of Leibniz, popularized his Leibnizian metaphysics in his *Lettres philosophiques,* published in 1729 in Geneva. Buffon is reputed to have broken off the discussion with Bourguet in 1734, because he took issue with Bourguet's arguments in support of preformation.[18]

Presumably already sensitized to Leibnizian ideas, Buffon responded to the discovery of the polyp. He took Trembley's description of regeneration of parts from cuttings as the paradigm for embryonic growth in general. In contrast to Trembley, who resisted drawing more general conclusions from his study of the polyp, Buffon used the discovery as one of the empirical bases for his philosophical system.

We percieve that a cube of sea-salt is composed of other cubes, and that an elm consists of other smaller elms, because by taking an end of a branch, or root, or a piece of the wood separated from the trunk, or a seed, they will alike produce a new tree. It is the same with respect to polyps, and some other kinds of animals, which we can multiply by cutting off and separating any of the different parts; and since our rule for judging in both is the same why should we judge differently of them?[19]

He argued that just as every particle that made up the body of the polyp was capable of becoming a complete being, all living beings must be composed of tiny organic particles that played a role in generation. Living particles, freed into the environment during decomposition of plants and animals, came together by chance to produce submicroscopic organized bodies similar to those visible in the macroscopic world.

Buffon rejected the idea of preexistence on mathematical grounds. Preexistence turned clear thinking into a "labyrinth of infinity." Both geometrical infinity and infinite progression (development) were abstractions that Buffon demonstrated could not exist in nature.[20] Rather, the seed or egg was capable of producing a new individual only because it contained the ubiquitous organic particles. No longer was the seed in plants, or the egg in animals the passive container for a successive development of preexisting beings, placed there by God's initial creative act. Egg and seed had a role in reproduction only within a wider context of the nutrition and growth of organisms. For Buffon the essential point was that the organic particles were living, and to an extent differentiated, prior to their incorporation into an organized being. They were drawn into the body of the plant or animal with other nutritive substances. There they encountered a *moule intérieur,* a matrix, mesh or mold, that provided the form that development took. Reproduction occurred only after the organism had reached maturity. Excess organic particles, no longer needed exclusively for growth, were sent to the reproductive organs. At conception these organic particles in the semina of male and female combined to produce a new organic being—a miniature that would expand in size prior to germination or birth.[21]

In thinking about reproduction within a context of nutrition and growth, Buffon was no doubt influenced by his study of Stephen Hales's *Vegetable Staticks,* which he translated in 1735. Buffon consciously compared the unexplainable action of the *moule* in governing growth and reproduction to Newtonian forces like that of gravity.[22]

Buffon's system was a rational construction, conceived before he attempted any experimental verification. Nevertheless, Buffon claimed to have created a system that was not only rationally compelling, but grounded in empirical fact. He insisted at more than one point in his "Initial Discourse" that at the base of natural history is the knowledge of facts. "The precise description and the accurate history of each thing is, as we have said, the sole end which ought to be proposed initially."[23] However, facts could not be ends in themselves; he also felt that it was important to make generalizations about nature, based on certain common sense notions. Buffon argued, for example, that it was intuitively clear that there was a division between the living and nonliving. Moreover, revealing one of the cherished presuppositions of the age, he confidently acknowledged that nature was arranged in a chain that stretched from the simplest to the most complex. In fact, Buffon's reaction upon hearing of Trembley's discovery of the polyp and of Bonnet's proof that aphids were capable of giving birth by parthenogenesis was that the discoveries demonstrated that "the whole of nature proceeds by gradations, and that all that can be is."[24] To arrive at knowledge about nature, it was equally permissible to reason inductively or deductively, either from the particular to the general, or from the general to the particular. "Both ways are good, and the choice of one or the other depends more on the bent of the author than on the nature of things, which always allows of being treated equally well by either method."[25]

In his experimental work, Buffon followed his own advice. The polyp, he reasoned, was the link between the plant and animal realms, since it shared properties common to each. In locomotion and feeding, it behaved like an animal, but its mode of reproduction was like a plant.[26] Through deduction, not observation, he concluded that if nature were arranged in a hierarchical ladder, organic particles must exist as the simplest unit of living matter. They must be the organic building blocks for all organized beings, either animal or plant. "I then began to suppose," Buffon reported, "that, by a microscope, I might be able to attain a discovery of the living organic particles, from which I thought every animal and vegetable drew their origin."[27] The microscope confirmed what he had anticipated. In collaboration with John Turberville Needham, Buffon studied various infusions under a microscope. Together they discovered microorganisms that they claimed were zoophytes—living organized bodies formed from the chance combination of organic particles.

Buffon had to refute Malpighi and Vallisnieri because Bourguet and others had relied on their empirical evidence and authority to argue for a theory of preexistence based on the egg. Thus, in addition to empirical confirmation of the link between organic particles and zoophytes, Buffon needed proof that male and female in higher animals contained organic particles from which the embryo formed. In the same year that he and Needham did their experimental work on infusions, Buffon claimed to have discovered an animated seminal fluid in females analogous to that of males. The main point of the paper, officially communicated to the Paris Academy of Sciences in May 1748, was that prior to impregnation the ovaries of viviparous females did not contain eggs. Regarding the term "ovaries" as a misnomer, Buffon examined what he referred to as the "glandular bodies or testicles" of female dogs in heat as well as other animals, brought to him immediately after slaughter. Buffon concluded that females produced tiny flagellated bodies, similar to Leeuwenhoek's spermatic animalcules. Buffon triumphantly reported to Cramer that his presentation to the Royal Academy of Sciences was extremely well received. "It is now very certain that eggs do not exist at all in viviparous females, and that they have, on the contrary, an animated seminal fluid containing, as does that of the male, bodies in movement, spermatic animals, which in each species of animal are absolutely similar both in form and movement to the spermatic animals contained in the fluid of the male."[28]

Buffon's observations supported his new theory of generation: the embryo is formed from the union of organic particles in the semen of males and females. Just as the "organic particles" in the experiments on infusions formed spontaneously to create zoophytes, so were nutritive particles drawn into the animal body transformed in the reproductive organs into flagellated bodies necessary for reproduction.

There is no question that Buffon had convinced himself of the existence of animated seminal fluid in females and the spontaneous generation of zoophytes from organic particles *before* he took the trouble to peer into a microscope. Both his microscopic work with Needham and the "discovery" of spermatic animals in females were two prongs born of rational necessity. Each provided empirical evidence to support his system. Buffon's lack of experience in anatomy had led him to imprecise descriptions of the glands that produced this supposed fluid. His possible observation of Brownian motion in cells of bacteria bred in sperm notwithstanding, the interpretation of his observations was flawed by an eye ready to yield to the imperialism of the mind. The experiments on infusions were equally problematic. Again, despite the fact that Needham may have seen flagellated zoospores produced by a type of water mold, Needham admitted that, like Buffon, he looked in nature for the validation of a previously conceived system.[29]

The discovery of animated seminal fluid in the glands of female dogs proved controversial almost at once. Buffon's contemporaries could not repeat these observations and quickly reached the conclusion that Buffon's commitment to his system was stronger than his empirical evidence. For example, Abraham Trembley called Buffon's new theory a "foolhardy hypothesis," and he objected to Buffon's mode of reasoning that, he argued, took rash conjectures and made them into first principles. He rightly saw that Buffon's new system was only tenuously based on observation. In a letter to Count Willem Bentinck in 1750, Trembley wrote:

> I have just read the three volumes of the Natural History of M. Buffon which appeared some time ago. They contain several curious facts; but not enough in proportion to their bulk. There are a very great number of conjectures, several of which are very rash. It often happens that M. de Buffon gives them first as conjectures, and then afterwards makes use of them as demonstrated Principles. . . . M. de Buffon claims to explain nearly everything about generation; but, I admit that I cannot help but consider his system as a foolhardy hypothesis. He tries to prove too much with the facts with which he has constructed it. It seems sometimes that he lets himself be carried away by his imagination. If his work is popular, I fear that he will do harm to Natural History in bringing back the taste for Hypotheses.[30]

Bonnet and Trembley, and their French patron René Antoine Réaumur, questioned the observations on which Buffon based his "hypotheses." Nevertheless, while they questioned the validity of the observations themselves, it is clear that they were motivated by a deeper opposition to the theological implications of Buffon's system. Chance did not operate in a universe controlled by God's providence. At Bonnet's urging, Lazzaro

Spallanzani, exposed the carelessness of Needham's infusion experiments. The infusions had not been sufficiently heated to destroy the "seeds" or "eggs" present in the water prior to sealing. Thus, microorganisms did not generate spontaneously from the supposed organic particles.[31] Rather than using theory as a heuristic device, Buffon and Needham had let their theory determine what they saw. Although Trembley, Bonnet and Réaumur were guided by the idea that the visible universe was the product of intelligent design, their observations were less liable to distortion by the lens of the mind. Surprisingly, those who practiced natural history within the frame of natural theology were left freer to explore with the outward eye. Trembley expressed his credo at the end of his book on the polyp when he wrote: "We cannot work to greater advantage to explain the Facts that we know than to strive to uncover new ones. Nature should be judged by nature, not by our own views.... The Beauty of Nature certainly appears better, when what we know of it, is not mixed with our Imaginings."[32] For Trembley, Buffon's imaginings pushed the facts too far.

Nevertheless, in their efforts to dispense with the elevated position of the egg, Buffon and Needham were grappling with questions that those who adhered to the preexistence theory could not ask. *Emboîtement,* or the encasement of germs, by definition, had to be taken on faith because no instrument could be conceived to make empirical confirmation of the infinitely small. Moreover, explanations based on encapsulation inadequately explained the seeming equal roles of male and female in the physical appearance of offspring. In carrying out their microscopical observations, Buffon and Needham had the courage to shine the torch of science on a corner of nature previously unexamined, and their efforts cleared the way for a naturalistic account of how animals reproduced. By attempting to see the *relationships* among biological phenomena, such as nutrition, growth, and reproduction, they began to expose the futility of understanding nature through the simple amassing of observational data. Generation was reduced to a special case of the natural processes of nutrition and growth at work in the organic world.[33] Buffon's attempt to formulate a new theory of generation was both appropriate and honest, if offensive to his more orthodox critics.

Bonnet's Response to Buffon

Trembley's polyp had unavoidable metaphysical implications and no one in the eighteenth century found any way to challenge his careful observations of budding and regeneration through cuttings. Regeneration and budding could not be explained on the old preexistence model. To adequately account for these perplexing phenomena, it was necessary to move from the outward eye to the eye of the mind.

After his first publication, the *Traîté d'Insectologie* in 1745, Bonnet struggled for several years with a loss of eyesight that might be considered both real and symbolic. The polyp, he wrote, "upset all my ideas and had put my brain in combustion."[34] The problems that the polyp raised for the theory of preexistence could not be solved on an empirical level. Increasingly, Bonnet turned to metaphysics. He now abandoned the art of observation, previously carried to near perfection with his discovery of parthenogenesis. In the winter of 1748 Bonnet read Leibniz's *Theodicy* for the first time, an event that he compared to the discovery of a new universe. While Leibniz's scholastic discussions and digressions bored him, he was impressed by his discussion of preexistence and souls (which Leibniz discussed in relation to monads).[35] Thus, Bonnet, like Buffon somewhat earlier, was drawn to Leibniz. Both used elements of Leibnizian thought to illuminate very different explanations of generation.

Though Bonnet's autobiography made much of his reading of the *Theodicy*, the unsettling effect of Buffon's work is noted in his letters. His reaction to Buffon was at least as important in the shaping of his philosophy as Leibniz's *Theodicy*. The letters of Bonnet and Réaumur after 1749 reflect ther criticism and distaste for Buffon's *Histoire naturelle*. Réaumur responded to a missing letter from Bonnet in 1751: "With the good and wise mind that I know you have, I am not at all surprised that you have not been seduced by the tone with which M. de Buffon exposed ideas that are too rash and which, if they are adopted, will throw us again into the shadows from which we have been brought by the works of so many great men."[36] At that time Réaumur and his friend the Abbé Joseph Adrien Lignac, had engaged a skilled microcopist, Mathurin-Jacques Brisson, to assist them in seeing where Buffon's and Needham's system had distorted their empirical data.[37]

While Réaumur attempted to expose Buffon's work by challenging the observations that Buffon had advanced in support his theory, Bonnet took a philosophical approach. In November 1751 he revealed in a letter to Réaumur that he had been working on his own system to account for generation.[38] He claimed to have conceived his system several years *prior* to the appearance of the *Histoire naturelle*. Whether this was in fact the case is not entirely clear, but there is little doubt that Buffon's work contributed to the urgency of his musings. The significance of his letter is that it fixes in time some of Bonnet's early thoughts on generation. The tone of the letter and Réaumur's careful response suggest that this may have been Bonnet's first rather full expression of his views on a subject which would continue to preoccupy him throughout his life.

In contrast to Buffon, the cornerstone of Bonnet's system is preexistence of germs. "All the organized bodies," Bonnet wrote in his letter, "were contained from the beginning in germs that represented them in miniature." Bonnet starts, not with organic particles as the basic unit from which living organized bodies arise, but at a relatively high level of organizaton, the germ. However, at this early stage in his thinking, he had not yet decided whether germs were encased one within the other or spread throughout creation. The idea of encasement of germs, Bonnet's first option, was clearly from Malebranche, whom Bonnet had read as a student. "The calculations that have been heaped up against the former [encasement] shocks only the imagination, and the imagination must keep silent in things which are the province of the understanding." However, he acknowledged Buffon's criticism in amending Malebranche to assume not the "actual division to infinity, but an indefinite division."[39]

Bonnet built his theory on the traditional idea that the seminal fluid provided nourishing saps appropriate for the growth of the germ or egg. Like Buffon, he saw a relationship between reproduction and nutrition that reveals the common influence of Stephen Hales's *Vegetable Staticks.* However, Buffon's translation had played down and in some cases eliminated, Hales's allusions to an intelligent planner in his introduction and preface to the *Vegetable Staticks.*[40] Bonnet, though he used Buffon's translation, took the argument from design as his starting point. Bonnet wrote to Réaumur that the organs of generation were "made with an art so marvellous that they are a sort of abstract of the whole body." They separated and worked the nutritive particles to stimulate the heart, thus beginning the circulation and growth within the germ. In explaining his earliest theory, Bonnet wrote to Réaumur:

> I consider the seminal fluid as a nourishing fluid, whose subtlety is proportional to the smallness of the parts of the Germ. These parts work this fluid, and incorporate it within themselves. It is carried to the heart and begins a circulation which only stops when life does. But as the solids of the Germ are not all of an equal consistency, and as they require nourishing saps of a relative diversity, I presume that the seminal liquor contains such saps, saps appropriate to each type of part. I imagine that the organs of generation are made with an art so marvellous that they are a sort of abstract of the whole body, in such a manner that they separate the particles analogous to the diverse parts, etc. Growth operates by the reception or the intermixing of the Alimentary Atoms in the meshes of the solids, which I presume are endowed with a great ductility. The relations that nature has established between the germ and the seminal liquid destined to procure the first growths determines the possibility of generation.[41]

In the same letter, he emphasized the predictability of his experiments on the regeneration of worms after sectioning—experiments inspired by Trembley's discovery of the polyp's amazing ability to regenerate from cuttings. Was this regularity not evidence of intelligent planning? He criticized Buffon in his letter to Réaumur for ignoring his discussion of this phenomenon published in 1745 in his *Traité d'Insectologie.* "There is something to which M. de Buffon has not paid any attention and which merits a great deal. This is what is offered to us in the Intermediate and Posterior portions of this species of long worm of which I spoke in my observations and which grows a Tail instead of a head. The frequency of this Phenomenon has a great deal that is surprising."[42]

Using the carefully controlled experimental approach of the *Vegetable Staticks* as a model, Bonnet had explored various aspects of growth of his worm segments. If he made his cutting either slightly ahead of, or slightly behind, the midpoint, the intermediate portions of the worm would regenerate whatever portion was missing. Regeneration occurred in these intermediate portions the most rapidly, though there was a limit to how many times a given worm could reproduce the missing part. He also discovered that the extreme anterior and posterior portions did not have the same ability to regenerate as the intermediate portions. If cuts were made too close to the head or to the tail, these severed parts were likely to die. One of his most important discoveries was that if he cut a worm exactly in half he could produce a monster. The posterior portion or tail section of his worm, instead of producing a head as he had first expected, grew a second tail.

What impressed him was that these experiments demonstrated the *regularity* or *predictability* of the regeneration of the segments. Even the creation of monsters could not be the result of chance combinations of particles. Though Bonnet could not entirely resolve the difficulty of accounting for monsters, he referred to the family in which several members were born with twenty-four digits, discussed by Réaumur in *L'art de faire eclorre et d'enlever en tout saison des oiseaux domestiques des toutes esspeces*. Bonnet suggested that germs stuck to one another, or grafted on to each other, might offer an explanation. Thus, the problem of monsters, whose philosophical implications would continue to perplex him, was grounded in his prior empirical work on worms.[43]

Réaumur responded to Bonnet's early attempt at the metaphysics of reproduction tactfully but firmly. While Bonnet's arguments for "the system of germs" appeared plausible, Réaumur was clearly uncomfortable with thoughts not entirely tethered to observed fact. In addition to deciding whether the germ was living or not prior to fertilization, Réaumur warned that Bonnet's description of the role of the male seminal fluid was not entirely satisfactory.

> The question is to determine if the germ lives, or if it is deprived of life; if the fluids circulate or do not circulate. . . . It is to employ a very natural use of the seminal fluid to use it to begin the circulation, the development. But it seems to me that you dismiss rather freely one of the greatest difficulties when you posit that in the seed, and in the saps provided for growth, some [saps] are appropriate for each type of part. Simpler ideas would please me on this point.[44]

Réaumur correctly realized that if Bonnet argued that the saps and seminal fluids were differentiated prior to incorporation with the germ, he seemed to presuppose a level of preorganization of matter that brought the theory uncomfortably close to materialism. Indirectly leveling criticism at both Bonnet and Buffon for leaving the solid ground of observation to speculate, Réaumur wrote: "a mind which is not happy except with ideas that are clear and well proven will have difficulty satisfying itself on a subject with respect to which our sense cannot give sufficient light." His eyes had served him well enough to see that "these microscopical observations from which M. de Buffon and Abbé Needham have drawn such strange conclusions are poorly done, often false, observations."[45]

If Buffon's system conformed imperfectly to facts revealed through observation, could a preexistence theory offer a more plausible alternative? Bonnet had already rejected the old simplistic idea of the germ as a miniature that merely swelled in size after conception. He also had to abandon the idea that the nutritive particles had a special preorganization, an idea that brought him too close to Buffon's organic particles. In his definition of the germ, Bonnet retreated from the nascent materialism of his early letter to Réaumur and his *Essai de Psychologie* (1754). In *Contemplation de la Nature* (1764) Bonnet defined the germ simply as "every preordination, every preformation of parts capable by itself of determining the existence of a Plant or an Animal."[46] Later in *La Palingénésie philosophique* (1769, 1783) he wrote that the germ, consisting of inorganic elements provided only a blueprint for development; the nutritive molecules, trapped in the germ's meshes, interacted with the elements within the germ to facilitate growth.

> I conceived the elements of the germ as the *primordial foundation*, on which the nutritive molecules went to work to increase in every direction the dimensions of the parts.
>
> I pictured the germ as a network, the elements of which formed the meshes. The nutritive molecules, incorporating themselves into these meshes, tended to enlarge them....
>
> Strictly speaking, I said (Art. 83, pp. 47, 48), the elements [inorganic] do not form organic bodies; they only develop them, and this is accomplished by nutrition. The primitive organization of the germs determines the arrangement which the nourishing atoms must take in order to become parts of the organic whole.[47]

In stipulating that generation, nutrition, and growth were all dependent on the interaction between the elements of the germ and the nutritive molecules, Bonnet attributed an activity to matter absent in previous preexistence theories. The embryo did not passively swell in size like a sponge filling with fluid. The primordial foundation provided no more than a stage for the reciprocal action of the internal and external particles. Like that of his rival, Bonnet's system went well beyond what could be verified through observation.

The significance of the hypotheses of Buffon and Bonnet lies in their complementary effort to integrate new observations in biology into a rational fabric. Both realized that, partly as a result of Trembley's discovery, the heaping up of empirical facts without the ordering of the mind's eye had reached a limit. While Trembley might complain that Buffon tried to "prove too much with the facts," the utility of collecting facts the better to reveal the wisdom of God's rational planning had passed. The systems of Bonnet and Buffon and the objections that neither could be fully justified by the observations reveal the dilemma of biologists of the middle of the eighteenth century. Though bitter rivals, Buffon and Bonnet were attempting to provide a new synthesis. Each brought different preconceptions to the problem, yet each had essentially the same empirical evidence to marshall in support of his system. Each was deeply indebted to the work of Stephen Hales and their theories had more in common than either would have cared to admit. Neither theory was entirely satisfactory. The hypotheses of Buffon and Bonnet reflected the efforts of two individuals to reweave the raveling skein of eighteenth-century biological thought. Though materialism took its place briefly in the waning years of the century, the resolution of the mysteries of generation had to await new observations with better instruments, as well as new preconceptions to be realized in the cell theory of the early nineteenth century.

Notes

1. Materialism is discussed in Shirley A. Roe, "John Turberville Needham and the Generation of Living Organisms," *Isis* 74 (1983): 159–84, and in "Anatomia animata: The Newtonian physiology of Albrecht von Haller" in Everett Mendelsohn, ed., *Transformation and Tradition in the Sciences,* (Cambridge University Press, 1984), 273–300, and in the introduction to Renato G. Mazzolini and Shirley A. Roe, *Science Against the Unbelievers: the Correspondence of Bonnet and Needham, 1760–1780,* vol. 243 of *Studies on Voltaire and the Eighteenth Century* (Oxford: Voltaire Foundation, 1986). For a discussion of the early Leiden empirical tradition, see Pierre Brunet, *Les Physiciens Hollandais et la méthode expérimentale en France au XVIIIᵉ Siècle* (Paris: Librairie Albert Blanchard, 1926); *L'Introduction des théories de Newton en France au XVIIIᵉ siècle. I: Avant 1738* (Paris: Blanchard, 1931). Robert E. Schofield also tackles some of the same themes from a different point of view in "An Evolutionary Taxonomy of Eighteenth-Century Newtonianisms," *Studies in Eighteenth-Century Culture* 7 (1978): 175–92. The "positivistic" designation is used by John Lyon and Phillip R. Sloan in *From Natural History to the History of Nature: Readings from Buffon and His Critics* (Notre Dame: University of Notre Dame Press, 1981), 19. For previous comparisons of Bonnet and Buffon, see Peter J. Bowler, "Bonnet and Buffon: Theories of Generation and the Problem of Species," *Journal of the History of Biology* 6 (1973): 259–81 and Carlo Castellani, "The Problem of Generation in Bonnet and in Buffon: A Critical Comparison," *Science, Medicine and Society in the Renaissance: Essays in Honor of Walter Pagel,* edited by Allen G. Debus, (London: Heinemann, 1972), 265–88.

2. Peter Gay's characterization of the march toward objectivity in *The Enlightenment: An Interpretation* (New York: Knopf, 1966–69), 2: 159 is challenged in P. M. Heimann and J. E. McGuire, "Newtonian Forces and Lockean Powers: Concepts of Matter in Eighteenth-Century Thought," *Historical Studies in the Physical Sciences* 3 (1971): 234.

3. A critical study of eighteenth-century microscopy is needed. A good general reference is *The Billings Microscope Collection* (Washington, D.C.: Armed Forces Institute of Pathology, 1974). See also S. Bradbury, *The Evolution of the Microscope,* (Oxford: Pergamon Press, 1967); Brian J. Ford, *Single Lens: The Story of the Simple Microscope* (New York: Harper & Row, 1985). New light is shed on Buffon's microscopy by Philip R. Sloan in *From Natural History,* 164–69, and "Organic Molecules Revisited," to be published in the Proceedings of the International Conference on Buffon, Paris, 1988.

4. Georges-Louis Le Clerc de Buffon, *Discours de la nature des animaux,* quoted by Jean Torlais, "Une rivalité célèbre: Réaumur et Buffon," *La Presse Médicale, Paris* 66 (1958): 1057–58.

5. Wolf Lepenies, "De l'histoire naturelle à l'histoire de la nature," *Dix-huitième siècle* 11 (1979): 175–84.

6. On the discovery of the polyp and its philosophical context, see Virginia P. Dawson, *Nature's Enigma: the Problem of the Polyp in the Letters of Bonnet, Trembley and Réaumur,*

(Philadelphia: American Philosophical Society, 1987); Aram Vartanian, "Trembley's Polyp, La Mettrie, and Eighteenth-Century French Materialism," *Journal of the History of Ideas* 11 1950: 259–86.

7. On the social and intellectual connections of the Bentinck family, see Margaret Jacob, *The Radical Enlightenment: Pantheists, Freemasons and Republicans* (London: George Allen & Unwin, 1981).

8. For an assessment of the precision of Trembley's experimental methodology by a contemporary hydra researcher, see Sylvia G. Lenhoff and Howard M. Lenhoff, *Hydra and the Birth of Experimental Biology—1744* (Pacific Grove, Calif.: Boxwood Press, 1986), 16–25.

9. In fact de Graaf did not identify the ovum, but only the follicles of the ovary from which the eggs emerged. See Peter Bowler, "Preformation and Pre-Existence in the Seventeenth Century: A Brief Analysis," *Journal of the History of Biology* 4 (1971): 221–44.

10. Quoted by Jacques Roger, *Les Sciences de la vie dans la pensée Française de XVIIIe Siècle* (Paris: Armand Colin, 1963), 336. See his entire discussion of preexistence, 325–84; also Paul Schrecker, "Malebranche et le préformisme biologique," *Revue Internationale de Philosophie* 1 (1938): 81.

11. Quoted by John W. Yolton, *Thinking Matter: Materialism in Eighteenth-Century Britain* (Minneapolis: University of Minnesota Press, 1983), 13, note 4.

12. Jacques Roger, *Les Sciences de la vie,* 418–27. Roger shows neo-Platonic influence on Georg-Ernst Stahl and Nicholas Hartsoeker.

13. A more extended discussion of this point can be found in Dawson, *Nature's Enigma,* 127–30.

14. Pierre Des Maizeaux, *Recueil de diverses pièces sur la Philosophie, la Religion naturelle, l'Histoire, les Mathématiques, &c.* (Amsterdam: Sauzet, 1720), vol. 1, lxx.

15. W. H. Barber, *Leibniz in France: From Arnauld to Voltaire* (Oxford: Clarendon Press, 1955), 92–99.

16. Lyon and Sloan, *From Natural History,* 22. Schofield in "An Evolutionary Taxonomy" calls this group "Leibnizian Newtonians." See also John L. Greenberg, "Mathematical Physics in Eighteenth-Century France," *Isis* 77 (1986): 59–78 for a discussion of the introduction of Leibnizian mathematics into France in the 1730s and 1740s.

17. "Je vous dirai que la longue lettre que vous m'avez écrite en defense de Leibnitz m'a occupé pendent quelques jours et que je voulois vous envoyer il y a près de trois ans une longue reponse...." in François Weil, "La Correspondance Buffon-Cramer," *Revue d'histoire des sciences* 14 (1961): 124.

18. Lesley Hanks, *Buffon, avant "l'Histoire Naturelle"* (Paris: Presses Universitaires de France, 1966), 70–71.

19. Buffon, "The Generation of Animals," in Lyon and Sloan *From Natural History,* 173.

20. Buffon discusses the concept of infinity in the preface to Isaac Newton's *Method of Fluxions and Infinite Series,* in Lyon and Sloan, *From Natural History,* 46–47.

21. Bowler makes the important point in "Bonnet and Buffon," 264, that the term epigenesis was not used by Buffon and that his theory was not epigenetic in the sense that Harvey used the term to describe the sequential formation of the parts of the embryo. For Buffon, a miniature was immediately created upon the mixing of the male and female semina. The miniature grew by expansion. Later opponents of Buffon, like Haller, used the term to designate any theory that denied preexistence.

22. On the Newtonian aspect of Buffon, see Robert Wohl, "Buffon and his Project for a New Science," *Isis* 51 (1960): 192. See Lesley Hanks, *Buffon,* for influence of Hales, 92.

23. Buffon, "Initial Discourse," Lyon and Sloan, *From Natural History,* 111.

24. Quoted by John R. Baker, *Abraham Trembley of Geneva* (London: Edwin Arnold, 1952), xvii. Buffon's letter to Martin Folkes 18 July 1741 was not published in the *Philosophical Transactions,* although it is alluded to in a note to "Extract of a Letter from J. F. Gronovius, November 1741," *Philosophical Transactions of the Royal Society* 42 (1742–43): 219.

25. Buffon, "Initial Discourse," Lyon and Sloan, *From Natural History,* 127.

26. Buffon, "Histoire naturelle, Mammifères—1," in *Oeuvres complètes de Buffon* (Paris: Bureau de Publications Illustrées, 1944), 3: 3.

27. Buffon, "Experiments on the Method of Generation," Lyon and Sloan, *From Natural History,* 187. Roger, *Les sciences de la vie,* 542, and Lesley Hanks, *Buffon,* 66, agree that the experiments of Buffon and Needham were made *after* Buffon had conceived his system.

28. " . . . il est maintenant très certain qu'il n'existe point d'oeufs dans les femelles vivipares, et qu'elles ont au contraire une liqueur seminale animée et qui contient comme celle du male des corps en mouvement des animaux spermatiques qui dans chaque espèce d'animal sont absolument semblables et pour la forme et pour le mouvement aux animaux spermatiques qui contient la liqueur du male." Buffon to Cramer, 14 December 1748, MS Cramer, Suppl. 384 fol. 120–121, Bibliothèque publique et universitaire de Genève, quoted in Weil, "Correspondence," 131. Buffon's experiments are reported in "Découverte de la liqueur seminale dans les femelles vivipares, et du réservoir qui la contient," *Première suite des mémoires de mathématique, et de physique tirés, des régistres de l'Académie Royale des Sciences, de l'année 1748* (Amsterdam: Schreuder et Pierre Mortier, 1756), 309–35.

29. Lyon and Sloan, *From Natural History,* 166–67, also 209, note 5, have explored the idea that Buffon observed Brownian motion. See also Roe in "John Turberville Needham," 163 who believes that Needham observed zoospores.

30. "Je viens de lire les 3 vol. de l'Hist. Nat. de Mr. Buffon qui ont paru depuis quelques tems. Ils renferment quelques faits curieux; mais, pas assés à proportion de leur grosseur. Il y a un très grand nombre de conjectures, dont plusieurs sont fort hardies. Il arrive souvent à Mr. de Buffon de les donner d'abord pour des conjectures, et puis de s'en servir comme de Principes demontrés. . . . Mr. de Buffon pretend presque tout expliquer sur la generation: mais, j'avouë que je ne puis considerer son systeme, que comme une hypothese hasardée. Il fait trop prouver aux faits sur les quels il l'a batit. Il semble quelques fois qu'il se laisse emporter à son imagination. Si son ouvrage est fort gouté, je crains qu'il ne fasse tort à l'Hist. Nat. en ramenant le gout des Hypotheses." Letter from Trembley to Willem Bentinck, 9/20 janvrier 1750, Ms Egerton 1726, British Library. Also quoted in M. Trembley, *Corresp. inédite,* note p. 330. I am indebted to Elaine Robson of the University of Manchester for obtaining a copy of this letter for me.

31. See Shirley A. Roe, "Needham's Controversy with Spallanzani: Can Animals Be Produced from Plants?" *Lazzaro Spallanzani e la biologia del settecento: Teorie, esperimenti, instituzioni scientifiche,* edited by G. Montalenti and P. Rossi (Florence: Leo s. Olschki, 1982), 295–303.

32. Abraham Trembley, *Mémoires, pour servir à l'histoire des polypes, à bras en forme des cornes* (Leiden: Verbeek, 1744), 312.

33. Lyon and Sloan discuss the limitations of the observational style of natural history in *From Natural History,* 14–27.

34. Raymond Savioz, ed., *Mémoires autobiographiques de Charles Bonnet de Genève,* (Paris: Vrin, 1948), 65. In his *Mémoires autobiographiques,* an autobiography consisting of letters to his closest friends, Bonnet describes a manuscript, *Contemplation sur l'Univers,* that he said he produced during this period. Though this work does not appear to be among his papers at the *Bibliothèque publique et universitaire* in Geneva, the introduction, consising of his early thoughts in response to Buffon's system, possibly conceived as early as 1749 but at least by 1751, was published in 1762 as the first five chapters of his *Considérations sur les Corps Organisés.* Bonnet, *Mémoires autobiographiques,* 101. See also Jacques Marx, *Charles Bonnet contre les Lumières, 1738–1850* (Oxford: The Voltaire Foundation, 1976) in vol. 156–57 *Studies on Voltaire and the Eighteenth Century,* 80 ff.

35. Bonnet, *Mémoires autobiographiques,* 101.

36. "[A]vec le bon et sage esprit que je vous connois, je ne suis point étonne que vous n'aiez point été seduit par le ton avec lequel M. de Buffon expose des idées trop hazardées et qui, si elles étoient adoptées, nous replongeroient dans les tenèbres dont nous avons été tirés par les travaux de tant de grandes hommes." Réaumur to Bonnet, 27 April 1751, Papers of Réaumur, (DB) Paris Academy of Sciences.

37. Lignac was preparing an anonymous attack on Buffon through his *Lettres à un Amériquain,* a project possibly conceived in Réaumur's company. [Joseph Adrien Lelarge de Lignac], *Lettres à un Amériquain sur l'Histoire Générale et particulière de M. de Buffon* (A Hambourg et se trouve à Paris chez Duchesne). *Lettres à un Amériquain* was probably printed near Paris. It is uncertain whether Réaumur collaborated with Lignac in the first three volumes, which appeared in 1751. Maurice Trembley established that Réaumur took part in Lignac's efforts to refute the infusion experiments of Needham and Buffon in the two volumes published in 1752. See *Les sciences de la vie,* 692 and M. Trembley, *Corresp. inédite,* note, p. 362–63.

38. Draft of letter from Bonnet to Réaumur, 25 November 1751. Ms. Bonnet 70 fol. 27v–28, Bibliothèque publique et universitaire de Genève. The part of the text of Bonnet's unpublished letter that pertains to his theory of generation is included here because it fixes in time an important

turning point in Bonnet's intellectual development. I have retained idiosyncratic punctuation and spelling.

Voicy Mr. très en Gros mon petit Systeme que je vous prie neanmoins de ne pas communiquer. Je l'avois imaginé plusiers années avant que l'Histoire naturelle de Mr de Buffon eut parû.

Je crois que touts les Corps Organises étoient contenus des le commencement dans les Germes qui les representoient en petit. Je ne diside point sy ces Germes etoient renfermés les uns dans les autres, ou s'ils étoient répendus partout. Ces deux Hipotheses me semblent également raisonnables. Les calquiels qu'on antasse contre la premier ne frayent que l'Imagination; et l'Imagination doit se taire dans les choses qui sont du ressort de l'entendement. L'emboitement des Germes les unes dans les autres ne supose pas la division actuelle de la matiere a l'infini mais une division indefinie. Le Grand et le petit sont-ils quelque chose dans la Nature [?] Seras ce en accumulant des zeros que nous fixerons tous la grosseur des dernieres Particules de la matiere [?] et par rapport a la proportion de vitesse dans l'accroissement des Germes qui fournissent aux Generations sucessives, pour quoy n'observeroit elles que les loix a nous connues [?] Je pense donc que les Germes contiennent toutes les parties solides qui constituent l'espece de la plante ou de l'animal. Mais je n'admets pas, ou du moins je ne le crois pas necessaire, que ces solides contiennent actuellemt des fluides. Je conjecture que ceux cy sont fournis originerement par l'acte de generation. Je regarde la liqur seminale comme un fluide nourissier, dont la subtilité est proportionnée a la petitesse des parties du Germe. Ces Parties travaillent ce fluide, et se l'incorporent. Il est porté dans le coeur, et y commence une circulation qui ne finira qu'avec la vie. Mais comme les solides du Germe ne sont pas tous d'une mème tisseure, et qu'ils demandent par consequent des sucs nouriciers d'une diversité relative. Je suppose que la liqueur seminale contient de tels sucs, des sucs apropriés a chaque genre de parties. J'imagine que les Organes de la Generation sont faits avec un art si merveilleux, qu'ils sont une Espece d'abrégé de tous le corps; en sorte qu'ils separent les particules analogues au diverses parties, &c. L'Acroissement sopere par la reception, ou l'intrmixtion [?] des Atomes Alimentaires dans les mailles des Solides, que je supose douées d'une grande ductilité les raports que la Nature a établis entre le Germe et la Liqueur Seminale destinée a procurer ces premiers accroissements determinent la posibilité de a Generation. S'ils peuvent setendre fort au dela de ce que nous pensons, et embrasser mème des classes differentes. De la les Mulets. S'ils n'en gendrent point, c'est en vertu d'une loi qui veut quils ne soit que la Liqueur Seminale propre a lespece qui puisse developer dans le Germe les Organes de la Generation. la raison de cette loi pouroit être assignée par diverces conjectures. Mais je [... ?] icy parce que j'aperçois que mes Idées sobscurcient. Vous voyés, Mr., a peu près quelles sont Mes Idées sur la Generation. vous en jugeriés mieux, et elles paretroient mieux aleurs [?] avantage, si lisiés le fil de mes méditations. Ce n'est pas que j'en sois fort satisfait je les regarde comme un Roman, et la famille aux 24 doigts ne contribuent pas peu a me les suspects. Daignés, Mr. mayder [m'aider] a debrouiller un peu ce caos en me faisant part de votre jugemens. Au reste, des Monstres a 24 doigts sont connus depuis un grand nombre de siecles. Ou en lit un Exemple dans un auteur de grands poids, dans le prophete Samuël. Des Germes collés les uns aux autres, ou greffés en tout ou en Partie &c. me fournissoient une solution physique des differentes Genres de Monstres.

Il en est un auquel Mr de Buffon n'a prété aucune attention et qui en merite beaucoup; C'est celuy que nous offrent les Portions Intermediares et posterieures de cette Espece de ver long dont j'ai parlé dans mes observations et qui poussent une Queuë au lieu d'une tête. Le frequence de ce Phenomene a bien de quoy surprendre.

39. See above, note 20.
40. Lesley Hanks, *Buffon,* 82–87. Bonnet used Hales as the model for his experimental study of the growth of worm segments, reported in his first book, *Traîté d'Insectologie* (1745), published as vol. 1 of *Oeuvres d'histoire naturelle et de philosophie* (Neuchâtel: S. Fauche, 1779–83).
41. Bonnet to Réaumur. 25 November 1751. For French text, see note 37 above.
42. Ibid. See note 37 above.
43. On polydactylism in Réaumur and Bonnet, see Bentley Glass, "Maupertuis, Pioneer of Genetics and Evolution," in *Forerunners of Darwin: 1745–1859,* edited by Bentley Glass, Owsei Temkin, William L. Straus, Jr. (Baltimore: The Johns Hopkins Press, 1959), 65.
44. Réaumur to Bonnet, #62, 10 December 1751, Papers of Réaumur, (DB) Paris Academy of Sciences. Capitalization and accents are lacking in the original.

[L]es idees dont vous m'avez fait part sont des plus vraisemblables qu'on puisse avoir sur une si obscure matiere et ont pour base le systeme des germes. a vous permis de croire que les germes ne contiennent que les parties solides de la machine animale; on ne vous demontrera jamais le contraire; comme je ne vois pas que vous puissies prouver que ces germes sont prives de liqueurs. la question est de scavoir si le germe vit, ou s'il est prive de vie, si les liqueurs y circulent ou n'y circulent pas. car quand vous en excludes les liqueurs c'est sans doute seulement leur circulation, des liqueurs se trouvent dans touts les corps solides quoiqu'inanimes. C'est faire un usage fort naturel de la liqueur seminale que de l'employer, a faire commencer la circulation; le developpement. mais il me semble que vous vous chargez assez gratuitement d'un des plus grandes difficultes lorsque vous voules que dans la semence, et dans les sucs fournis a l'accroissement, il y en ait d'appropries a chaque genre de parties. des idees plus simples me contentent sur cet article. le suc fourni par la terre fait croitre et developper toutes les parties de cette plante; quelque difference qu'il ait entr'elles de tant d'aliments differents l'estomach scait faire un chyle assez semblable.

45. Ibid. Also quoted in M. Trembley's partial transcription of this letter in *Corresp. inédite,* note 362–63.

46. Quoted by Bentley Glass, "Heredity and Variation in the Eighteenth Century Concept of Species," *Forerunners of Darwin,* 166.

47. Quoted by Glass, "Heredity and Variation," 167. Translation by C. O. Whitman, "The Palingenesia and the Germ Doctrine," in vol. 3 of *Biological Lectures Delivered at the Marine Biological Laboratory at Woods Holl in the Summer of 1894* (Boston: Ginn, 1896), 246. Whitman objected that when Bonnet modified the preexistence theory to account for the regeneration of hydra, his definition of the germ came so close to epigenesis that it revealed "a fatal inconsistency that sinks the whole speculative fabric below the dignity of a 'romance.'" Glass has responded that Bonnet's theory, in fact, reconciled preformation and epigenesis. Bonnet's definition of the germ bears a striking similarity to the concept of the *genome* shared by modern geneticists and developmental biologists. In the *genome,* DNA molecules are preorganized in such a way as to control the development of an embryo or a bud. "The particles of matter, the genes and the chromosomes," writes Dr. Glass, "are preformed and duplicated in every part of the plant's or animal's body; but the development that ensues is the reproduction of form, the differentiation of parts, de novo." Personal communication to the author, 3 February 1988.

The Influence of Samuel George Morton on American Geology

PATSY GERSTNER

Late in 1827 Samuel George Morton (1799–1851), a Philadelphia physician, read the first part of a paper on the Secondary formation of the east coast of the United States and its characteristic fossils at a meeting of the Academy of Natural Sciences of Philadelphia. It was the first of many papers by Morton in which he developed the concept of a Cretaceous formation in the United States.[1] Morton successfully used fossils to identify and trace the formation from New Jersey to Virginia, the Carolinas, Georgia, Alabama, and thence north and west toward the Mississippi. He also correlated the American Cretaceous with that of Europe on the basis of fossils. Morton's paleontological approach to stratigraphy was still new in the United States and was the first extensive attempt to use fossils, rather than the structure and mineral content of the rock, as a means of identifying geological formations. Therefore, it constitutes an important and direct contribution to the growth of the geological sciences in this country, a contribution that Leonard Wilson astutely summarized a few years ago in the context of the growth of American geology.[2] It is the purpose of this paper to suggest that Morton's influence on the development of geology was even more extensive than Wilson or others have supposed.

Morton graduated from the University of Pennsylvania in 1820. Intent on further study in Europe, as were many Americans interested in medicine at the time, he attended the University of Edinburgh from which he graduated in 1823. Morton had been interested in natural history since childhood, an interest stimulated principally by his stepfather Thomas Rogers, and while in Edinburgh he pursued this interest by attending the lectures of the Wernerian geologist, Robert Jameson.[3] Although what Morton learned in these lectures cannot be demonstrated, key changes were taking place in geology in England and France at the time, changes that emphasized the importance of fossils in stratigraphy. In the late eighteenth century geologists divided the rocks of the earth's surface into several major groups on the basis of their structure and content. From the oldest to the youngest they were called Primitive, Transition, Secondary, Tertiary, and Alluvial. Individual formations within these broad categories were ordered on the same bases. Early nineteenth century studies by William Smith in England, George Cuvier and Alexandre

Brongniart in France, and others demonstrated that fossils were more reliable guides, however. Among other things the use of fossils resulted in defining many specific formations such as the Cretaceous (the upper part of the Secondary), and provided a sound basis for correlating widely separated and structurally distinct beds.

Because Jameson approved of many of Cuvier's views as logical extensions of Werner's theories, and applauded Cuvier's use of fossils in stratigraphy as an extension of Werner's methods, it is likely that he discussed the use of fossils in stratigraphy in his lectures. If he spoke of Cuvier, he may also have spoken of Brongniart, who played an important role in describing the Cretaceous formations of Europe. Brongniart demonstrated that although white chalk was the most distinctive part of the Cretaceous, fossils indicated that beds of very different composition were also part of the Cretaceous. Since the white chalk is not present in America, Morton argued in a similar vein.[4]

Reflecting both old and new interests, Morton began, soon after his return, to collect fossils from New Jersey and Delaware where the construction of the Delaware and Chesapeake Canal was exposing new areas.[5] Morton had joined the Academy of Natural Sciences of Philadelphia before going to Edinburgh and became very active in this organization after his return. His reputation as an expert in the identification of fossils, especially invertebrates, seems to have grown quickly, and he began arranging the Academy's "whole series of European fossils . . . (upward of two thousand Species from all sorts of formations)."[6]

Although the use of fossils in stratigraphy advanced rapidly in Europe in the first quarter of the century, this was not the case in the United States. On the basis of their composition, the deposits along the Atlantic coast had been identified as Alluvial and/or Tertiary by some of America's pioneering geologists, including William Maclure and Amos Eaton. It was not until 1823 that Peter Finch, a visitor from England, suggested that fossils indicated that the beds were both Tertiary and Secondary. His limited study was followed by the more detailed work of Lardner Vanuxem who was skilled in the use of fossils in stratigraphy by virtue of an education at the École des Mines in Paris where he had studied with Brongniart.[7] A trip to Mexico interrupted Vanuxem's plans to publish his studies, so he asked Morton to arrange his notes for publication. Because of his extensive knowledge of fossils and possibly because of shared familiarity with the work of Brongniart, or at least of the use of fossils as stratigraphical indices, it is not surprising that Vanuxem asked Morton to arrange his notes. In fact, Morton may have been the only other person in the United States qualified to do so.

In addition to preparing Vanuxem's notes for publication, Morton substantially added to them by preparing a study of the fossils characteristic of the Secondary deposits of New Jersey and Delaware to accompany them. It was part of this paper that he read to the Academy late in 1827 and completed at another session in early 1828.[8]

Morton's influence on geology beyond his publications has received little attention, and it is the aim of the rest of this paper to point out that geology in the United States owes much more to Morton and that his influence was widespread. This suggestion is based primarily on a sampling of his correspondence, major collections of which are preserved at the Pennsylvania Historical Society and the American Philosophical Society, while smaller important collections are found in such diverse places as the Marietta (Ohio) College Library and the Alexander Turnbull Library in Wellington, New Zealand. The correspondence shows that, in addition to arguing for the use of fossils in stratigraphy and encouraging others to follow his lead, Morton was considered one of the nation's leading experts on invertebrate fossils. He tried to give his help to all those who requested it and in any way that he could. In this, and in tireless efforts to build up useful collections of fossils for study at the Academy of Natural Sciences of Philadelphia, Morton helped in the advance of geology in the United States as surely as did his papers on the Cretaceous. The following are a few examples selected from Morton's correspondence to illustrate the point. No effort is made to be comprehensive or to do more than suggest by example the importance of this aspect of Morton's labor.

Although a seemingly unagressive person, Morton was so convinced of the legitimacy of using fossils that he was prepared from the beginning to do whatever he could to encourage the acceptance of their use, even if it meant challenging the establishment in the person of Amos Eaton. Eaton's continued insistence on calling the Atlantic formations Tertiary in spite of the evidence given by Morton for their Secondary nature led to an exchange of correspondence between the two in 1829. Morton sent Eaton several publications on the Atlantic formations and emphatically told him that if the fossils found there were referable to Tertiary deposits

> then certainly so much modern Geology as is based on Fossil conchology, must fall to the ground—and the laws which have been established or adopted by Cuvier, Brongniart, Conybeare, Phillips & others must be abandoned entirely.[9]

This brash attack from the young Philadelphian elicited an answer from Eaton in which he demanded some assurance that Morton had indeed qualified himself with adequate examination of the rocks in question. Had Morton visited the New Jersey marl pits in person and was he familiar with the standard references on fossil concholgy?[10] Morton lost little time in sending assurances that he had extensive first hand acquaintance with these things, that he lived on the very margin of the New Jersey formation, and that he was very familiar with European fossils. He informed Eaton of his studies of the extensive series of European fossils at the Academy of Natural Sciences of Philadelphia and noted that he had personally collected many fossils from the American formation for comparison. He told Eaton that he had visited the Chesapeake and Delaware Canal excavation at six different periods of excavation. "In fact—my *personal* exertions in this particular have been extreme,—of which a handsome acknowledgment has recently been made by the [American] Philos[ophical] Society, in submitting to my examination (as chairman of a committee) a series of official documents and a fine collection of Fossils recently received by the society from one of the Engineers of the Ches. and Del. Canal."[11] Eaton yielded on the possible significance of fossils but was not readily convinced that the fossils were Secondary instead of Tertiary, and the debate went on in the pages of the *American Journal of Science*.[12] Nevertheless, Eaton's opinion of Morton grew. He kept an open mind to Morton's theories on the Secondary nature of the deposits and on their correlation with European deposits, and he was eventually to be convinced of the correctness of Morton's views.[13]

While Morton did not hesitate to nudge the old-line geologists, his main importance in the use of fossils in stratigraphic studies lies in the assistance given to a newer generation of geologists, particularly those associated with state geological surveys. In no case is this more apparent than in the encouragement and direction given to Timothy Abbot Conrad. While Morton was working on the Cretaceous, Conrad had begun an intensive study of the fossils characteristic of the southern Tertiary formation that resulted in his *Fossil Shells of the Tertiary Formations of North America* (1832–1833). Conrad and Morton, both Philadelphians and members of the Academy of Natural Sciences, were close friends and together had "resolved that the geology of this country shall be thoroughly explored."[14] A steady flow of letters between Morton and Conrad characterized the duration of Conrad's survey. There is something of the student-to-master relationship in many of Conrad's letters to Morton and the conclusion that Conrad owed much of his ability and interest to Morton is inescapable. There is evidence of Morton's hand in Conrad's initial venture into the Tertiary study, for Conrad's publication in 1832 prompted Morton to write to his English friend, Gideon Mantell, that he had "long ago projected such a work."[15] There seems to be a sentiment in this letter suggesting that Conrad was carrying out Morton's idea and perhaps doing so because of Morton's encouragement. That this was so is clearer in a subsequent letter to Mantell in which Morton says that "Mr. C. undertook this long journey [to study the Tertiary] solely at my suggestion and under my exclusive patronage."[16]

Morton recommended Conrad as director of the New Jersey Geological Survey that began in 1835, but this much-sought-after position went to Henry Darwin Rogers.[17] A short time later Morton recommended him for a post with the New York Geological Survey, and Conrad was appointed to head the geological reconnaissance of the third district of New York when the survey of that state began in 1836.[18] A year later he became paleontologist for the entire survey. At the time of his appointment, he was one of two geologists on the Survey competent to use fossils as geological indices; the other was Vanuxem. Eaton, also a geologist for the survey, may have felt by then more comfortable in the use of fossils as stratigraphical indices, but this is not apparent in his publications about the survey. Of the several geologists working on the New York Survey, Conrad was the leader in using fossils to identify, arrange, and correlate strata. His influence is seen throughout the early years of the Survey as he worked to establish this basis for the work in New York. Although not the earliest of the state surveys, New York's was the most influential, and the successful use of fossils there assured interest in their use elsewhere. Thus, through Conrad, Morton's influence was felt widely in America. Vanuxem might also have influenced geology in a similar manner, but he was never as outspoken on the issue as was Conrad.[19]

Studies of fossil invertebrates in America were limited when Morton began his work in the 1820s, and he became a source of information and help in identifying specimens, whether they were to be used in stratigraphy or not. Morton received dozens of letters in the late 1820s and early 1830s requesting such assistance. One of Morton's early correspondents was James E. DeKay in New York.[20] DeKay had proposed a work on American fossils and asked Morton to help with the invertebrates. Others followed, including William Cooper of the Lyceum of Natural History in New York and conchologist Edmund Ravenel. Two series of letters, one with Edward Hitchcock of Massachusetts and the other with Samuel Hildreth of Ohio, give a good sense of Morton's help.

Hitchcock was appointed the state geologist of Massachusetts in 1830. Unfamiliar in the early years of the survey with the invertebrates, he turned to Morton for help and assistance in naming and describing the organic remains he encountered in his survey. A letter written to Morton on 11 June, 1832 tells Hitchcock's story:

> I have been for some time engaged in the examination of the geology of Massachusetts under the direction of the government and have recently found a number of fossils on Martha's Vineyard and Nantucket in a formation which I suppose to be a continuation of that of N. Jersey so ably described by you. . . . They consist of bones and teeth in conglomerate & bones shells & zoophytes in green and ferruginous sand; also vegetable remains in red & argillaceous oxide of iron. As I am but little acquainted with fossil geology & have scarcely no means in N. England for instituting a comparison between those which I have found & specimens from other parts of the country & Europe I write you at this time to enquire whether you can conveniently render me some assistance in this matter. I wish if possible in the course of a few weeks to come to Philadelphia with the specimens provided you can assist me in their examination. Or if it should not be in my power to go could you, sir, look over the specimens & name them provided I should send them to You?[21]

Hitchcock never got to visit Philadelphia and in July shipped the specimens to Morton. Subsequent shipments were sent in August and in October. Among these were specimens that Hitchcock did not think were from formations continuous with those of New Jersey. While it might be expected that people would consult Morton about Secondary, or especially Cretaceous fossils, the

Hitchcock correspondence demonstrates what is seen many times over in the Morton correspondence, that people turned to him for general assistance with invertebrates and did not limit their requests to the period Morton knew best.

When he sent the specimens in October, he had not yet received any word from Morton about the earlier shipments, and Hitchcock wrote that "I shall be greatly obliged to you if you can ere long give me your opinion as to these remains and those forwarded some time since as I am anxious to give some description of them in my report to the government . . . " and, almost with a note of apology, asked Morton to "excuse this request in view of the difficulties with which I am surrounded in respect to the study of this part of geology."[22] Shortly thereafter Morton sent the commentary on the fossils and apparently remarked to Hitchcock that he (Hitchcock) had overrated Morton's abilities.[23] Morton liked to tell his correspondents that he was competent only with certain groups of invertebrates from one specific formation, but, like most to whom he made this comment, Hitchcock was not convinced of its complete truth. "You think I have overrated your attainments in this department of knowledge but I assure you I should like to receive more commentary of the same kind. I could not expect indeed that you would be as familiar with the organic remains of the older rocks as with those to which you have given special attention with a view to publication. Still if you can give me only the generic names I shall be quite thankful."[24]

In the mid-1830s Morton was collecting skulls for craniological studies and in the hope of adding North American Indian skulls from Ohio to his growing collection, he started to correspond with Hildreth, who lived in an area of Ohio rich in Indian burial sites. Hildreth would later serve as the first assistant to W. W. Mather on the Ohio Geological Survey, but at the time of contact with Morton he was writing a paper on the coal deposits of Ohio. He felt uncomfortable with the fossil invertebrates and plant fossils that he found in association with the coal and asked Morton for help. In the spring of 1835 Hildreth sent the fossils to Morton, who responded that he was "willing to try what I can do with them, but I have been accustomed to organic remains in a state so much more perfect (generally speaking) that I feel some doubts of my competence to describe these."[25] But even with these reservations Hildreth was eager to have Morton assist him, and Morton's description of the organic remains accompanied Hildreth's publication in January, 1836.[26] When Hildreth first contacted him, Morton thought that Hildreth wanted his help only with the shells from the deposits and discovered only shortly before the article was to be published that Hildreth expected him to help with the plants also. He did the best he could and his descriptions were included in the publication. However, he told Hildreth that if he would leave these fossil impressions with him for a while longer, he would describe them more fully so that the descriptions could be published in a subsequent issue of the *American Journal of Science*.[27] But this was one of the few times that Morton had a change of heart, writing to Hildreth that he was too uninformed on the plants of the Ohio region and knew of no one competent to handle them.[28]

Even as Americans grew in their familiarity and use of invertebrate fossils, many still turned to him for help or encouragement. James Hall, who joined the New York Survey in 1837 and became the state paleontologist in 1843 when Conrad left because of poor health, is an example. Although the author of a multi-volume study of the state's fossils, the first volume of which appeared in 1847, he seemed unsure of himself when he became the paleontologist.[29] Looking on Morton as "the pioneer in American Paleontology" he asked for his "opinion concerning this matter of paleontology." Morton's encouragement offered him the support he needed. "If I could believe," he told Morton, "I merit half the praise you bestow my days would be brighter and my dreams pleasanter, but I toil on hoping only to accomplish my task and feeling still that what I do may not be favorably received or perfectly done. Such a letter as yours gives me hope and encourages and makes me happier and better."[30]

Morton recognized the necessity of having geological and paleontological specimens for study and with which to compare other specimens, and he consequently devoted himself to developing collections at the institutions of Philadelphia, especially the Academy of Natural Sciences. As early as 1829 he tried to arrange for the Academy to purchase a large collection from Joseph Dorfeuille of the Western Museum in Cincinnati. Failing that, he tried to arrange a joint purchase between the Academy and the American Philosophical Society or between the Academy and the Lyceum of Natural History in New York.[31] Although unsuccessful, this may have been one of the first efforts in the United States at a cooperative purchase between scientific societies. As curator of the Academy from December 1830 to December 1832, as corresponding secretary from November 1831 to May 1840, as vice president from May 1840 to December 1849, and as president until May 1851, Morton worked tirelessly to build the collections and, as noted above, to organize them for further use. An 1835 curators report at the Academy noting shortcomings in the condition of the collections emphasized that the authors did "not speak in reference to the departments of geology, mineralogy, or conchology, the arrangement of which, the work of our fellow-member, Dr. Morton, has so long been a subject of admiration."[32] An especially interesting component of Morton's quest for fossil collections is the attention given to acquiring European specimens, a reflection of Morton's own reliance on European specimens for comparison and his belief in the correlation of formations between the two continents. His acquisitions were almost always part of an exchange so that he was able to add significantly to collections in Europe while building those at home, and his correspondence with Europeans is extensive. Among the earliest letters are those exchanged in 1828 with James De Carle Sowerby, mineralogist and paleontologist. In return for specimens from the United States, Sowerby declared that he would send as many of the fossils Morton wanted as possible and expressed interest in a continuing exchange of specimens with Morton. Sowerby had been looking for a convenient channel for exchanging fossils with the United States and Morton could provide it. He also expected Morton to be able to clarify the identification of several fossils already in his collection.[33]

With Henry Galiotti of Brussels, Morton arranged an exchange of fossils from Mexico, Belgium, and Italy.[34] By frequent exchanges of specimens between Morton and Gideon Mantell in England, who became through correspondence one of Morton's strongest allies and closest friends in geological pursuits, both were able to add significantly to local collections. For example, in one box Morton found a helpful series of shells showing a relation between "ancient tertiary" and the present strata as well as fossil plants and a variety of other specimens.[35] With the assistance of Morton, Mantell strove to establish the best American collection in London. One of the first shipments of fossils from Morton prompted Mantell to remark that the English "were suddenly in the possession of all the treasures of North America."[36]

Morton's devotion to the Academy is reflected in his last will and testament in which he wrote:

I much regret that my large family, and the serious losses I have sustained during the past few years, will not permit me to leave a monied legacy to the Academy of Natural Sciences; an Institution which I have endeavored to serve, with my time and other resources, during a period of twenty nine years. I trust in this instance 'the will may be taken for the deed,' and that my past services may suffice to associate my memory with this, my favourite institution."[37]

The foregoing are only a few examples of Morton's influence on American geologists. His correspondence offers much more, covering a wide range of efforts to advance American geology. For example, he willingly loaned his manuscript of the *Synopsis of the Organic Remains of the Cretaceous Group* before publication to the young Pennsylvania geologist Henry Darwin Rogers who wanted to include some of Morton's information in a paper on American geology he was preparing for the British Association for the Advancement of Science.[38] He corresponded with many state geologists in addition to those already mentioned, including Owen of Indiana and Troost of Tennessee, who like the others, recognized Morton as one of the most esteemed members of the geological community. In addition to the collecting he did for the Academy, he tried to get other important collections to the United States, one of which was Mantell's own which was offered for sale in 1838.[39] His correspondence with Mantell and other European geologists created a ready access to information for him and his colleagues, and the correspondence provides a generous look at the many things going on in geology in the United States in the late 1820s and 1830s.

Morton's influence in establishing the Cretaceous in the United States and correlating it with the European Cretaceous was direct. That he made a less direct, but no less important, contribution through his many contacts and his work at the Academy seems clear. Nevertheless, his influence has remained in the background. This may be due in part to the fact that Morton, by temperament, preferred to remain in the background. Another reason, and one perhaps historically more significant, was his interest in craniology. Morton's interest in craniology developed when he could not find specimens to illustrate an introductory lecture in anatomy in 1830 on "The Different Forms of the Skull as Exhibited in the Five Races of Men."[40] From about 1832 there is a distinct shifting of emphasis in his correspondence from geology to craniology, and peers began to associate him more closely with that study. Ultimately his craniological studies became an important basis for the racist arguments of Nott, Gliddon, and others, and Morton has not been treated kindly by modern authors who examine only this aspect of his work. His influence on geology beyond his work on the Cretaceous clearly deserves more attention.

Notes

1. Morton's paper was read in December 1827 and January 1828 and was published as "Description of the Fossil Shells Which Characterize the Atlantic Secondary Formation of New Jersey and Delaware; Including Four New Species," *Journal of the Academy of Natural Sciences* 6 (1829): 72–100. Morton's other publications on the Cretaceous include: "Additional Observations on the Geology and Organic Remains of New Jersey and Delaware," *Journal of the Academy of Natural Sciences* 6 (1830): 189–203; "Synopsis of the Organic Remains of the Ferruginous Sand Formation of the United States; with Geological Remarks," *American Journal of Science* 17 (1830): 274–94 and 18 (1830): 243–250; "On the Analogy Which Exists Between the Marl of New Jersey, etc. and the Chalk Formation of Europe," *American Journal of Science* 22 (1832): 90–95; "Supplement to the 'Synopsis of the Organic Remains of the Ferruginous Sand Formation of the United States,' Contained in Vols. XVII and XVIII of This Journal," *American Journal of Science* 23 (1833): 288–94 and 24 (1833): 128–32; *Synopsis of the Organic Remains of the Cretaceous Group of the United States,* (Philadelphia: Key and Biddle, 1834); "Description of Some New Species of Organic Remains of the Cretaceous Group of the United States: With a Tabular View of the Fossils Hitherto Discovered in the Formation," *Journal of the Academy of Natural Sciences* 8 (1842): 205–27.

2. Leonard G. Wilson, "The Emergence of Geology as a Science in the United States," *Cahiers D'Histoire Mondiale* 10 (1967): 416–37.

3. Morton's early interest in natural history and his acquaintance with Jameson is discussed by George B. Wood, *A Biographical Memoir of Samuel George Morton, M.D.,* (Philadelphia: T. K. and P. G. Collins, 1853), 5; Charles D. Meigs, *A Memoir of Samuel George Morton, M.D., Late President of the Academy of Natural Sciences of Philadelphia,* (Philadelphia: T. K. and P. G. Collins, 1851), 15; Henry S. Patterson "Memoir of the Life and Scientific Labors of Samuel George Morton," in J. C. Nott and George B. Gliddon, *Types of Mankind,* (Philadelphia: J. B. Lippincott and Co., 1871), xvii–lvii.

4. Wilson, "Emergence," 423 and 426–27. Brongniart's principal work on the Cretaceous is "Sur les caractéres zoologiques des formations, avec l'application de ces caractéres à la détermination de quelques terrains de craie," *Annales des Mines* 6 (1821): 537–72.

5. In an 1828 letter to Benjamin Silliman, Morton said he had been collecting the fossils "for several years past." 17 August, 1828, Boston, Mass. Public Library.

6. Morton mentioned this activity to Amos Eaton in a letter of 30 September, 1829, New York Historical Society (New York City), Misc. Mss. M.

7. For a brief sketch of Vanuxem see the article by John Wells in C. C. Gillispie, ed., *Dictionary of Scientific Biography* (New York: Scribners, 1972). Vanuxem, born in Philadelphia, taught at South Carolina until 1826. He travelled extensively up and down the coast to outline the formations.

8. Vanuxem's notes were incorporated in a paper read in January 1828 and published as "Geological Observations on the Secondary, Tertiary, and Alluvial Formations of the Atlantic

Coast of the United States of America. Arranged from the Notes of Lardner Vanuxem," *Journal of the Academy of the Natural Sciences* 6 (1829): 60–71.

9. Morton to Eaton, 16 September, 1829, Pennsylvania Historical Society (Philadelphia), Gratz Collection.

10. Eaton to Morton, 26 September, 1829, American Philosophical Society (Philadelphia), Morton Papers.

11. Morton to Eaton, 30 September, 1829, New York Historical Society (New York City), Misc. Mss. M.

12. Morton, "Synopsis of the Organic Remains of the Ferruginous Sand Formation of the United States; with Geological Remarks," *American Journal of Science* 17 (1830): 274–95 and Amos Eaton, "Geological Prodromus," *American Journal of Science* 17 (1830): 63–69.

13. Eaton to W. L. Marcy, 2 April, 1836, quoted in Ethel M. McAllister, *Amos Eaton Scientist and Educator 1776–1842,* (Philadelphia: University of Pennsylvania Press, 1941), 325–26. Original letter in the Gratz Collection of the Pennsylvania Historical Society (Philadelphia).

14. Morton to Gideon Mantell, 16 September, 1832, Alexander Turnbull Library (Wellington, New Zealand), Mantell Collection.

15. Ibid.

16. Morton to Mantell, 1 May, 1833, Alexander Turnbull Library (Wellington, New Zealand), Mantell Collection. Past suggestions indicate a belief that Conrad's trip was financed by at least three people, Morton, Isaac Lea, and Charles A. Poulson. See Harry Edgar Wheeler, *Timothy Abbott Conrad With Particular Reference to His Work in Alabama One Hundred Years Ago* (Ithaca, N.Y.: Paleontological Research Institute, 1935). It would seem from Morton's comment to Mantell that this may not have been the case.

17. Morton to P. D. Vroom, governor of New Jersey, March 1835, Pennsylvania Historical Society (Philadelphia). Among those contending for this position in addition to Rogers were Thomas Green Clemson, James Pierce, and possibly Jacob Green. Joseph Henry recommended Rogers as did Richard Stockton Field, member of the New Jersey Assembly. Joseph Henry to Rogers, 9 March, 1835, Massachusetts Institute of Technology (Boston), Rogers Papers; Nathan Reingold, ed., *The Papers of Joseph Henry, November 1832–December 1835, The Princeton Years,* vol. 2 (Washington, D.C.: Smithsonian Institution Press, 1975), 364.

18. Morton to William L. Marcy, Governor of New York, 6 May, 1836, Pennsylvania Historical Society (Philadelphia).

19. Conrad, for example, attacked Henry Rogers, the director of the Pennsylvania Geological Survey in the pages of the *American Journal of Science* in 1839 over his failure to use fossils in the identification of the Silurian formations in Pennsylvania. Conrad, "Observations on Characteristic Fossils, and Upon a Fall of Temperature in Different Geological Epochs," *American Journal of Science* 35 (1839): 243–44. Editorial remarks on Rogers opposing view are in the same issue, 250–51. A good sense of Conrad's work and influence in New York comes from his annual reports of the survey such as the Second and Fifth: State of New York Assembly, Documents 275, 27 February, 1839 and 150, 17 February, 1841.

20. James E. DeKay to Morton, 25 August, 1827 American Philosophical Society (Philadelphia), Morton Papers.

21. Edward Hitchocok to Morton, 11 June, 1832, American Philosophical Society (Philadelphia), Morton Papers.

22. Hitchcock to Morton, 1 October, 1832, American Philosophical Society (Philadelphia), Morton Papers.

23. Morton to Hitchcock, 28 October, 1832, American Philosophical Society (Philadelphia), Morton Papers.

24. Ibid. Morton's assistance in identifying invertebrate fossils does not necessarily imply that he convinced each person of the principal importance of fossils as stratigraphical indices. Hitchcock is a case in point. When Hitchcock published his *Final Report on the Geology of Massachusetts* (Northhampton, Mass: J. H. Butler, 1841) he noted that it did not appear that organic remains alone were sufficient to identify the chronological position of rocks widely separated from each other. However, fossils did "show an approximate identity as to period of their deposition; and in regard to rocks in a limited district, it will show complete identity." (p. 745).

25. Morton to Hildreth, 22 July, 1835, Marietta College Library (Marietta, Ohio), Hildreth Collection.

26. Hildreth to Morton, 18 August, 1835, American Philosophical Society (Philadelphia),

Morton Papers and Hildreth, "Observations on the Bituminous Coal Deposits of the Valley of the Ohio, and the Accompanying Rock Strata; With Notices of the Fossil Organic Remains, and the Relics of Vegetable and Animal Bodies, Illustrated by a Geological Map, By Numerous Drawings of Plants and Shells, and By Views of Interesting Scenery," *American Journal of Science* 29 (1836): 1–154. Morton's identifications were contained in an appendix entitled "Being a Notice and Description of the Organic Remains Embraced in the Preceding Paper," pp. 149–54.

27. Morton to Hildreth, 18 November, 1835, Marietta College Library (Marietta, Ohio), Hildreth Papers.

28. Ibid., 3 March, 1836, Marietta College Library (Marietta, Ohio), Hildreth Papers.

29. James Hall, *Paleontology of New York* (Albany: Charles Van Benthuysen, 1847–1894).

30. See letters from Hall to Morton, n.d. (but probably late summer 1844), 29 August, 1844, 10 September, 1844, and 30 October, 1844, Pennsylvania Historical Society (Philadelphia), Library Company of Philadelphia Collection-Morton Papers.

31. Morton to Dorfeuille, 25 July, 1829, Pennsylvania Historical Society (Philadelphia), Dreer Collection-Physicians.

32. Academy of Natural Sciences of Philadelphia. Collection 86, Curators Report, 1835, signed by Charles Pickering, Walter R. Johnson, and W. McEwen.

33. Specimens and accompanying letter from Morton to Sowerby were transmitted by Thomas Hodgkins, a correspondent of Morton. Sowerby to Hodgkins, 28 February, 1828, American Philosophical Society (Philadelphia), Morton Papers.

34. Galiotti to Morton, 27 December, 1840 and 24 February, 1842, Pennsylvania Historical Society (Philadelphia), Library Company of Philadelphia Collection-Morton Papers.

35. Mantell to Morton, 27 September, 1832, American Philosophical Society (Philadelphia), Morton Papers.

36. Ibid., 16 January. 1836, American Philosophical Society (Philadelphia), Morton Papers.

37. Last Will and Testament of Samuel George Morton. Will No. 142, 1851, County of Philadelphia, Pennsylvania.

38. Morton to Mantell, 31 January, 1835, Alexander Turnbull Library (Wellington, New Zealand), Mantell Papers.

39. Silliman to Mantell, 22 March, 1838, Yale University Library (New Haven, Connecticut), Silliman-Mantell Correspondence. Although it is not clear how Morton tried to help, it may have been in regard to Mantell's hope that Girard College might be interested in his collection. Mantell thought Girard was one of the few institutions in America that could afford his asking price of £3,000–5,000. The collection went eventually to the British Museum.

40. A summary of the problems he encountered is in Samuel George Morton, "An Account of a Craniological Collection With Remarks on the Classification of Some Families of the Human Race," *Transactions of the American Ethnological Society* 2 (1848): 215. Morton's major works in craniology are: *Crania Americana; or a Comparative View of the Skulls of Various Aboriginal Nations of North and South America* (Philadelphia: J. Dobson, and London: Simpkin, Marshall & Co., 1839) and *Crania Aegyptica; or, Observations on Egyptian Ethnography Derived from Anatomy, History, and the Monuments* (Philadelphia: John Penington, London: Madden & Co., 1844).

Methodology and Its Rhetoric in Nineteenth-Century Chemistry: Induction versus Hypothesis

ALAN J. ROCKE

By way of introducing a critique of the atomic theory of John Dalton, Jacob Berzelius commented:

> There is a very essential difference between the researches of Mr. Dalton and myself. Mr. Dalton has chosen the method of an inventor, by setting out from a first principle, from which he endeavors to deduce the experimental facts. For my own part, I have been obliged to take the road of an ordinary man, collecting together a number of experiments, from which I have endeavored to mount from experiment towards the first principle; while Mr. Dalton descends from that principle to experiment. It is certainly a great homage to the speculations of Mr. Dalton if we meet each other on the road.[1]

Through the superficial appearance of self-effacement and respect towards his scientific adversary may be discerned a certain derogatory character in this passage. Berzelius was imbued with the Baconian inductivist ideals so cherished by the leading figures of the Enlightenment, and scorned Dalton's unbridled use of hypothesis—or, to repeat his more negative term, speculation.[2]

Of course Berzelius did have Dalton appropriately pegged as an instinctive and untroubled hypothetico-deductivist. Dalton's methodological style was of a piece with his relatively straightforward materialist metaphysics and his adherence to the textbook and lecture tradition of popular Newtonianism.[3] Unfortunately for Dalton, this set of methodological and metaphysical commitments was perceptibly out of step with most of the scientific world at the turn of the century and became even more so as the years went by. Despite the respect accorded to him by the British scientific community, especially in his later years, his atomic theory was subjected to repeated attacks, ranging from wholesale alteration to total rejection. Many critics echoed Berzelius by decrying the overly hypothetical character of the theory. On rare occasions Dalton conceded his use of hypotheses but defended his assumptions as both inherently probable and consistent with post hoc experiments. But Dalton was fighting a rear-guard action; Berzelius was speaking for his age.

Hypothetico-deductivism or, more simply, the method of hypothesis came

back into fashion during the course of the nineteenth century: Larry Laudan
has looked at this shift in general terms, and Kenneth Caneva has suggested
how it came about in German physics.[4] I wish to examine here both the
rhetoric and the reality of inductivism and deductivism in nineteenth-century
European chemistry. I will argue that, after some delay, chemists followed the
trend established by physicists toward describing their work in hypothetico-
deductive terms, but also that there was a remarkable degree of continuity in
actual chemical practice—as distinct from rhetoric. That reality was much
closer to the method of hypothesis than it was to classic Baconian enumerative
induction.

It has long been recognized that a sharp distinction needs to be made
between "private" and "public" science, between the "context of discovery"
and the "context of justification."[5] Scientific writing rarely describes either
the methodology, the chronology, or often even the logic of the actual process
of scientific discovery and verification followed by the writer in the laboratory
on a day-to-day basis—nor, perhaps, is such historical accuracy necessary or
desirable in such a context. This disparity may be self-conscious and deliberate
on the part of the author of a scientific paper, or there may be a significant
degree of self-delusion.[6]

Demonstrating a disparity between a scientist's professed and actual
methodology is far easier than describing the latter in detail. On one level, in
this paper I assume a simple dichotomy between inductive and hypothetico-
deductive methods. In examining the nineteenth-century rhetoric of scientific
methodology, this assumption is unproblematical, but the methodological
reality is certainly much more complex than this simple dichotomy would
imply. I regard scientific method, of a piece with scientific creativity, as to a
large degree *sui generis* with every individual scientist, and it is probably safe
to suggest that no one ever produces good science without in some way
combining inductive and deductive procedures—both in discovery and in
verification. The complexity of scientific methods in practice does not,
however, negate the utility of seeking common general patterns that may
indeed be abstractable from the flux of particularities, or of seeking genera-
tional changes in substantive as well as in rhetorical fashion; it merely teaches
caution against the temptations to overgeneralize. Consequently, this essay is
intended to be more suggestive than definitive.

1

Hypotheses were by no means unknown in Enlightenment science: Laudan
has identified George LeSage, David Hartley, and Rudjer Bošković as
examples of open and fruitful users of hypothetico-deductivism in the mid-
eighteenth century, and many more examples could be cited. But for their
pains these thinkers encountered widespread criticism and gathered few
disciples.[7] In the same spirit, the philosophes, and ultimately Aufklärer as

well, believed Lavoisier's professions of empiricism in preference to the phlogistic conjectures of Stahl. Although Lavoisier's virtuosity in the laboratory obviously played a crucial role, the suddenness with which the chemical revolution captivated Europe was likely aided by its apparent concordance with accepted modes of procedure. The strongly hypothetical orientation of Lavoisier's actual practice—for instance, his continuing faith in an imponderable caloric fluid—went unrecognized.

Count Rumford and his youthful employee, Humphry Davy, at London's Royal Institution were also acting in the spirit of the new age when they attempted, separately, to discredit the hypothesis of the caloric fluid and to substitute for it a dynamical theory of heat. Davy was also known as an outspoken critic of Dalton's atomic theory. When Dalton first spoke to him about the theory in 1803, Davy told Dalton that he regarded the ideas as "more ingenious than correct"; eight years later he expressed the same sentiment in a letter to Berzelius. And when Davy, as president of the Royal Society, presented the first Royal Medal to Dalton in 1826, he noted carefully that his praise of his countryman's theory referred to "the fundamental principle, and not the details, as they are found in Mr. Dalton's system of chemical philosophy."[8]

The respected London chemist William Wollaston had, it would seem, a similar reaction to Dalton's theory. Expressing doubts as to the reliability of Dalton's conjectural assignment of molecular formulas—a prerequisite for the determination of atomic weights—Wollaston concluded that

> since the decision of these questions is purely theoretical, and by no means necessary to the formation of a table adapted to most practical purposes, I have not been desirous of warping my numbers according to an atomic theory, but have endeavored to make practical convenience my sole guide.[9]

Davy thought that Wollaston's table of chemical equivalents "separates the practical part of the doctrine from the atomic or hypothetical part," and many of his contemporaries concurred. According to this widespread view, "equivalent weights" (or, in Davy's terminology, "proportions") had the proper empirical and inductivist cachet, in contrast to the hypothetical "atomic weights" of John Dalton.[10]

A similar pattern worked itself out on the Continent. J. L. Gay-Lussac followed Davy in calculating "nombres proportionnels" or equivalents for many elements, and his numbers were adopted by many later French chemists. In Germany, too, the influential chemist and journal editor J. S. C. Schweigger criticized Dalton's predilection for hypotheses, celebrated Wollaston's "avoidance of all atomistic sophistry," and advocated the new equivalents.[11]

Other chemists besides Dalton who openly utilized hypothetico-deductivism were treated similarly. The most obvious example is the rejection of the molecular hypotheses of Amedeo Avogadro. This notorious case study has recently been reexamined, and it is now clear that there were some excellent reasons for the unfavorable reception, given the context of early nineteenth-century science, other than the blind antagonism of the elite towards an unknown. However, the unfashionable status of the method of hypothesis in science undoubtedly made the reception of Avogadro's ideas even colder than might otherwise have been the case.[12]

Another celebrated morality play from the history of chemistry in the first half of the nineteenth century involves the unhappy fates of the close friends and scientific radicals Auguste Laurent and Charles Gerhardt. Following a magisterial dictum of their contemporary Adolphe Wurtz, most commentators have inappropriately conflated the contributions and styles of these two chemists. In fact, whereas Gerhardt professed to reject all hypotheses and made a positivistic commitment his hallmark, Laurent's work was distinguished by the prominent place accorded to hypotheses and theories of all types. In his posthumous manifesto, he argued that even false theories are useful, and even necessary, since all theories "may perhaps be nothing but a perpetual mirage destined to lead us incessantly to the exploration of new countries."[13] Of course, the reforms of both men were rejected; Laurent in particular was singled out for criticism regarding his use of hypotheses. France was probably the most positivistically oriented country of Europe at this time.

Yet another example of a hypothetico-deductivist whose work was finally rejected may seem surprising considering the passage cited at the beginning of this paper: Jacob Berzelius. It is unclear whether in responding to Dalton Berzelius was being cynically disingenuous, or simply self-deluded, but virtually all who have studied Berzelius have noted his predilection for hypotheses and the extent to which they guided his work. His ardent defense and elaboration of both chemical and physical atomism is the most representative symbol of this methodological commitment, but there are many other examples to choose from. His good friend Alexandre Marcet wrote to him in 1813: "How you love chemical theory! Shall I tell you the fault that I find in your writings? They cover too much at one time. . . . Your writings thus have a richness that dazzles ordinary philosophers."[14]

Throughout the second decade of the nineteenth century, many of the leading French chemical theorists hesitated to accept Berzelius's atomistic rules of combining proportions since they appeared to be too conjectural. However, by the 1820s Berzelius had become enormously influential throughout Europe, and by 1830 he had persuaded leading chemists, especially the Germans, to adopt his set of atomic weights in preference to the rival equivalents of Wollaston and Leopold Gmelin. His hegemony was, however, only temporary, and by 1845 Continental chemists were returning decisively to equivalents (British loyalties had never faltered).

The leading protagonist in this episode was the dynamic chemical empire-builder Justus Liebig. Following a brief youthful flirtation with *Naturphilosophie* and a thorough chemical apprenticeship with Gay-Lussac, Liebig settled down at the University of Giessen and began to develop a model teaching and research laboratory. His conversion to the Berzelian orthodoxy in 1830 proved decisive for the direction of the German chemical community. However, the violent quarrels over competing conceptions of atoms, radicals and molecules in which he engaged, especially with Jean-Baptiste Dumas, led by 1840 to a profound apostasy. Not only did he now reject Berzelian atomic weights in preference to the supposedly more empirical equivalents, but he claimed to turn resolutely from theory altogether, and he began to pursue agricultural, physiological, and technological chemistry.[15] About the same time he warned his former student Gerhardt, then in Paris, that Gerhardt's tendency to propound theories threatened his youthful career.[16]

Although on the personal level Liebig's shift was doubtless more due to fatigue, illness, and bitterness than any fundamental transformation of his instinctive values, his public rejection of deductive theorization seems to have struck a responsive chord. Berzelius himself was embittered in turn by these events, all the more so since his hypotheses regarding the theory of organic radicals were experiencing increasing resistance during the 1840s. Ironically, Liebig was a (witting or unwitting) crypto-deductivist, after 1840 as well as before; in this regard he proved to be remarkably similar to his erstwhile model Berzelius. In particular, the development of Liebig's ideas on plant and animal nutrition follows classical hypothetico-deductive lines.[17] However, this fact is surely clearer to latter-day historians than it must have been to many of Liebig's own contemporaries (or to Liebig?), who, themselves imbued with Baconian inductivist ideals, failed to recognize the hypothetical content of Liebig's method.

There is a much broader area in which this same methodological myopia kept nineteenth-century chemists from a proper appreciation of the nature of their own procedures. The inductivists were not aware that in embracing the supposedly empirical "chemical equivalents," they were simply adopting a competing set of atomic weights, no less theoretical than Dalton's or Berzelius's numbers. By starting with *any* assumed set of fixed relative weights for the elements, they were *implicitly* proposing a hypothesis from which molecular weights and formulas were deduced. One could then manipulate these formulas in comparison and interaction with experiment and ultimately come to conclusions as to whether the resulting theoretical edifice was more or less simple, elegant, heuristic, consistent, intuitively reasonable, and so on, than a competing system constructed upon a different hypothetical base. Once again, this is clearly hypothetico-deductive procedure, but it went unrecognized under the guise of a basis of "inductive" and "empirical" chemical equivalents.[18]

One final example of methodological practice needs brief mention. John Brooke has shown that Berzelius and Gerhardt (both ardent inductivists in rhetoric), as well as the hypothetico-deductivist Laurent, based much of their logic of discovery on skillful use of analogical reasoning. Analogy is in fact often a rich source of testable hypotheses. For instance, Berzelius attempted to build a system of organic chemistry based solidly on principles of inorganic chemistry, especially the model of electrochemical dualism that had proven so successful in the latter field.[19] The dualistic approach toward organic chemistry, despite its well-advertised weaknesses, did in fact suggest over the course of several decades a profusion of often highly fruitful experiments.

2

Although this essay is focused on developments within chemistry, a few glances at the wider context in philosophy and the other sciences will be helpful here. Laudan has traced the fate of the method of hypothesis from its (very limited) defense by Bacon himself, through its heyday in the seventeenth century and its eclipse in the eighteenth.[20] I have already alluded to the attacks on phlogiston and caloric as manifestations, to some degree, of the distrust in hypothetico-deduction at the beginning of the nineteenth century. Another example that might be cited is the critique of Newton's corpuscular theory of light by Thomas Young, a colleague of Davy's at Rumford's Royal Institution. However, in the hands of Young's French competitor Augustin Fresnel, the new wave theory of light led to a surprising twist. From the mathematics of Fresnel's essay on diffraction written for the Paris Academy prize competition of 1819, S. D. Poisson deduced that the theory would predict a bright spot in the center of the shadow of a small–disc diffractor, a result that seemed to contradict both the corpuscular theory favored by Poisson, and common sense. However, the experiment subsequently performed to demolish Fresnel's theory instead strongly supported it, since the unexpected prediction was verified.[21] Laudan has plausibly argued that this event represented an important turning point, not only providing a satisfying example of the success of hypothetico-deduction, but also leading to a new epistemic criterion for hypothesis evaluation: verification of novel and unexpected phenomena deduced from the hypothesis.[22]

This new criterion was discussed by a number of scientists and philosophers of science between 1830 and 1843—especially J. F. W. Herschel, Auguste Comte, William Whewell, and John Stuart Mill—in the context of a resurgence of interest in and respect for the method of hypothesis in general.[23] The subject of the heuristic evaluation of hypotheses and theories now began to appear not just in explicitly philosophical treatises, but in purely scientific works as well. In a textbook on the mathematical theory of diffraction published in 1835, F. M. Schwerd wrote that "the wave theory predicts the phenomena of diffraction just as reliably as the theory of gravitation predicts the movements of the heavenly bodies."[24] And in a lecture delivered a few months later, J. B. Dumas opined, in a more relativistic vein:

Theories have always been regarded as things quite different from truth; for this reason theories have long been accorded an importance proportional to the service that they render.... In chemistry, our theories are crutches; to show that they are good, they must be used to walk.... A theory established by twenty facts must explain thirty, and lead to the discovery of ten more; but nearly always it is modified or succumbs to ten new facts added to all that went before.[25]

Caneva has shown that it was just about this time that German physicists working in electromagnetism—such as Karl Gauss, Wilhelm Weber, Gustav Fechner, H. F. E. Lenz, and Franz Ernest Neumann—began to embrace the method of hypothesis, in contrast to the qualitative, concrete, and strongly inductive style of their immediate predecessors.[26] By using the new method successfully and fruitfully, these physicists must have increased its scientific visibility and popularity. However, the explicit and untroubled application of the method of hypothesis to the microcosmic world was as slow to arrive in nineteenth-century physics as it was in chemistry. Not as many physicists were interested in atomic-molecular questions, and those that were experienced difficulties in gaining a hearing in their peer community.[27]

The first to achieve success in this regard was Rudolph Clausius, who revived the kinetic theory of gases in 1857 after the failed efforts of John Herapath, J. J. Waterston, and A. Krönig. An objection from the Dutch meteorologist C. H. D. Buys-Ballot led Clausius to further refine his theory the following year, by introducing the concept of mean free paths of his hypothetical molecules. This last article brought the theory to the attention of James Clerk Maxwell, who read an English translation in the February 1859 number of the *Philosophical Magazine*. Maxwell was doubtful, but sufficiently intrigued to reach a number of novel results "as an exercise in mechanics," communicating them privately to his friend G. G. Stokes on 30 May and publicly to the British Association on 21 September. A major two-part article by Maxwell appeared the following year. Maxwell was the first kineticist to employ a statistical distribution of molecular velocities rather than a single mean speed, and he deduced several predictions of gaseous transport processes from the theory.[28]

One of the most interesting of these predictions was that the viscosity of a gas ought to be independent of its density, a result that seemed to contradict both common sense and some earlier rather equivocal experiments on the question. But new experiments carried out in the 1860s by Maxwell and others confirmed the prediction; the unexpected character of the phenomenon gave significant support to the kinetic hypotheses used to deduce it.

During the third quarter of the century, the kinetic theory made rapid strides toward general acceptance. Such major physicists as Ludwig Boltzmann, Lord Kelvin, J. D. van der Waals, Thomas Graham, and John Tyndall contributed to formulating and propagating the new theory, and Clausius and Maxwell continued to improve their own versions, learning from each other's critiques. A number of estimates of the sizes of molecules inferred from experimental data emerged in the late 1860s and early 1870s, some of them independent of kinetic assumptions and all of them reasonably consistent among themselves. The new ability to specify molecular weights in grams and molecular volumes in milliliters lent concreteness and credibility to the theory, even though the figures remained highly approximate.

In the meantime, Maxwell was fruitfully exploiting hypothetico-deductive method in another field as well, electromagnetism. His work on this subject is a classic instance of imaginative use of heuristic models and analogies, and his professed image of his own methodology is unusually sophisticated and surprisingly accurate.[29]

Whether due to examples of the successful use of hypothetico-deductive method in the wave theory of light, the kinetic theory of gases, and electromagnetic theory, or the altered climate of opinion among philosophers of science, or some other combination of factors,[30] by mid-century Baconian induction began to lose much of the bloom that it had at the beginning of the century. In 1847 an American commentator attempted to deflate "the lofty pretensions of the Baconian philosophy."[31] The English historian H. T. Buckle wrote twelve years later that "it might easily be shown, and indeed will hardly be disputed, that during the last fifty years an opinion has been gaining ground, that the Baconian system has been overrated...."[32] W. Stanley Jevons, in a fervent and detailed defense of the method of hypothesis first published in 1874, referred to "the gradual reaction which has taken place in recent times against the purely empirical or Baconian theory of induction."[33] Two years later, a generally positive article on Bacon in the *Encyclopaedia Brittanica* concluded:

> The inductive formation of axioms by a gradually ascending scale is a route which no science has ever followed, and by which no science could ever make progress. The true scientific procedure is by hypothesis followed up and tested by verification; the most powerful instrument is the deductive method, which Bacon can hardly be said to have recognized.[34]

Another witness for this change in attitude is Charles Darwin. Darwin read John Herschel's *Preliminary Discourse on the Study of Natural Philosophy* early in 1831 while still an undergraduate at Cambridge and was extremely enthusiastic; he was also much influenced by the brilliant professor of mineralogy William Whewell.[35] Darwin was obviously receptive to these advocates of the method of hypothesis, and he learned his lessons well. In a letter to Charles Lyell he later remarked: "... without the making of theories I am convinced there would be no observation." To the economist Henry Fawcett he expressed his frustration against those who did not understand that "all observation must be for or against some view if it is to be of any service!" He added that he recalled "much talk thirty years ago [ca. 1831]" about the necessity of observing with no hypotheses in mind, but he regarded this opinion as both foolish and passe.[36]

Darwin's strong feelings in this regard were conditioned not only by his education at Cambridge, but also by the reception accorded his *Origin of Species,* a work distinguished by its hypothetico-deductive character.[37] Samuel Wilberforce wrote a blistering anonymous review, published a month before his celebrated encounter with T. H. Huxley, in which (*inter alia*) he repeatedly indicted Darwin before the bar of the "stern Baconian law of the observation of facts"; the venerable Adam Sedgwick fired off a letter within days of receiving the *Origin,* in which he mixed avuncular kindnesses with the severest opprobrium against Darwin's desertion of the inductive philosophy.[38] Fawcett and Huxley both recalled how very general was the complaint, "Oh, but this is contrary to the Baconian method!"[39] Darwin was chastened but unrepentant. To a young naturalist he advised: "... *let theory guide your observations,* but till your reputation is established, be sparing of publishing theory. It makes persons doubt your observations."[40]

Of course Darwin had supporters as well as detractors of his views on the method of hypothesis. Fawcett, like Darwin, thought the critics were guilty of specious rhetoric based on a shallow understanding of science; he reported to Darwin a conversation with John Stuart Mill in which Mill expressed strong approval of Darwin's logic of discovery.[41] The experience with the reception of the *Origin* led Huxley (according to his son and biographer) to "a settled disbelief" in Baconian induction. In the 1860s Huxley began to speak out regarding his detestation of "the pseudo-scientific cant which is talked about the Baconian philosophy"; in 1878 he detailed his belief in the indispensability of hypotheses in general. With unconcealed irritation he commented, "To be accused of departing from the canons of the Baconian philosophy is almost as bad as to be charged with forgetting your aspirates...." He wrote to a friend concerning this speech, "I have been oppressed by the humbug of the 'Baconian induction' all my life, and at last *the worm has turned.*"[42]

3

Chemistry was slower to adopt the model of hypothetico-deduction than physics. At the same time that new versions of induction were being delineated by Herschel, Auguste Comte, William Whewell and John Stuart Mill, each of whom accorded the method of hypothesis much greater importance than in classical Baconian enumerative induction, and at the same time that respect for the method was beginning to penetrate the physics community, chemists were aggressively defending the autonomy and inductive purity of their discipline. Daltonian physical atomism continued to draw criticism; Avogadro's hypotheses continued to experience rejection in part because they were physical and deductive; the shift from Berzelian atomism to equivalents was fundamentally motivated by the nonchemical and presumed hypothetical elements in Berzelius' system. The effort in the early 1840s by Laurent and Gerhardt to reform Berzelius's system of atomic weights and molecular formulas was promptly derailed by this self-conscious (but self-deluded) resort to inductivism and empiricism.[43] In short, inductivism dominated chemistry throughout the first half of the nineteenth century—despite limited and desultory defenses of the method of hypothesis by such masters as Dumas and Liebig.[44]

About the time when Laurent and Gerhardt suffered their tragically early deaths (1853 and 1856 respectively), their reforms began to be championed by such newcomers as Adolphe Wurtz, Alexander Williamson, William Odling, and August Kekulé. This success was occasioned by the circumstance that these reformers were now able to adduce more convincing chemical evidence to support the new ideas. The fact that the reform was largely synonymous with a return to Berzelius's system—a system derived hypothetico-deductively from physical as well as chemical data—was merely fortuitous, since Laurent and Gerhardt, like their peers, eschewed physical data, and Gerhardt, at least, explicitly eschewed the method of hypothesis.

Just when the movement appeared to be gaining ascendancy in the European chemical community, a new face with a different approach appeared on the scene. Stanislao Cannizzaro published a "Sketch of a Course in Chemical Philosophy," given at the University of Genoa, in which he outlined a system of atomic weights and molecular formulas based on a conviction of the truth of Avogadro's hypotheses. Using physical laws such as the Dulong-Petit law of atomic heats and Mitscherlich's law of isomorphism, Cannizzaro showed that all apparent anomalies could be eliminated—if one also made some additional assumptions such as gaseous dissociation of certain labile compounds and variable submolecularity in molecules of elements. His system was, of course, virtually identical to the Laurent-Gerhardt reform (and to the modern system).

Cannizzaro knew how to dress his discussion in convent⁚ ⁻ⁿally fashionable clothing, so that the hypothetico-deductive nature of the enterprise was not immediately apparent. For instance, he termed his central operational rule for atomic weight determinations the "law of atoms," despite the fact that it concealed Avogadro's hypotheses within itself. This sleight of hand no doubt deceived some of Cannizzaro's contemporaries, as it has at least one of our own.[45] But there is no denying the self-conscious methodological innovation that Cannizzaro was pursuing. When Cannizzaro averred that the coherence of the chemical system built on the assumption of Avogadro's hypotheses provided another argument in favor of that assumption, he was following classical hypothetico-deductive procedure.

An example will further illustrate the point. Despite Berzelius's fruitful use of hypotheses and of physical evidence, he accorded the principle of chemical analogy the premier heuristic position; he used many of the same physical laws as Cannizzaro, but refused to accept their full generality, especially when they contradicted chemical analogy. The same was true of Gerhardt and even of Laurent. Cannizzaro, on the other hand, accepted the dictates deduced from the physical evidence with little regard for chemical analogy. For example, despite the evidence from atomic heats and isomorphism, it simply did not seem reasonable to most chemists to place the alkalies in a different class from the alkaline earths. Whereas from 1826 Berzelius erred in classifying basic metal oxides in the ferrous oxide (MO) group, Gerhardt put them in the alkali metal (M_2O) group. Hence, Berzelius proposed atomic weights for the monovalent metals that are twice the modern values. Gerhardt's "correction" worked well for these monovalent metals, but halved Berzelius's good divalent metal atomic weights. Cannizzaro simply rejected the argument from chemical analogy and accepted the implication from physics to place both groups in the same category.

What factors led Cannizzaro to this new approach at this particular time? Fisher has cogently argued that it was a "teacher's reform" and must be viewed in the context of Cannizzaro's pedagogical brilliance.[46] I have elsewhere argued for a direct, but largely unprovable, influence from the emerging kinetic theory of gases, from which Avogadro's hypotheses could be, and were in fact regularly, deduced.[47] In the larger context supplied by this essay, one might also suggest, even if the demonstration cannot be provided, that the "time was ripe" for such a change in chemical methodology.

Cannizzaro's essay was published in 1858 in an obscure Italian journal, but his ideas received considerably wider exposure at the Karlsruhe Congress two year later. The congress was an outgrowth of the reform movement then headed by such men as Wurtz, Kekulé, and A. W. Williamson. It represented an attempt by these chemists to confer official orthodoxy on the new ideas. Cannizzaro was, of course, in the reformist camp. But Kekulé's speeches at the congress exemplify the hesitancy even among the progressives to admit physical data on an equal level with chemical data. He disagreed with Cannizzaro in refusing to identify the molecules that take part in chemical reactions with the physical molecules of the kinetic theory of gases. His hesitation seemed warranted because dissociation in the vapor phase was just then becoming widely recognized; this phenomenon suggested that some physical molecules consist of varying numbers of chemical molecules depending on the conditions. Cannizzaro held strongly to the view that the chemical need not be distinguished from the physical molecule; all observers need to distinguish are the physical gas molecules—whose relative weights are given directly by vapor densities—from the chemical *atoms* that form those gas molecules and that enter into the composition of chemical compounds.

The evidence suggests that Cannizzaro really did make a remarkable impression at the congress, as the conventional mythology asserts. His most celebrated convert was the physical chemist Lothar Meyer, who later described how the "scales fell from" his eyes upon reading the offprint at the end of the meeting. Significantly, Meyer had been trained as a physicist, under the early hypothetico-deductivist Neumann. Meyer immediately became an enthusiastic disciple for Cannizzaro's views. In a friendly critique of a paper published by Kekulé five years after Karlsruhe, Meyer wrote:

> Perhaps my honored friend will now decide to concede what he has so often contested in our private conversations and correspondence: that the fundamental hypotheses of chemistry should be derived not just from purely chemical data in a narrow sense, but rather, as in every investigation of the most intimate nature of matter, *all* scienticfic aids must be applied.[48]

Although Kekulé never did concede this point, he too began to embrace the method of hypothesis ever more openly. In the same year as Meyer's critique, 1865, he published a theory of aromatic compounds that well exemplifies hypothetico-deductive precepts.[49] Kekulé's benzene theory has often been depicted as one of those anomalous cases where a sudden inspiration, even perhaps as a result of a dream or hallucination, dramatically revolutionized a scientific field. Closer examination of the episode tends to transform it into a more conventional story of the continuity, gradualism, and rationalism of scientific progress, and of the important role of experiment in the adjustment of theory. But it also represents at least as much a case study of the reverse process, the role of hypothesis (and deductions therefrom) in guiding the selection and interpretation of experiment.

For instance, Kekulé proposed his theory at a time when at least some organic chemists were convinced of the existence of isomers of benzene and of benzoic acid, and of the existence of aromatic compounds with less than six carbon atoms. Such substances would violate the predictions of the theory. Kekulé ignored them as probable experimental errors, and within a year or two after publication of the theory they were generally recognized as such. In a broader sense, it was Kekulé's full confidence in the validity of the principles of his *structure* theory that provided the psychological and conceptual prerequisites for the creation of the *benzene* theory. Although he proceeded cautiously, Kekulé seems never to have seriously doubted the verity of the cyclic, six-carbon, triple doubly bonded structure once he had conceived it. It seemed simply too "elegant" not to be true.

The hypothesis provided yeoman service in suggesting, and helping to interpret, experiment. If the hypothesis were true, the hydrogen atoms should all be chemically equivalent. Were they? There should be no isomers or lower homologues of benzene or of monosubstituted benzene. Did such really not exist? There should be three isomers of every di-, tri-, and tetraderivative. Could these be identified? And so on. In the first flush of enthusiasm for his theory, Kekulé wrote to a friend: "A great deal is in the works; the plans are unlimited, for the aromatic theory is an inexhaustible treasure-trove. Now when 'German youths' need dissertation topics, they will find plenty of them here."[50] This last prediction was not the least accurate of all those spawned by Kekulé's hypothesis on the structure of aromatic substances.

Near the end of the century, the English chemist Francis Japp averred in his "Kekulé Memorial Lecture":

> The accuracy of Kekulé's predictions has done more to inspire a belief in the utility

of legitimate hypotheses in chemistry, and has therefore done more for the deductive side of the science, than that of almost any other investigator. His work stands preeminent as an example of the power of ideas.[51]

A generous discount for the hyperbole of an obituarist still leaves a remarkably strong statement and one which I would characterize as accurate provided the name of Cannizzaro is added to that of Kekulé. For my purposes here, I would underline the circumstances that by this time the fashions of rhetoric in scientific method had been so transformed that Kekulé could actually be praised rather than excoriated for his use of hypothesis. The same change of fashion explains why Kekulé himself was eager publicly to proclaim his famous dream anecdotes in 1890, and not earlier—as shown below.

4

During the period when Kekulé was first beginning to explore the implications of his aromatic hypothesis, his mentor Justus Liebig published a violent attack on Francis Bacon and the inductive philosophy. Always a man easily roused by criticism and quick to express his strong opinions, Liebig had been wounded by the adverse reception in Britain of his ideas on plant nutrition. The resistance by British—and American—agriculturalists was clearly based on a Baconian preference for the sort of practical experimentation conducted by J. B. Lawes and J. H. Gilbert, which seemed to contradict the more strongly theory-based conceptions of Liebig.

As president of the Bavarian Academy of Sciences, Liebig was compelled to present semiannual addresses of general interest. In the summer of 1862 he began to study Bacon and his times as well as the works in philosophy of science by Whewell and Mill. He later related that he quickly recognized that "there was a kind of demon, who was my worst enemy, concealed under the words *induction, principles,* and *axioms* of Bacon."[52] The research was summarized in his speech of 28 March 1863, entitled "Francis Bacon von Verulam und die Geschichte der Naturwissenschaften."[53] To his close friend Friedrich Wöhler he wrote:

My essay on Bacon will appear very soon in the [Augsburger] Allgemeine Zeitung. You will be amazed what a swindler this man was. . . . He is the hero of dilettantes, but I believe his adoration will end. . . . [I]t is also appearing in English, and I am curious what John Bull will say about it.[54]

The speech was published as a short monograph later that year, with an introduction in which Liebig railed at the obstinacy of the English farmers, making clear his original motivation for the work. The source of opposition to his ideas, he candidly opined, "lay in a peculiar direction of the English mind . . ." illustrated particularly clearly in the works of Francis Bacon, which were, "if not the source, at all events the ideal, of the methods of experiment and conclusion in use among the dilettanti in science.'' The reaction was so strong against this polemical introduction that Liebig was induced to delete it from the numerous republications and translations of the work.

Liebig had carefully studied Macaulay's celebrated essay on Bacon, and his biographical portrait was even more negative than the Englishman's. Bacon was depicted not only as a man justly convicted of bribery, but as a toadying sycophant who betrayed his friend and benefactor Essex for personal advancement. But unlike Macaulay, Liebig found Bacon's philosophy of science to be no more useful or admirable than his character. Enumerative induction, Liebig asserted, is scientifically worthless, and is never utilized by the scientist.

> Bacon's method is the method of many instances, and since every single unexplained instance is a zero, and thousands of zeros accumulated in whatever order you like cannot result in any [positive] number, it is clear that his entire process of induction consists in a farrago of inexact sensual perceptions. *The result at which one arrives by following his method must always be zero.*[55]

If the inductive method was useless to the scientist, it was the hypothetico-deductive process that constituted the only real and fruitful scientific methodology.

> The true method of science . . . is diametrically opposed to that of Bacon. . . . [A]ll scientific research is deductive or aprioristic; the experiment is only an aid for the thinking process, like calculation; thinking must precede [experimentation] necessarily and in all cases, if the latter is to have any significance at all. Empirical scientific research in the customary sense simply does not exist.[56]

This is a remarkably strong statement coming from a man who in 1840 had ostentatiously rejected chemical theory altogether and who four years later had turned resolutely from ''hypothetical'' Berzelian atomic weights to the empirical haven of equivalents. That this inconsistency was due more to Liebig's self-delusion in the 1840s than a transformation of his values by 1863 is clear from passages in the first edition of his *Familiar Letters on Chemistry* (1843), where he expressed his conviction that theories ''scarcely ever fail'' to produce discoveries, even if the theory itself fails to survive the experiments unmodified.[57]

The storm of protest that greeted Liebig's essay makes clear that Baconian induction still had strong defenders in 1863 (as Huxley and Darwin had also discovered in 1859). Among the Germans, C. Siegwart and H. Böhmer penned lengthy rebuttals, to the first of which Liebig responded in kind.[58] An anonymous English reply in *Fraser's Magazine* referred to Liebig's ''angry and uncompromising onslaught,'' displaying ''blind animosity'' and ''chop-logic with a vengeance.'' The writer accused Liebig of seizing upon Bacon ''in searching for the means of depreciating [Bacon's] countrymen.''[59] Liebig wrote Wöhler that he could not let this last charge go unanswered, since ''I am on the contrary an admirer of this remarkable people, and truly love the English.''[60] His lengthy reply was published in the magzine, but its force was considerably weakened by detailed, often sarcastic footnotes written by the same anonymous author.[61]

5

Despite the angry reaction to Liebig's speech, the climate of opinion on this issue was perceptibly different in the 1860s than it had been thirty years earlier. Although Liebig's address was surely influential in making the method of hypothesis more respectable, the developments discussed above suggest that the essay was more a symptom than a cause of this change. At a commemorative celebration in his honor in 1890, Liebig's student August Kekulé reminisced about his experiences during this period. As Japp commented a few years afterward, there are few better examples of an imaginative user of the hypothetico-deductive method; Kekulé described how two of his most celebrated ideas came to him in dreams and were only later sufficiently supported by the empirical evidence to be regarded as well verified. Significantly, in both cases he hesitated to publish until, as he put it, both the theory and the time were ripe.[62] In both cases he omitted all reference to dreams; in the first case, the structure theory, he even disguised the hypothetico-deductive nature of his methodology. But what might have been considered professionally dangerous in the 1850s became less so by the 1860s, and the second theory, the benzene theory, was proposed and developed with candid reference to hypothetico-deduction.

The cases of Kekulé and Liebig, as well as other examples cited here, support the unsurprising conclusion that the character of a scientist's methodology cannot be decided with reference to his explicit claims on the subject in published papers alone, for his public words are often chosen with more regard to peer pressure and professional expectations than neutral description;[63] moreover, the scientist is not always the best observer of his own approaches to scientific creativity and problem-solving. What I have sketched here is not so much an alteration in actually methodology, than a sea change in what was *considered* proper and respectable for the careful scientist. I have sought to establish that there was a certain convergence between the rhetoric and practice of chemical methodology during the course of the nineteenth century; in other words, that the rhetoric began more accurately to reflect reality.

Still, whether in substance or form, a sea change it was, and a few concluding conjectures on the causes of the shift are in order. Other than in German electrodynamics, where there is evidence of transition as early as the 1830s,[64] the critical period for most of the sciences seems to have been the 1850s and 1860s. This was the period that saw Cannizzaro's return to Avogadro's hypotheses, Kekulé's theories on chemical structure and aromatic compounds, and the emergence of Darwin's theory of evolution and Clausius's and Maxwell's development of kinetic theory (not to mention Liebig's celebrated attack on Bacon). Although all of these hypothetico-deductive theories were controversial, they were immediately recognized by most serious scientists as highly important and on the whole experienced more acceptance than rejection by the elite in each field. It is reasonable to suppose that a certain self-reinforcing cycle began to take hold during this period in elevating the method of hypothesis to greater visibility and respect: nothing succeeds like success. But it is also reasonable to believe that the cycle may have owed its initial impulse to the work a generation earlier of respected scientist-philosophers such as Herschel and Whewell, and logicians such as Mill. I have noted the clear influence of Herschel and Whewell on Darwin, and of Whewell and Mill on Liebig. Other, more diffuse influences may also have been operating on other of our protagonists.[65]

Notes

I wish to express my thanks to Elizabeth Garber, whose helpful comments materially improved this paper.

1. Jacob Berzelius, "An Address to Those Chemists who Wish to Examine the Laws of Chemical Proportions, and the Theory of Chemistry in General," *Annals of Philosophy* 5 (1815): 122–31, 122.

2. For a general discussion of the vogue of Baconian induction in the Enlightenment and the early nineteenth century, see Larry Laudan, *Science and Hypothesis: Historical Essays on Scientific Methodology* (Boston: Reidel, 1981); 9–15, 72–127. A wide-ranging review of many of the issues discussed here is J. A. Schuster and R. R. Yeo, eds., *The Politics and Rhetoric of Scientific Method* (Dordrecht: Reidel, 1986). George Daniels discusses the hegemony of Bacon in early nineteenth-century America in *American Science in the Age of Jackson* (New York: Columbia University Press, 1968), 63–85. Susan Cannon doubts the meaningfulness of the label "Baconian" in *Science in Culture: The Early Victorian Period* (New York: Science History, 1978). Exceedingly valuable for nineteenth-century British views of Bacon is Harold Stolerman, "Francis Bacon and the Victorians, 1830–1885," (Ph.D. diss., New York University, 1969); also Richard Yeo, "An Idol of the Marketplace: Baconianism in Nineteenth Century Britain," *History of Science* 23 (1985): 251–98. The inductive character of nineteenth-century chemistry in particular has been emphasized by Nicholas Fisher, "Avogadro, the Chemists, and Historians of Chemistry," *History of Science* 20 (1982): 90–92, and by Mary Jo Nye, "Berthelot's Anti-Atomism: A 'Matter of Taste'?" *Annals of Science* 38 (1981): 585–90. See also J. H. Brooke, "Methods and Methodology in the Development of Organic Chemistry," *Ambix* 34 (1987): 147–55, and A. J. Rocke, "Convention Versus Ontology in Nineteenth-Century Organic Chemistry," in J. G. Traynham, ed., *Essays on the History of Organic Chemistry* (Baton Rouge: Louisiana State University Press, 1987), 1–20.

3. The eighteenth-century background on these issues is treated in Robert Schofield, *Mechanism and Materialism: British Natural Philosophy in an Age of Reason* (Princeton: Princeton University Press, 1970), and in Arnold Thackray, *Atoms and Powers: An Essay on Newtonian Matter Theory and the Development of Chemistry* (Cambridge: Harvard University Press, 1970).

4. Laudan, *Science and Hypothesis;* Kenneth Caneva, "From Galvanism to Electrodynamics: The Transformation of German Physics and Its Social Context," *Historical Studies in the Physical Sciences* 9 (1978): 63–159, especially 95–122. A major contention of Laudan's book is that scentific methodology (and its rhetoric) is in general influenced much more by practicing scientists or scientist-philosophers than by philosophers per se. My case study supports and amplifies this contention. David Brewster was one of the first to suggest that the value of Baconian induction was more a hobby horse of philosophers than scientists: see [Margaret M.] Gordon, *The*

Home Life of Sir David Brewster, 2d ed. (Edinburgh: Edmonton and Douglas, 1870), 128, 130.

5. See Hans Reichenbach, *The Rise of Scientific Philosophy* (Berkeley: University of California Press, 1951), 231.

6. A superb case study that gives instances of both of these situations is Frederic L. Holmes, *Lavoisier and the Chemistry of Life: An Exploration of Scientific Creativity* (Madison: University of Wisconsin Press, 1985); other examples can be found in Richard S. Westfall, *Never at Rest: A Biography of Isaac Newton* (Cambridge: Cambridge University Press, 1980).

7. Laudan, *Science and Hypothesis,* 9–15, and Laudan, *Science and Values* (Berkeley: University of California Press, 1984), 54–60.

8. Arnold Thackray, *John Dalton: Critical Assessments of His Life and Science* (Cambridge: Harvard University Press, 1972), 92; H. G. Söderbaum, ed., *Jac. Berzelius Bref* (Uppsala: Almqvist & Wiksell, 1912–61), vol. 1, pt. 2, 23 (Davy to Berzelius, 24 March 1811); J. Davy, ed., *Collected Works of Sir Humphry Davy* (London: Smith, Elder and Co., 1839–40), 6: 97.

9. W. H. Wollaston, "A Synoptic Scale of Chemical Equivalents," *Philosophical Transactions of the Royal Society of London* 104 (1814): 1–22, 7.

10. D. C. Goodman, "Wollaston and the Atomic Theory of Dalton," *Historical Studies in the Physical Sciences* 1 (1969): 37–59.

11. J. S. C. Schweigger, "Ueber stöchiometrische Scale . . . ," *Journal für Chemie und Physik* 15 (1815): 496.

12. John Brooke, "Avogadro's Hypothesis and Its Fate: A Case Study in the Failure of Case Studies," *History of Science* 19 (1981): 235–73; Fisher, "Avogadro."

13. Auguste Laurent, *Méthode de chimie* (Paris: Mallet-Bachelier, 1854), xiii; *Chemical Method* (London, 1855), xvi.

14. Söderbaum, ed., *Berzelius Bref,* vol. 1, pt. 3, 43 (Marcet to Berzelius, 5 May 1813).

15. For a description of these events with full literature citations, see Rocke, *Chemical Atomism in the Nineteenth Century* (Columbus: Ohio State University Press, 1984), 171–82.

16. Edouard Grimaux and Charles Gerhardt, Jr., *Charles Gerhardt: Sa vie, son oeuvre, sa correspondance* (Paris: Masson et Cie, 1900), 42.

17. For which, see e.g. Frederic L. Holmes, "Justus von Liebig," *Dictionary of Scientific Biography* (New York: Scribner, 1973), 8: 329–50, and works cited therein.

18. This paragraph summarizes a major theme of my *Chemical Atomism in the Nineteenth Century.*

19. J. H. Brooke, "Organic Synthesis and the Unification of Chemistry—A Reappraisal," *British Journal for the History of Science* 5 (1971): 363–92; Brooke, "Chlorine Substitution and the Future of Organic Chemistry: Methodological Issues in the Laurent-Berzelius Correspondence (1843–1844)," *Studies in History and Philosophy of Science* 4 (1973): 47–94; Brooke, "Laurent, Gerhardt, and the Philosophy of Chemistry," *Historical Studies in the Physical Sciences* 6 (1975): 405–29; and Brooke, "Methods and Methodology."

20. Laudan, *Science and Hypothesis.*

21. D. H. Arnold, "The Mécanique Physique of Siméon-Denis Poisson: The Evolution and Isolation in France of his Approach to Physical Theory (1800–1840). V. Fresnel and the Circular Screen," *Archive for History of Exact Sciences* 28 (1983): 321–42.

22. Laudan, *Science and Hypothesis,* 128–29. Elizabeth Garber has suggested to me in a private communication that his case has been misinterpreted. A final judgment must await the publication of her work on Poisson now in progress. [Editor's note, see p. 162.]

23. See Laudan, *Science and Hypothesis,* 127–80; also C. J. Ducasse, in E. H. Madden, ed., *Theories of Scientific Method: The Renaissance through the Nineteenth Century* (Seattle: University of Washington Press, 1960), 153–232. Laudan, 10–18, notes that there were several attempts to resurrect the method of hypothesis before Herschel, Comte, Whewell, and Mill. For instance, Dugald Stewart praised the heuristic value of hypotheses in his *Elements of the Philosophy of the Human Mind* (London, printed for A. Strahan and T. Cadell 1792–1827), and David Brewster missed no opportunity to criticize Baconian induction and illustrate the importance of hypothetical reasoning, most notably in his *Life of Sir Isaac Newton* (London: J. Murray 1831). In 1824 Brewster wrote to a correspondent: "The opinion so prevalent during the last thirty years, that Lord Bacon introduced the art of experimental inquiry on physical subjects, and that he devised and published a method of discovering scientific truth, called the method of induction, appears to me to be without foundation, and perfectly inconsistent with the history of science." Gordon, *Home Life,* 128.

154   METHODOLOGY IN NINETEENTH-CENTURY CHEMISTRY

24. F. M. Schwerd, *Die Beugungserscheinungen aus den Fundamentalgesetzen der Undulationstheorie analytisch entwickelt* (Mannheim: Schwann und Goetz, 1835), ix–x.

25. J. B. Dumas, *Leçons sur la philosophie chimique,* 2d ed. (Paris: Gauthier–Villars, 1878), 66–67 (lecture delivered on 23 April 1836). A year earlier Dumas had written: "Sciences are based on facts, but they do not become sciences until these facts are grouped by concepts that are sure, each taking its place in a system that leaves spaces to fill, which revealing the ideas and predictions which flow out of such a methodical arrangement." Dumas, *Traité de chimie appliquée aux arts* (Paris: Béchet jeune, 1828–46), 5: 72; cited by Leo Klosterman, "A Research School of Chemistry in the Nineteenth Century: Jean Baptiste Dumas and his Research Students," *Annals of Science* 42 (1985): 42.

26. Caneva, "Transformation of German Physics," 95–122.

27. Elizabeth Garber, "Molecular Science in Late-Nineteenth-Century Britain," *Historical Studies in the Physical Sciences* 9 (1978): 265–97, especially 265. Not only was it necessary to use the method of hypothesis in investigating this area, but transductive procedures were also required, which worked to make the field even more controversial.

28. For material in this and the following paragraphs, see especially Stephen G. Brush, *The Kind of Motion We Call Heat: A History of the Kinetic Theory of Gases,* 2 vols. (Amsterdam: North-Holland, 1976).

29. See C. W. F. Everitt, "James Clerk Maxwell," *Dictionary of Scientific Biography* 9 (New York, Scribners, 1974), 198–230, especially 204–17, and works cited therein.

30. Caneva argues that a changing institutional and social context for German physics was the major factor in the shift of styles toward abstract, mathematical, and hypothetico-deductive methodology. Caneva, "Transformation of German Physics," 123–59.

31. "Bacon's Philosophy," *Methodist Quarterly Review* 31 (1847): 22, cited in Daniels, *American Science,* 67.

32. H. T. Buckle, "Mill on Liberty," *Fraser's Magazine,* 59 (1859): 522.

33. W. Stanley Jevons, *The Principles of Science: A Treatise on Logic and Scientific Method,* 2d ed. (London: Macmillan, 1892), 506 (first edition published in 1874).

34. R[obert] Ad[amson], "Francis Bacon," *Encyclopaedia Britannica* 9th ed. (American reprint, Philadelphia: M. Stodort, 1876), 3: 187.

35. N. Barlow, ed., *The Autobiography of Charles Darwin* (New York: Harcourt-Brace, 1958), 66–68; Michael Ruse, *The Darwinian Revolution* (Chicago: University of Chicago Press, 1979), 61. Ruse discusses the wider context of Darwin's philosophy of science, 56–63, 174–80, and 234–39. Michael Faraday was equally influenced by Herschel's book. On 10 November 1832 he wrote Herschel: "When your work on the study of Nat. Phil. came out, I read it as all others did with delight. I took it as a school book for philosophers and I feel that it has made me a better reasoner & even experimenter and has altogether heightened my character and made me if I may be permitted to say so a better philosopher." L. Pearce Williams, ed., *Selected Correspondence of Michael Faraday* (Cambridge: Cambridge University Press, 1971), 1: 235.

36. Francis Darwin, *The Life and Letters of Charles Darwin* (New York: Appleton, 1887), 2: 108 (Darwin to Lyell, [1 June 1860]); Francis Darwin, *More Letters of Charles Darwin* (New York: Appleton, 1903), 1: 195 (Darwin to Fawcett, 18 September [1861]). Also see Darwin, *Autobiography* (New York: Schuman, 1950), 141, for his admitted lifelong addiction to hypotheses.

37. For which see Ruse, *Darwinian Revolution;* Michael Ghiselin, *The Triumph of the Darwinian Method* (Chicago: University of Chicago Press, 1984); and especially David L. Hull, *Darwin and His Critics: The Reception of Darwin's Theory of Evolution by the Scientific Community* (Cambridge: Harvard University Press, 1973), 3–77.

38. [S. Wilberforce,] *"On the Origin of Species . . . by Charles Darwin,"* *Quarterly Review* 108 (1860): 225–64, especially 231, 239, 249; Darwin, *Life and Letters,* 2: 43 (Sedgwick to Darwin, [November 1859]).

39. Darwin, *More Letters,* 1: 189 (Fawcett to Darwin, 16 July [1861]); Leonard Huxley, *Life and Letters of Thomas Henry Huxley* (London: Macmillan, 1900): 1: 485–86.

40. Darwin, *More Letters,* 2, 323 (Darwin to J. Scott, 6 June 1863).

41. Ibid., 1: 189 (Fawcett to Darwin, 16 July [1861]).

42. L. Huxley, *Life and Letters of Thomas Henry Huxley,* 1: 485–86. In 1880 he referred in a letter to "that sneak Bacon," 2: 14.

43. The following discussion of Laurent, Gerhardt, and Cannizzaro is summarized largely from my *Chemical Atomism in the Nineteenth Century,* 200–311.

44. See above, note 25 (Dumas), and below, note 57 (Liebig).

45. G. S. Morrison, "Cannizzaro's Atom–Free Stoichiometry," *Journal of Chemical Education* 53 (1976): 723.

46. Fisher, "Avogadro," 217–18. Fisher also emphasized the innovative—hypothetico-deductive—nature of Cannizzaro's methodology: *Ibid.,* 90–92, 218–23.

47. Rocke, *Chemical Atomism,* 290–91, 295.

48. L. Meyer, "Kritik einer Abhandlung von A. Kekulé...," *Zeitschrift für Chemie, 8* (1865): 253–54.

49. For which see Rocke, "Hypothesis and Experiment in the Early Development of Kekulé's Benzene Theory," *Annals of Science* 42 (1985): 355–81, and Rocke, "Kekulé's Benzene Theory and the Appraisal of Scientific Theories," in A. Donovan, L. Laudan, and R. Laudan, eds., *Scrutinizing Science* (Dordrecht: Reidel, 1988), 145–61.

50. Kekulé to Adolph Baeyer, 10 April 1865; letter preserved in the Kekulé Sammlung, Institut für Organische Chemie, Technische Hochschule, Darmstadt, FRG. I thank Professor K. Hafner for permission to publish this extract.

51. F. Japp, "Kekulé Memorial Lecture," *Journal of the Chemical Society* 73 (1898): 138.

52. J. Liebig, *Fraser's Magazine* 75 (1867): 483.

53. Liebig, "Francis Bacon von Verulam und die Geschichte der Naturwissenschaften," *Augsburger Allgemeine Zeitung,* nos. 100–105 (1863); it appeared as an offprint later that year, and was frequently republished, most accessibly in Liebig's collected *Reden und Abhandlungen* (Leipzig: C. F. Winter, 1874), 200–254. Otto Sonntag's excellent article "Liebig on Francis Bacon and the Utility of Science," *Annals of Science* 31 (1974): 373–86 deals only with Liebig's views on Baconian utilitarianism.

54. A. W. Hofmann, ed., *Aus Justus Liebig's und Friedrich Wöhler's Briefwechsel* (Brunswick, 1888), 2: 133, 138 (Liebig to Wöhler, 9 April and 6 June 1863).

55. Liebig, *Reden und Abhandlungen,* 248.

56. Ibid., 248–49.

57. Liebig, *Familiar Letters on Chemistry,* first American edition (New York: D. Appleton and Co., 1843), 13–14.

58. C. Siegwart, "Ein Philosoph und ein Naturforscher über Francis Bacon von Verulam," *Preussische Jahrbücher* 12 (August, 1863); H. Böhmer, *Ueber Franzis Bacon von Verulam und die Verbindung der Philosophie mit der Naturwissenschaft* (Erlangen: F. Ente, 1864); Liebig, *Reden und Abhandlungen,* 255–95.

59. Anonymous, "Was Lord Bacon an Imposter?" *Fraser's Magazine* 74 (1866): 719, 730, 736, 739.

60. Hofmann, ed., *Briefwechsel,* 2: 227 (Liebig to Wöhler, 23 December 1866). On the same day he wrote a letter to Dumas, expressing similar sentiments (Archive de l'Académie des Sciences, Paris, Fonds Dumas).

61. See above, note 52.

62. A. Kekulé, "Benzolfest: Rede," *Berichte der Deutschen Chemischen Gesellschaft* 23 (1890): 1302–11. See Rocke, "Subatomic Speculation and the Origin of Structure Theory," *Ambix* 30 (1983): 1–18; and Rocke, "Kekulé's Benzene Theory."

63. For discussion of the intersection of personal and career relations with the pursuit of science, see e.g. Caneva, "German Physics," 123–59, or several of the papers in the volume edited by Schuster and Yeo, *Politics and Rhetoric.*

64. Caneva, "German Physics."

65. It is interesting, and even relevant, to note that the debate between Whewell and Mill on the precise character of scientific induction was still attracting widespread attention in the critical decade of the 1850s. Regarding this debate, see, for example, H. T. Walsh, "Whewell and Mill on Induction," *Philosophy of Science* 29 (1962): 279–84.

Siméon-Denis Poisson: Mathematics versus Physics in Early Nineteenth-Century France

ELIZABETH GARBER

Historians of physics and mathematics do not usually concern themselves with the historical relationship between the disciplines they study. However, such an examination can illuminate at least the history of physics in ways not visible by looking at the conceptual changes historians assume mark the development of the modern discipline. This is an attempt to examine the historical relationships between mathematics and physics that, through the work of Siméon-Denis Poisson throws light upon unexplored presuppositions in the history of physics.

One assumption that is obvious from just a brief examination of the histories of the technical development of physics is that the fusion of mathematics and physics, albeit of great significance, is seen as a natural evolution. Equally obvious is the attitude that, although described by physicists as their language, mathematics is treated by historians of physics as a tool. It is a passive instrument taken up and used at will, not an active developing problematic facet of the intellectual struggle for understanding certain natural phenomena.

Mathematics, therefore, remains in the background of their accounts, an unproblematic component of the development of modern physics. Its content and methods might change, but this has had no impact on how physicists have used mathematics; it remains a tool. Historians usually locate the integration of mathematics into physics in the work of Galileo Galilei and have seen no reason to reassess the relationship analysed there. The problem is that historians assume that the relation postulated in those works of Galileo remain the same in the next three centuries. They have failed to think of the implications for the mathematics-physics connection of the transformation of both disciplines in that same period. Nor have they examined the intellectual origins and language used in considering mathematics as the "tool" of the sciences. In fact they seem to expect physicists since Isaac Newton to use mathematics in the ways with which philosophers say that mathematics is used in the twentieth century. They also seem deaf to the explanations of physicists themselves, that mathematics is their language, and that using mathematics to solve a problem is neither obvious nor easy.

This is understandable, given that the focus of attention is the technical history of physics, but leaving these assumptions unexamined has introduced difficulties into the historical narrative. In assuming a constancy of the relationship over time, historians have introduced physical interpretations into an author's mathematical arguments or introduced extended mathematical analysis when reconstructing an author's physical reasoning where none exists in the original.

Neither process addresses an author's intentions nor his difficulties with mathematical language or physical imagery. More importantly, these approaches do not take into account the radically different nature of physics and mathematics, say, in the middle of the eighteenth century from the disciplines of physics and mathematics of a century later that affect their relationship to one another. In particular, when they redefined the calculus in the early nineteenth century, mathematicians made foundational issues the primary focus of their research, and these issues replaced the older concentration on solving problems. Physical problems ceased to be the springboard into investigations of the calculus. Mathematicians no longer saw the origins of their calculus in the real-world problems of physics. Physics was abandoned as the guarantor of the validity of their analysis or the existence of a solution to their differential equations. In tearing physical reality and the calculus apart, new ties had to be forged that helped define the mathematical, theoretical physics of the late nineteenth and twentieth centuries.

The new relationships that emerged around the middle of the nineteenth century were not reached easily. As part of the effort to understand the changes, historians have to reevaluate what "physique mathematique" meant at the beginning of the nineteenth century. Specifically it might be useful to look at the "géométre" most closely associated with the origins of "mathematical physics", Siméon-Denis Poisson. In so doing, they will see that more than the simple coming together of experiment and mathematics was needed to create the language and methods of late nineteenth-century physics. Equally, we shall see that backward extrapolations of a general notion of a mathematical, theoretical physics, as historians understand it, is insufficient to comprehend the goals and methods of the practitioners of the earlier discipline.

Assessments of Siméon-Denis Poisson's contributions to physics and mathematics have ranged from the highly appreciative to the brusquely dismissive. This sharp division of opinion started during Poisson's lifetime and reflected both the intellectual upheavals occurring in the disciplines of physics and mathematics and the politics within the community of their French practitioners.[1] In the generations following his own Poisson was appreciated for his mathematics not his physics. James Clerk Maxwell was silent on the physical content of Poisson's work in electrostatics or magnetism and discussed only his mathematical results.[2] He even attributes, correctly, the idea of the potential to others. Maxwell was acutely aware of the history of the research areas in which he worked, yet he treated Poisson and others of his generation as mathematicians rather than physicists, citing their useful mathematical relationships for physics without attributing to them any physical interpretations.

Edmund Whittaker also recorded his appreciation of Poisson's work in mathematical terms even while assigning some of Poisson's success to the state of development of the calculus.

> The rapidity with which in a single memoir Poisson passed from the barest elements of the subject to such recondite problems as those just mentioned [attraction between charged spheroids] may well excite admiration. His success is, no doubt, partly explained by the high state of development to which analysis had been advanced by the great mathematicians of the eighteenth century; but even after allowance has been made for what is due his predecessors, Poisson's investigations must be accounted a splendid memorial of his genius.[3]

This has also been the evaluation of others, mostly engineers and physicists, delving into the history of their disciplines.[4] One such evaluation found his work in elasticity useless because of the lack of attention to the details of the physical process.[5]

Recent historical assessments of Poisson's work focus on his importance for the history of physics, although there is disagreement about what in Poisson's work makes it important for that history. Poisson's basic disinterest in physical processes has been seen as the source of his importance for the development of "early modern physics." Poisson concentrated upon mathematics that brought with it clarity and precision and neglected the vagueness of mechanisms. He was content to develop a mathematical analysis that was compatible with experimental results.[6] Similarly, without any detailed analysis of the paper's content, Poisson's 1811 electrostatics paper has been seen as a turning point both for electrostatics (by narrowing the problem under discussion) and physics (by putting its theory on a new, mathematical footing).[7] The main thrust of the argument is that the social circumstances of Poisson's life led him to the study of physical problems rather than mathematics—and hence to change physics.

Robert Fox has detailed the physical model used by Poisson without considering its relationship to analysis. However, his goal in examining Poisson and Laplace was to explore the social dynamics of early nineteenth-century French science rather than trace the intellectual development of physics as a discipline.[8]

The most recent examination of Poisson's work is a narrative, detailed account of the development of Poisson's mathematical physics. This account uses the assumption that physical problems quickly became the focus of Poisson's attention and mathematics was an instrument in this design. Mathematical criteria were laid aside or reworked in light of the needs of Poisson's physical model of matter and Poisson's solutions to physical problems.[9] The difficulty with seeing Poisson's work as significant for the development of physics is, as the author states, that Poisson's physical model was inadequate, and he largely worked in increasing isolation from his colleagues. His physical ideas had little or no impact on the historical development of physics.

Despite this range of approaches, the above historians share certain assumptions. They assume a certain continuity in the relationship between mathematics and physics in Poisson's work to that in the theoretical physics of the late nineteenth and twentieth centuries. In fact, John Heilbron makes this assumption explicit.[10] Historians extrapolate this relationship back to the early nineteenth century (although they disagree on the nature of that relationship). This method of fusing mathematical analysis and physical imagery continued from the eighteenth into the nineteenth century, where it naturally spread from mechanics into the disciplines of electrostatics and theories of light. Thus, historians can examine Poisson's work and that of his contemporaries with the expectation of seeing the same mix of mathematics and physical imagery they see in twentieth century physics; mathematics serving the same function for Poisson as it does for later physicists—to investigate nature.

The particular aspects of theoretical physics presumed to be present in Poisson's work are that Poisson's investigations begin in a physical problem or process, and that the understanding of that process was the goal of his investigation. Thus historians expect Poisson to begin in physical imagery that is transformed into a mathematical formulation of the problem. Instead of the most general, mathematical solution of the analytical problem posed, they would expect the physical conditions of the problem to delineate and guide his mathematical analysis.[11] While the calculus constitutes the language in which the problem is stated, the lines of development of this language are limited by the physical images embedded symbolically in the initial mathematical relationships. Unlike the mathematician who strives for the most general analytical solution, the theoretical physicist looks for the solution from which he can squeeze the most physical information. Much of the possible mathematical analysis and development of this language is superfluous for the physicist's purposes, and presumably Poisson's.

In general, and in the particular case of Poisson, knowing what aspects of the analysis can be dispensed with depends on understanding what in the mathematical analysis represents plausible physical processes, given the particular imagery used to set up the problem. The dialetic between physical metaphor and mathematical language continues as the analysis develops. The mathematics is taken along lines that reveal physically meaningful results, although not necessarily mathematically interesting ones. Each aspect, physical ideas and analysis, guide and control the conceptual and logical development and consequences inherent in the initial imagery. This means that from the point of view of the mathematics both at the beginning, middle, and at the end, physical considerations may intrude, both to curtail and direct it. The physical significance of any stage in the analytical development can be made explicit and justifies modifications of it.

Reciprocally, results derived from analysis can change physicists' interpretations of the imagery that they used to set up the problem. Known experimental results are integrated into or deduced from the mathematical analysis and given a physical interpretation in terms of the processes initially postulated or reinterpreted. Beyond this, the new insights into physical processes offered by mathematics drive theoreticians beyond known analogies and metaphors into the unknown to predict new physical outcomes of the extended analytical process. (Prediction, however difficult or rare, is taken for granted as an aspect of physics in the past 130 years.) It is the venture into the physically unknown rather than the replication of the known that historians can also take for granted as a core aspect of modern physics.

All this may not be accomplished by one theoretical physicist alone, but the goals of the practitioners within the discipline remain well defined by these terms. Also, theoretical physicists of the last 130 years or so were conscious of the different goals of mathematicians. Physicists contributed to mathematics in separate journals, knowing that they were addressing an audience with different expectations.[12]

The integration of physical imagery, mathematical language, and experimental method defines modern physics. It is not often seen in the work of a single physicist, especially in the twentieth century. But the members of the discipline as a whole work to understand nature, not to develop mathematics, even when, as in certain domains of high energy physics today, "physics is mathematics." The physical model or analogy may be absent but the physical context of the problem is not forgotten.[13] The historical question is whether this applies in Poisson's work.

Poisson claimed to have developed a "mathematical-physics," a particular model of matter from which a vast array of disparate phenomena could be deduced through the application of analysis, that is, the calculus. He derided the deductive nature of Lagrange's mechanics because it was not based upon a physical model of matter. Poisson thought he had produced an equally deductive system, based upon a mechanical model. Lagrange had taken the abstract method as far as was possible by replacing the physical connections between bodies by equations between coordinates of points. But Lagrange's approach did not lead to the solution of certain kinds of problems, such as the instantaneous collisions of perfectly hard bodies.[14]

The deductive nature of Poisson's system is explicit in the title to his treatise on heat and his explanation of that title. Poisson proposed to deduce

rigorously from the calculus the consequences of a general hypothesis, in this case on the communication of heat between molecules in a body.[15]

It appears as though at the bottom of Poisson's mathematical physics is a purely mechanical model of matter and the belief that all physical phenomena can be traced back to the interactions of the molecules of matter through the forces acting between them.[16] Therefore mechanics, both dynamics and statics based upon the concept of force, was fundamental to Poisson's research. This in itself meant a complete reformulation of mechanics.[17] Dynamics became more fundamental than statics, which was reduced to a special case of dynamics. Throughout the eighteenth century, culminating in the work of Lagrange, the equations of motion of a system were deduced from its equilibrium equations using virtual displacement and velocity governed by minimum principles.[18]

Even with this stated, historians still have to examine whether Poisson was able to develop a mathematical physics, a theoretical physics based upon a particular mechanical model by applying the calculus. Or, rather, were his results just deductions from the calculus, not from his mechanical model. Was Poisson an early example of the theoretical physicist historians are familiar with, or was he something quite different—indeed an example of a practitioner of a discipline that shares only a name in common with its twentieth-century descendant. Implicit in all this is the question of whether the grounds upon which Poisson's work has been judged in the past are adequate.

In short, the problem is that most of the crucial characteristics of theoretical physics developed later are absent from Poisson's work. And this is true of the papers of his contemporaries, save Fourier's, addressed to physical problems. The physical implications or results of Poisson's papers are not brought out by himself or others. The physical significance that historians see and that are cited as in Poisson's work are read back into his papers from later accounts. More importantly, historians can use Poisson's work to examine how mathematics and physics in France in the first third of the nineteenth century differed from later disciplines of the same name. The problem, then, is to see what Poisson was doing and how his work differed from the theoretical physics of the latter half of the nineteenth century.[19]

We shall find Poisson's discipline was not a mathematical physics but an attempt to capture the emerging subfields of electrostatics and elasticity from the experimental discipline of physics and annex them for mathematics. Poisson was an eighteenth-century mathematician, not a modern theoretical physicist. Mathematicians in the eighteenth century used physical problems not to explore the physical world but to establish mathematical relationships that were then solved as mathematical problems. Since mathematics was viewed as reflecting the structure of nature, physical problems could lead directly into a extended exploration of the calculus, even to pioneering studies in new areas of analysis. However, the aim was the exploration of the calculus, not the solution of physical problems.[20]

Poisson not only worked within this eighteenth-century tradition but with

a calculus already becoming obsolete. He was fighting to maintain a mathematical tradition that already was being abandoned. Therefore, historians
must see his papers that are ostensibly on physical subjects within the context
of a deepening crisis within the discipline of mathematics. This crisis is
related to the reformulation of the foundations of the calculus and to the
conduct of mathematics itself—the kinds of problems mathematicians should
pursue—not to the evolution of modern, theoretical physics.

The transparency of the arguments in Poisson's mathematical papers versus
the convoluted nature of those in physics is a small indication of his
mathematical versus physical abilities. Poisson even as a student was able to
explain difficult mathematical concepts.[21] Evidence for Poisson the physicist
also rests on his "prediction" of the central bright spot in Fresnel's interference patterns whether the fringes were formed by an object or the edges of
a hole. However, as Arago narrates the incident, Poisson saw likenesses in the
mathematics of two physical cases. Fresnel's paper included a calculation for
the intensity of light at positions across the screen at the distances of full and
half wave lengths. (The physical foundations of this calculations are more than
obscure, to say the least, but Fresnel deduced an integral expression for the
brightness of the fringes.) Poisson then noted that this integral could easily be
deduced for the center for either case. He found that these *integrals* resembled
"that of reflected light in the phenomena of colored rings." Thus Poisson
drew the analogy with an already known physical result through the
similarities in the *mathematical* forms of the integrals. He did not predict a
new physical result from an understanding of the physical case itself. This is
definitely not a plunge into the physical unknown but a reaching back to
physical cases whose mathematical structures he already knew.[22]

All of Poisson's papers on electrostatics, magnetism, and elasticity
(including wave motions and light) begin with a more or less elaborate
statement of a physical model of matter.[23] Even the two electric fluids of his
early electrostatics papers were made up of molecules between which were
forces of attraction or repulsion. Poisson inherited this model from Laplace,
and from their point of view the model was empirically grounded.

Laplace had analytically deduced the law of gravitation from Kepler's laws,
which he took as observational. He extended the law to all particles of matter
in all heavenly bodies. Laplace thus removed the slur of hypothesis from
celestial mechanics. Newton's assumption of unviersal gravitation was also
deduced from a combination of analytical results and observation. (However,
the argument depends also on Newton's third law, and thus physical
hypothesis reenters the discussion.) Laplace mistrusted observations deeply
because they were prone to error, and he minimized his dependence on them,
relying instead on sometimes impenetrable tangles of analysis. Poisson
followed suit. Hypothesis—that is, physical assumptions about processes,
mechanisms, or structure—had no place in *mathematical* physics.

If, however, one trusted the apparently purely analytical deductions of

Laplace, the assumption that all matter is composed of molecules between which central forces acted was not a hypothesis. Laplace and Poisson remained blind to the idea that this model could not be extended to short-range intermolecular forces of unknown form and still maintain it was empirically grounded. In their work they reiterated the empirical origins of their model which was, therefore, of a different order than the hypotheses of physicists— hence Poisson's animus against physical imagery and his confidence in the truth of his model and its legitimacy in a discipline that, for him, admitted only the certainties of the calculus and observation.

Poisson's insistence on the reality of his model and his qualitative descriptions of it become longer and more dogged with time. Essentially, however, they do not change. All of his investigations into the multiplicity of physical phenomena had no impact upon his physical model. He does not appear to feed back the results of his detailed analysis into the interpretation of the physical system.

Poisson began his analysis of the equilibrium state and of the motions of elastic bodies with the model of matter as molecules between which were attractive and repulsive forces. If historians trace how Poisson transformed his initial physical model into mathematical realtionships, it is difficult to see how he used the above model to construct his mathematical functions that represent his forces. One clue to the difficulties Poisson experienced in putting his physical model into analytical form is that any discussion of how he constructs his equations of equilibrium and motion can proceed without any reference to his physical model.[24] The force acting on a molecule is represented by the summation of all the actions of countless individual molecules whose sphere of activity includes the molecule under consideration. This sum is in turn represented by a generalized function of the coordinates. This is purely mathematical. No statement follows that this is an arbitrary function whose form is at present unknown.[25]

Yet it is in these papers that Poisson claims that his physical model of matter can account for the observed macrophenomena of elastic bodies, pressure of fluids, tension in strings as well as complex wave motions. His hope was that in tracing the equilibrium and motions of bodies back to the internal molecular forces,

we will no longer need special hypotheses when we wish to apply the general rules of mechanics to particular questions.[26]

Thus, presumably, the tension in a flexible cord could be seen as the immediate result of the mutual actions of its molecules, and the states of lamina, fluids, and solids could be traced to the same source, although this is never stated by Poisson himself.

However, Poisson was also animated by another principle that springs from mathematics. As well as developing a complete, deductive physical system, the solution to the analytical problem had to be the most general mathematical solution possible. And in his papers this general solution was pursued wherever possible. From this general solution special cases were deduced, and from these he launched comparisons with experimental results. Giving, initially, a particular form to these functions (dictated by the physical conditions of the case) could lead only to particular solutions of the partial differential equations of motions and equilibrium.

The point is that in Poisson's work the mathematical relationships used to define the problem analytically bear no special relationship to the particular model used to introduce them. Poisson's particular model of matter cannot lead to the general mathematical solution he was searching for. Conversely, the general mathematical solutions he chose to pursue could not be deduced from the particularities of the molecular model Poisson insisted upon using. And, in Poisson's case, physics was always expendable.

The problem of the relationship between Poisson's model and his mathematical results was spotted very early by Navier. He pointed out that Poisson's results for vibrating plates had already been deduced by Lagrange from the ideas of Sophie Germain, where the forces that resisted flexure were inversely proportional to the principle radius of curvature of the plate. As Navier pointed out this is a geometrical condition, and

it remains to be shown how this geometrical idea is the result of the physical nature of molecular actions.[27]

Poisson could only insist that Lagrange and Germain's assumption was not legitimate in solving physical problems, but he did not address Navier's point.[28] His two standards are mutually exclusive, one arguing for the physical particularity that would restrict mathematical development, the other arguing for complete freedom to follow the analysis along its own mathematical course.

In generating expressions for the internal equilibrium of an elastic body, Poisson mathematically constructed the forces and moments acting on a parallelepiped of material from the rest of the body without going back into considerations of the forces from the molecules, that is, using only the general functions representing their actions. These are balanced by the forces within the parallelepiped on the rest of the body, again represented without returning to the actions of the individual molecules. In equilibrium these mathematical expressions must be equal; but here this could only be accomplished through much truncation of series and other mathematical manipulations. Mathematical necessity dictated when and how these series were terminated, powers of variables discarded and coefficients ignored. There is no reference to what physical conditions those variables, terms, or coefficients might represent.

At the end of the problem, when he had deduced a special case and matched it to known, experimental results, Poisson did not explore what these particular results could tell him about the forces between individual molecules, or even the internal state of matter in this special case. Neither numerical nor mathematical results are retranslated into terms that lead back into an exploration of the physical model. In addition, Poisson does not justify the appropriateness of the many approximations necessary to obtain his numerical results. The final outcome is that his model and the results of the analysis remain on two separate levels, isolated from one another.

This isolation of analysis and physics is present throughout the development of the calculation. As soon as the physical entities are defined mathematically they remain mathematical. They are regarded and discussed as functions, arbitrary constants, or coefficients, and Poisson is interested only in their mathematical characteristics. Conditions of equilibrium and equations of motion are mathematical equations, not statements of physical circumstances.

When Poisson established his initial equations, he restricted himself to those concepts available in his revised mechanics. The results of all his work in elasticity, electricity, and the theory of heat were minimal. While Poisson generalized Navier's equations of equilibrium and motion for elastic bodies and Laplace's on the attraction of spheroids and addressed what he considered the deficiencies of Fourier's mathematics in his theory of heat, he did not develop elasticity, electrostatics or heat theory conceptually.[29] And in his papers on elasticity Poisson did not carry out integrations even in particular cases where the integrals can be evaluated. The solution to the motions of a vibrating string are left as,

$$y = f(x + at) + F(x - at)$$

where y is the displacement, x the axis along which the cord lies when at rest, t is the time, and f and F are arbitrary functions of x and t. The above "is the complete solution to the problem," which it is, mathematically.

While we immediately use Fourier series to represent the arbitrary functions, this avenue of approach was closed to Poisson. In his paper on Fourier's unpublished 1807 paper on heat, Poisson had critiqued Fourier's use of infinite series of sines and cosines.[30] From the point of view of traditional calculus, Poisson was correct; the series involved did not converge in ways recognized by eighteenth-century mathematicians. They were not legitimate functions. And Fourier did not prove all the theorems required to use these series, that is, showing that they were convergent. Poisson, however, developed an alternative (or rather elaborated Lagrange's earlier mathematical analysis of wave motion) that led to results of the same form as Fourier's while denying that sine or cosine series could represent the general solution to the motion of a vibrating string.[31] Lagrange was seeking a functional solution to the wave equation. Poisson followed the same approach and transformed integrals of the offending series into integrals of exponential functions.[32]

More importantly, in his elasticity papers all the physical results Poisson discusses are deduced from the above "complete" solution. In the case of the vibrating string, rods, and membranes, he deduced the ratios of the frequencies of the various of possible vibrations—transverse and longitudinal in the case of the string; transverse, longitudinal, and torsional in the case of thin rods—then compared them with experimental results.[33] Considering a membrane, Poisson could solve only the special case of the membrane under constant tension around its boundary. (This was for mathematical reasons; the integrals were indeterminate.) Poisson then analysed the case of a free plate, ignoring all external forces, and then compared his results with Chladini's experiments. It is difficult to see how these can be compared, even ignoring gravity, as Chladini's plates were all supported somewhere.[34]

All these theoretically deduced frequencies depend on factors that were initially defined in terms of intermolecular forces. Poisson, however, bypasses the opportunity to explore the internal state of the vibrating body. Poisson had already insisted that the actual form of these forces could never be known so that their physical function is, in his case, problematic.

In this 1828 paper on elasticity, Poisson had set himself the task of setting up then solving the general equations of motion and equilibrium for elastic bodies. What he actually achieved was a generalization of Navier's equations for an isotropic elastic solid and examined some special cases. Conceptually he had not gone beyond Navier. However, his analysis is not altogether satisfactory in its own terms and he makes some unjustified assumptions that lead to less than valid conclusions. He deduced a ratio of $1:4$ for the axial elongation to the lateral contraction in a thin bar. This ratio is actually the result of mathematical approximations that are not valid in general. Poisson presented it as a valid, mathematically general result, and as such it was accepted.[35] Poisson was to return again to this problem in an attempt to encompass a mathematically more general analysis.[36]

Mathematically Poisson had also backed himself into a corner. He insisted that the mathematics previously used for the treatment of elastic bodies made up of molecules, that is, the methods of both Cauchy and Navier were wrong. The force acting at a point is the summation of forces from "indefinitely large" numbers of "disjoint" molecules, and it was illegal to transform these summations into definite integrals.[37] Poisson showed that if he transformed the summations that represented his molecular force into definite integrals, having previously assumed that the forces would be zero at the limits of the integration, when the external forces on the body changed and the distance between the molecules changed these integrals remained zero, which was absurd.[38] Obviously the result is true for the particular force law Poisson implicitly assumed but was not a demonstration of a general mathematical result.[39] Poisson then went on to take a specific example of a particular force law, but to obtain results at all he was forced to transform his summations into integrals.[40]

Yet the results he did obtain in specific cases were the same as those previously deduced using "illegal" methods. Poisson insisted on the above technique in his later papers and developed an argument to show that the summation of a series involved in the expressions for the forces between molecules was not the same when the coefficients of the terms were obtained as a sum or as an integral. Again Poisson made a number of particular assumptions that were arbitrary mathematically and vitiated the image of the molecules being disjoint.[41]

And in cases where he used specific force laws to illustrate these mathematical methods, Poisson used definite integrals, reverting to "illegal" methods. The physical content of Poisson's papers is confined, therefore, to their introductory sections in the shape of a specific model that may well have been necessary for him to enter into the mathematical process. Once entered into, however, the solution to the problems were those of a mathematician not the mathematical solutions of a physicist.

One is forced to note that this particular pattern was established very early in Poisson's mathematical physics in his papers on electrostatics. While insisting that the two electric fluids were composed of mutually repulsing molecules, Poisson does not use this model to set up or to solve the mathematical problem.[42] The mathematical problems actually solved and generalized are from two found in Laplace's *Mécanique Céleste,* namely, the attraction of two spheroids when there is no force within the spheroids and the depth of a fluid spread over such a spheroid.[43] The language Poisson uses in the statement of the physical problem is of forces and equilibrium in general rather than being specifically electrostatic. Indeed he refers explicitly to Laplace's problem in this connection.[44] What Poisson actually accomplished was the generalization of Laplace's mathematical method to any spheroid, not just those very close to a sphere. The problem is not merely analogous to the gravitational problem, it is the gravitational problem, reworked and mathematically generalized.

His only concession to electrostatics was to propose two expressions for force, one for repulsion, the other for attraction. The form in which the problem is couched and the methods used are taken over from Laplace's work, including a function V defined through the force acting at a point that represents

> the sum of the molecules of the body divided by their respective distances at the given point; denoting the sum by V with respect to the fluid layer which we are considering and we are looking for the values of V as a function of the coordinates of an arbitrary point in space.[45]

The potential then is taken from Laplace, and defined mathematically. The potential is a function V, "equal to the sum of all the particles of the spheroid, divided by their respective distances from the attracted point."[46]

The potential was introduced to solve the mathematical problem of the attraction of a spheroid on an external point. After its definition it remained, to both Poisson and Laplace, a mathematical function. In their work it carries no physical significance. Because Poisson does not explore the physical significance of the potential, some of his results are obtained only through laborious analytical methods that he does not connect to physical processes.[47] Neither does Poisson use physical imagery to shorten or simplify the analytical process. The physical characteristics of electric force or potential do not impinge on the mathematical generality of the analytical process. Because he relied on analysis, Poisson is forced to deduce arduously many of Charles Coulomb's experimental results, which he could have reached more easily with the judicious use of the physical ideas inherent in his model. Instead his physical comments are restricted to matching known experimental results with those of his analysis. These are, here and elsewhere, accompanied by remarks on the intrinsic inaccuracies of experimentation, often invoked to explain differences between the calculated and experimental numbers. Implicit here and explicit elsewhere is Poisson's belief in the absolute accuracy of analysis, an idea common among eighteenth-century mathematicians.

Recently historians have analyzed Poisson's work in both elasticity and electrostatics for their importance for the history of physics. However, on reading Poisson's paper one must ask where is the physics in all this analysis, much of which is so complex and seems designed to obscure the physics rather than clarify. At least no clarification of physical processes are obvious in the papers. This leads one to challenge a fondly held assumption that the mathematization of a problem necessarily elucidates the physical issues. Sometimes it may well muddy the waters.

Historians are also left with the question of what was Poisson doing if they cannot connect his work structurally with that of later mathematical, theoretical physicists? This has been a continuing dilemma. Most historians of mathematics consign his work to the history of physics. Yet it does not fit easily into the development of physics. Clearly, his work seems peripheral to both, particularly when contrasted to the brilliant work of his rivals, Fourier and Cauchy. As one historian has put it, Poisson was unable to match Cauchy's mathematical insight or Fourier's originality in physics.[48] Yet while we have more difficulty fitting Poisson into our backward extrapolations of mathematics and physics, neither Cauchy nor Fourier quite fit the mold either.[49]

Poisson followed early eighteenth-century mathematicians and grounded the calculus in physical reality, a reality that Lagrange's "geometrical" model denied. He was trying to undo Leonard Euler's and Lagrange's attempts to wrest the foundations of the calculus from physics and establish it in algebra. Lagrange had replaced forces and motions with coordinates denying the physical basis for mathematics and indeed reversing that relationship in his *Mécanique Analytique*.[50]

Poisson saw his method as the true approach to the study of physical phenomena, which accounts for his habit of beginning each new problem anew. This included definitions of elementary concepts, descriptions of basic observations and experiments, and the rederivation of well-known results. Consequently, most of his papers are not-so-small treatises on the problems he addresses. They are not research papers. Even given the relative loquaciousness of the era, he is long-winded and repeats much that has no bearing on the aspects of the problem on which he claims to shed new light. And in his papers Poisson reached mathematically more general partial differential equations and more general solutions to these equations rather than new physical insights. The problem for Poisson was to use a concrete physical model to set up the general mathematical solution and from these deduce particular results. The latter indeed was the goal of some eighteenth-century mathematicians who believed that mathematics represented reality. Mathematics was sufficient to itself to solve physical problems without recourse to logically problematic physical hypotheses. And yet Poisson's goal of using a concrete model to establish the most general mathematical solution was something he did not and could not accomplish.

The form then of Poisson's papers, while always following the same design, are deceptive. The problems begin after he has set up his mathematical expressions for the functions representing physical forces. Because these mathematical expressions are so general the particularities of the physical model are immediately lost. And Poisson always chose the path to the goal of a general, mathematical solution. And yet he insisted on the need for particular physical models.[51] It was not enough to reproduce the laws of cooling; the physical process of cooling itself had to be modeled. However, the process of radiation between molecules plays no part in the analytical argument.

It begins to make more historical sense to see Poisson's work as an attempt to reestablish a form of the calculus, which was already discarded by some mathematicians, and lead mathematicians back to the patterns of problem solving of the eighteenth century. His critique of Lagrange was based on the belief that algebra was an insufficient foundation for the calculus. To be true, the calculus had to be grounded in physical reality.

Together with this critique we can see all of Poisson's mathematical physics papers and his text on the theory of heat as a mathematical tirade against the methods represented by Lagrange's calculus and analytical mechanics and those contained in Fourier's work on heat. This calculus was based on assumptions that Poisson saw as threatening the foundations of the discipline, intellectually and socially. Poisson returned to the problem of heat over a period of thirty years and repeatedly explored methods to replace Fourier's series with functional solutions, thus denying the mathematical foundations from which Fourier's results sprang.[52] Poisson began by elaborating Lagrange's alternative functional solution to the wave equation. However, he went on to develop his own transformations for integrals of sine and cosine series, that is, reworking even the basis for this functional solution offered by Lagrange.[53]

Similarly, Poisson returned to the problems of elasticity, which again brought him into direct conflict with Cauchy and the new analysis that Cauchy developed in the decade of the 1820s. In this regard it is important to note that in the early decades of the nineteenth century, while Poisson did not explore the physical signficance of his work, neither did his contemporaries. The works that historians assume to belong to physics were discussed by the French in the early nineteenth century as if they were mathematical ones. Mathematical not physical criteria were used to judge the cogency of arguments. In Fourier's theory of heat the physical concepts seem to historians to leap from the page. Yet Fourier's contemporaries did not take note of the new physical concept of heat flux. What they did discuss were the shortcomings of his mathematics, in particular the legitimacy and generality of his solutions to his partial differential equations.

From the point of view of the available versions of the calculus, Fourier's use of definite integrals was also suspect.[54] Equally, all his critics missed Fourier's contention that physical criteria delineate the depth and direction of mathematical solution for any physical problem. Fourier did not claim to be solving a mathematical, but a physical problem. Fourier hedged the analysis about with the needs of the solution of the physical problem. He also limited the mathematical problems he considered to those of physical significance. In his solutions physical conditions controlled the depth of mathematical development. His work challenged the traditional relationship between mathematics and physics. As if to emphasis this break, Fourier did not prove all the theorems he needed until pressed by his critics to do so. That those theorems, when proved, also challenged many assumptions held about the foundations of the calculus only served to further undermine the validity of his work. And yet Fourier was content to meet his critics on their terms. He did not emphasize what historians can see as the crucial difference between the mathematician's and the physicist's solution to the same problem.

By the 1820s the foundations of the calculus were being transformed largely through the work of Augustin Cauchy. In the eighteenth century, mathematicians seem almost too busy extending and enjoying the rich mathematical domain of the calculus to worry overly much about the soundness of its basic concepts. Sheer exuberance in the power of analysis and technical brilliance displayed in solving problems were the hallmarks of the eighteenth-century mathematician. The goal was to see who was technically the most innovative, could push the calculus furthest, and open up yet more problems for mathematicians to solve. The starting points for many mathematicians were physical problems. They assumed that mathematical solutions to physical problems always existed, and the reliability of the calculus was guaranteed by the reality of mechanics. This was also extended to a belief that mathematics directly expressed the structure of reality and therefore, to solve the mathematical was to solve the physical problem.

Toward the end of the century, thoughtful mathematicians like Lagrange had begun to scrutinize the foundations of the calculus. Because he was unable to define function, differential, or integral satisfactorily for his students, Lagrange reexamined these fundamental concepts more closely. In so doing he reopened a new kind of research problem for mathematicians. This exploration of the foundations of the calculus was a challenge to the notion that mathematicians solved problems. Clearly those solutions had to be based upon foundations that were well defined, self-consistent, and logically defensible.

Lagrange himself critiqued current foundations of the calculus and put forward an alternative.[55] More importantly for the history of mathematics, he joined the definitions of basic concepts to proofs of specific theorems.[56] Others followed suit. By the time Poisson began his career, the content and direction of traditional mathematical research had changed.

In light of these remarks historians can reevaluate both Lagrange's and Laplace's claims that mathematics was a dying field. One has to ask, What mathematics were they referring to? Certainly Lagrange's career points to a revitalization of a discipline not to its imminent demise. He was probably referring to traditional analysis and the discipline focused on problem solving. So also was Laplace. Both searched for ways of escaping from the dilemma, Lagrange by concentrating on fundamentals, Laplace and then Poisson by reaffirming the traditional foundations and methods of the calculus. This was reinforced by annexing the newer fields of physics and solving problems based in physical reality with solutions that resided wholly in this older mathematics.

Both Laplace and Poisson questioned the new foundations developed by Lagrange and like-minded mathematicians. They criticized the new, algebraic foundation for the calculus that separated the physical grounds from which the calculus initially sprang. They based their calculus on the idea of the infinitely small without being able to connect this to their methods or validation of them. Implicit in Laplace's and Poisson's mathematics was the question of whether such a purely algebraic calculus was adequate to solve physical problems. Yet both Laplace and Poisson always worked with the mathematician's goal of developing the most general solution to any mathematical problem that presented itself.

In their opinion, while Lagrange removed mathematics from its basis in reality, Fourier's mathematics was simply wrong. In addition Fourier was explicit in his claims that the laws of heat could not be reduced to those of mechanics. This claim obviously required some answer. Poisson expended a great deal of effort in vain trying to demonstrate that they were so reducible. All he could do was reproduce Fourier's partial differential equations for heat flow, which were not clearly connected to the mechanical model he insisted on, then deduce his alternative functional solutions to those same partial differential equations.

While Fourier subtly defied the mathematics of Poisson and Laplace, Cauchy's challenged it directly. Cauchy based his new calculus on the derivative and the variable, not the differential and the function. He also established the definition of the derivative as a limit and integration as the limit of the summation process. The basis of both Poisson's mathematics and physics was under attack, especially the strict separation he claimed he maintained between the processes of summation and integration and the relationship of these mathematical processes to the physical processes they represented.

The anti-Cauchy group within the French mathematical community was small and seems to have consisted of only the politically powerful Poisson and Laplace. Both Fourier and Cauchy together redefined the relationship between mathematics and physics and the direction of research within mathematics itself and both were influential in changing the directions of these disciplines in the next generations of mathematicians and theoretical physicists—but not in France. The foundations of Cauchy's calculus were accepted, but the discipline retained its focus on the solution to problems. Also, the limits of the disciplines remained defined as in the eighteenth century, and analysis and experiment remained the twin pillars for the solution of physical problems. The solution to the mathematical problem was by definition also the solution to the physical one. Physics was still defined by the method that had defined it in the eighteenth century, experiment. In this scheme there was no place for physical imagery, the very domain that was to lead to the opening up of new areas of research for physicists in Germany and Britain.

In these countries theoreticians began to use physical models and processes to define the limits of the mathematical problems that were necessary for them to solve. Physical results were the only outcomes that were interesting in this redefined disciplinary matrix. Mathematical implications were published elsewhere.

Therefore, rather than seeing a "decline" of French physics in the nineteenth century, it continued with mathematics along disciplinary lines laid out in the late eighteenth century. This would account for the puzzlement and distaste the French expressed for the physics of the very model-prone British. This is nowhere more pronounced than in the reactions of Pierre Duhem to Maxwell's theory of electromagnetism and his equal enthusiasm for thermodynamics, shorn of its mechanical origins.

Duhem, like Poisson before him, was heir to a tradition in French science in which only the generalities of analysis and the results of experiment were legitimate. In this scheme there was no place for physical hypotheses joined to the analysis but limiting its directions of development. Experiment was integrated into theory by not merely matching results already obtained with the results of analysis but through the physical imagery used to extend the analysis into previously unexplored areas.

In Poisson's work his particular model did not enter into his analysis, nor did he draw out of the results of this analysis any physical significance. Indeed his physical ideas never changed over the course of a career in which a few important physical problems were reexamined several times. What did change was the analysis, in the direction of generalizing the mathematical solutions.

This and other aspects of his papers force historians to consider Poisson's work in the history of mathematics rather than physics. At the same time Poisson insisted upon a certain form for the calculus, and much of his effort on physical problems was to preserve a mathematics fast being repudiated and to remain a practitioner in a discipline that was being transformed beyond recognition. Conversely, theoretical physics did not develop through the simple mathematization of the experimental discipline of the eighteenth century through the work of such men as Poisson. Nor was Poisson successful in joining physical imagery and mathematical analysis into one new scheme. His mathematical physics cannot be linked directly to the discipline of mathematical, theoretical physics easily recognizable later in the century. He was working in an entirely different discipline. It is one defined by eighteenth-century mathematical criteria rather than those of nineteenth-century theoretical physics. He was not a new kind of physicist at all, but rather his life was an attempt to retain the goals and methods of the traditions of the eighteenth-century mathematician. Thus, later, his mathematical results could be useful while his analytical methods were abandoned. In addition his physical imagery was useless for experimental physicists since it could not be followed through the mathematical maze to emerge as suggestive ideas for future work. And for the same reason it was not a model for later generations of physicists in whose work we see the emergence of modern, theoretical physics.

Notes

1. In this paper I will be assessing Poisson's place in the intellectual history of physics through his research papers. As Poisson well knew and explicitly stated, there are pedagogical constraints in presenting ideas in textbooks. I am assuming that his research papers represent his methods of using and developing arguments in physics and mathematics better than his textbooks. I am also not directly concerned with his career and behavior towards colleagues. Thus his treatment of Fourier is less important than how he responded to Fourier's theory of heat and what those responses reveal about the ways in which he thought about problems in mathematics and physics.

2. James Clerk Maxwell, *Treatise on Electricity and Magnetism* (New York: Dover reprint of 1891 edition, 1954), 1: 89. See also E. Garber, S. G. Brush, and C. F. W. Everitt eds., *Maxwell on Molecules and Gases* (Cambridge: MIT Press, 1986), 174.

3. Edmund Whittaker, *A History of the Theories of Aether and Electricity* (New York: Thomas Nelson and Sons, 1962), vol. 1, *The Classical Theories,* 62. The paper to which Whittaker refers is Siméon-Denis Poisson, "Mémoire sur la Distribution de l'Électricité à la surface des Corps conducteurs," *Mémoires de l'Institut de France, Classe des Sciences Mathématiques et Physiques* (1811): 1–92, cited as Poisson, "Électricité."

4. Stephen Timoshenko, *History of the Strength of Materials,* with a brief account of the history of the theory of elasticity and theory of structures (New York: Dover reprint of 1953 edition, 1983), 114.

5. Louis L. Bucciarelli, "Poisson and the Mechanics of Elastic Surfaces," in Michel Metevier, Pierre Costabel, and Pierre Dugas, eds., *Siméon-Denis Poisson et la science de son temps* (Paris: École Polytechnique, 1981). In this connection it is interesting to note that Pierre Costabel, "Siméon-Denis Poisson," *Dictionary of Scientific Biography,* edited by Charles Coulston Gillispie (New York: Scribners, 1971) supplementary vol., 480–90, describes the details of his papers on electricity and elasticity without assessing their value for the development of physics.

6. John Heilbron, *Electricity in the Seventeenth and Eighteenth Centuries* (Berkeley: University of California Press, 1979), 499–500.

7. R. W. Home, "Poisson's Memoirs on Electricity: Academic Politics and a New Style in Physics," *British Journal for the History of Science* 16 (1983): 239–60.

8. Robert Fox, "The Rise and Fall of Laplacian Physics," *Historical Studies in the Physical Sciences* 4 (1974): 89–136.

9. D. H. Arnold, "The Mécanique Physique of Siméon-Denis Poisson: the Evolution and Isolation in France of his Approach to Physical Theory," *Archive for the History of the Exact Sciences* 28 (1983): 255–367; 29 (1984): 37–94, 287–307.

10. Heilbron, *Electricity,* 500.

11. In this paper "analysis" has two meanings. When referring to Poisson it is used in its eighteenth-century meaning, a synonym for the calculus. In general it refers to the unfolding mathematical argument.

12. In addition, chapters on mathematical methods appear in texts separate from the chapters on physics.

13. An early example of physics being mathematics is Ludwig Boltzmann's statistical description of a system's approach to thermodynamic equilibrium. His search for a physical, molecular, mechanical model for this statistical process ended in failure. He finally interpreted entropy in terms of fluctuations from an equilibrium value. See Boltzmann, "On Certain Questions of the Theory of Gases," *Nature* 51 (1895): 413–15, 581, and "Ueber die sogenannte H-curve," *Mathematischen Annalen* 50 (1898): 325–32, and Boltzmann and G. H. Bryan, "Ueber die mechanischen Analogie des Wärmegleichgewichtes zweier sich beruhrender Körper," *Sitzungsberichte Kaiserliche Akademie der Wissenschaften, Wien. Mathematischen-Naturwissenschaftliche Klasse* 103 (1894): 1125–34, and *London Physical Society, Proceedings* 13 (1895): 485–93.

14. Siméon-Denis Poisson," "Mémoire sur l'équilibre et mouvement des corps élastiques," *Mémoires de l'Académie des Sciences, Paris* 8 (1828): 357–570, 623–27, 360, cited as Poisson, "Corps Élastique."

15. Poisson, *Theorie mathématique de la chaleur* (Paris: Bachelier, 1835), 5

16. See Fox, "Rise and Fall," for details of the Laplacian model of matter.

17. Laplace gave a Newtonian version of dynamics in the first volume of his *Mécanique Céleste,* but, it was on an introductory level and he only touched on those aspects necessary for celestial mechanics. See Laplace, *Celestial Mechanics,* translated by Nathaniel Bowditch (New York: Chelsea Publishing Co., reprint of 1829 ed., 1966) vol. book 1.

18. For a general history of mechanics in the eighteenth century, see René Dugas, *Histoire de la mécanique* (Paris: Editions Dunod, 1950).

19. This paper is largely a presentation of negative results. The historian is at a disadvantage as there were no quotations forthcoming, only the silences of the lacuna. I have chosen to concentrate on one paper, albeit an important one in the theory of elasticity, Poisson, "Corps Elastiques'," where all the problems and shortcomings of Poisson's work converge.

20. This will be discussed further in E. Garber, *Mathematics as Language.*

21. See Arnold, "Poisson," 28: 283.

22. Francois Arago, "Rapport a l'Académie au sujet de le prix sur le diffraction," *Annales de chimie et physique* 11 (1819): 5; see also A. Fresnel, *Oeuvres completes,* edited by Henri de Senarmont, Emile Verdet, and Leonor Fresnel (New York: Johnson Reprint of 1864 edition, 1965), 1: 245.

23. Thus I must disagree with Heilbron that Poisson was indifferent to physical images or mechanisms. His repetition of the same molecular mechanisms to explain ranges of mechanical and nonmechanical phenomena were obviously fundamental to all his physique mathematique. See, Heilbron, *Electricity,* 499.

24. See Isaac Todhunter, *A History of the Theory of Elasticity and the Strength of Materials,* vol 1., *From Galileo to Saint Venant (1639–1860)* completed by Karl Pearson, ed. (New York: Dover Reprint of 1886 edition, 1960) 222–23. In his discussion of Poisson's most important paper in elasticity, Poisson, "Corps Élastiques," Todhunter gives a description of Poisson's model and the mathematics necessary if it is used, then ignores it, stating Poisson's general equation for equilibrium without any reference as to how to deduce the forces in the equations.

25. This could be done in 1828. Navier, whose paper inspired Poisson's 1828 paper had made exactly this statement at this point in his argument. See, C. L. M. H. Navier, "Mémoires sur les lois de l'équilibre at du mouvement des corps solides élastiques," *Mémoires de l'Académie des Sciences, Paris* 18 (1827): 375–93, 379.

26. Poisson, "Corps Élastiques," 360.

27. Navier, "Note relative a l'article intutile: Mémoire sur l'équilibre et le mouvement des corps élastiques, insére page 337 du tome précédent," *Annales de chimie* 38 (1828); 304–14. Navier also complained quite rightly, of Poisson's neglect of his work, wishing Poisson had "given the same attention to the living as the dead" in his account of the history of elasticity. This complaint would also cover Poisson's omission of the contributions of Sophie Germain and Lagrange. The paper to which Navier is referring is Poisson, "Mémoire sur l'équilibre et le mouvement des corps élastique," *Annales de chimie* 37 (1828): 337–55, which is an abstract of Poisson, "Corps Élastiques," much of it is in the introduction of the longer paper. As this controversy developed, Navier asked whether Poisson could show that his mutually interacting molecules could model the behavior of an elastic body. See Navier, "Remarque sur l'article de M. Poisson insére dans le Cahier d'auot," *Annales de chimie* 39 (1828): 145.

28. Poisson, "Response d'une Note de M. Navier insére dans le dernier Cahier de ce Journal," *Annales de chimie* 38 (1828): 435–40. Poisson's arguments against Navier are purely mathematical.

29. See his definition of the potential in Poisson, "Électricité." For a clearer, yet equally mathematical generalization of Navier's theory of elasticity based on point centers of force, see Augustin-Louis Cauchy, "Sur l'équilibre et le mouvement d'un système de points matériels sollicité par des forces d'attraction ou de repulsion mutuelle," *Exercices de mathématiques* (Paris: De Bure, frères, 1828) 3eme annee, 188–212. Cauchy's is the mathematicians' generalized solution to the problem, and he attaches no more physical significance to his results than Poisson. He considers the displacements of the molecules rather than the forces acting on them. His mathematics is simpler, more rigorous, and easier to follow than Poisson's.

30. Poisson, "Mémoire sur la propagation de la chaleur dans les corps solides," *Bulletin Societé Philomatique* (1807): 112–15. Ivor Grattan-Guiness sees this review as "denigrating" Fourier's 1807 paper and singularly myopic. See I. Grattan-Guiness in collaboration with J. R. Ravetz, *Joseph Fourier (1768–1830)* (Cambridge: MIT, 1972), 442–43. See also, John Herivel, *Joseph Fourier: The Man and the Physicist* (Oxford: Clarendon Press, 1975). Arnold, "Poisson," 28: 289, sees it as a fair review of the contents of Fourier's paper. However, Poisson made no comments upon Fourier's introduction of the concept of heat flux, the single most important physical innovation. The review describes the physical problem only in the most general terms. His detailed remarks are confined strictly to Fourier's mathematics, which Poisson critiqued both for its lack of generality and rigor. For details of this mathematical critique, see Grattan-Guinness, above, and, Herivel and Pierre Costabel, *Joseph Fourier face aux objections contre sa théorie de la chaleur, lettres inédites, 1808–1816* (Paris: Bibliotheque Nationale, 1980).

31. Jean Louis Lagrange "Recherches sur la nature, et la propagation du son," *Miscellanae Taurinensia* 1 (1754) [pub. 1759]: 1–112 reprinted in *Oeuvres de Lagrange* (Paris: Gauthier–Villars, 1867), 1:39–148. A discussion of the mathematical differences between the derivation of Lagrange's solution to the wave equation and Fourier analysis is in Grattan-Guinness and Ravetz, *Fourier,* 246–49.

32. Poisson, "Sur les integrales definiés," *Journal d'École Polytechnique* 11 (1820): 295–341, and , "Sur la maniere d'exprimer les fonctions par des séries de quantitiés périodiques, et sur l'usage de cette transformation dans le résolution se différentes problèmes," *Polytechnique* 11 (1820): 417–89. One of the problems Poisson addresses is that of the vibrating string. Again he transforms the series into a functional solution.

33. Poisson, "Corps Élastiques," 438–41.

34. Poisson, Ibid, 456.

35. Poisson, Ibid, 451. See Timoshenko, *History of Strength,* 112.

36. Poisson, "Mémoire sur l'équilibre et le mouvement des corps élastiques," *Mémoires de l'Académie des Sciences, Paris* 8 (1829): 357–570, 623–27, and, "Mémoire sur la propagation du mouvement dans les milieux élastiques," *Mémoires de l'Académie des Sciences, Paris* 10 (1831): 549–606.

37. This was an attack on both the mathematical methods of Navier and on the new calculus being developed by Cauchy and already in use, implicitly, in Fourier's theory of heat. I will return to this later.

38. Poisson "Corps Élastiques," 398–99.

39. Cauchy, "Mouvement d'un système," insisted on the same distinctions between summations and integrals and showed that Poisson's methods led to false results.

40. Poisson, "Corps Élastiques," 378.

41. Poisson, "Mémoire sur les Équations générales de l'Équilibre et du Mouvement des Corps solides élastiques et des Fluides," *Journal de l'École Polytechnique* 13 (1831): 1–174. Poisson had earlier argued that the sum and the integral of a function between the same limits might depend on the path. This was a sharp critique of Cauchy's emerging calculus. Poisson, "Sur les integrales definiés," 329–30.

42. Ivor Grattan-Guinness was the first modern historian to examine early nineteenth-century French mathematical physics in detail. Yet he sees the mathematization of physics as an evolutionary process, seamless and natural. He has explicitly analyzed French mathematical physics as an aspect of the history of mathematics without exploring the implications of his treatment for the history of physics. See I. Grattan-Guinness, " Mathematical Physics in France, 1800–1835," in H. N. Jancke and M. Otto, eds., *Epistemological and Social Problems in the Sciences in the Early Nineteenth Century* (Boston: Reidel, 1979) 349–70, and "Mathematical Physics in France, 1800–1840: Knowledge, Activity and Historiography," in J. Dauben, ed., *Mathematical Perspectives* (New York: Academic Press, 1981).

43. Laplace, *Mécanique Céleste* 2, 1–359. The section on the depth of fluid is in 241–359.

44. Poisson, "Électricité," 4–5.

45. Poisson, Ibid, 15.

46. Laplace, *Mécanique Céleste* 1, 45.

47. See Poisson's treatment of Gauss's law, the physical significance of which eluded him, Poisson, "Électricité," 52.

48. Grattan-Guinness and Ravetz, *Fourier*, 462–63.

49. While Fourier clearly separated the solution of the physical problem from the general mathematical one, he was content to defend his work on the ground dictated by his critics, i.e., his mathematics.

50. The sketch that follows of eighteenth-century calculus is taken from several sources, the most important of which are Grattan-Guinness, *The Development of the Foundations of mathematical Analysis from Euler to Riemann* (Cambridge: Cambridge University Press, 1970); H. J. M. Bos, "Differentials, Higher Order Differentials and the Derivative in the Leibnizian Calculus," *Archive for the History of the Exact Sciences* 14 (1974): 1–90; A. P. Youschkevitch; "The Concept of the Function up to the Middle of the Nineteenth Century,"*Archive for the History of the Exact Sciences* 16 (1974): 37–85; Judith Grabiner, *The Origins of Cauchy's Rigorous Calculus* (Cambridge: MIT Press, 1981).

51. In this respect it is interesting to note that Poisson did not develop the mathematical language to deal with the mechanics of a body made up of centers of force. This was done using extensions of Lagrange's methods by William Rowan Hamilton and their mathematical generalization by Carl Gustav Jacob Jacobi.

52. Poisson, "Sur la distibution de la chaleur dans les corps solides," *Mémoires de l'Académie des Sciences, Paris* 1 (1816): 1–70.

53. Ibid.

54. S. F. Lacroix's *Traité du Calcul Différentiel et Calcul Integrel,* 3 vols. (Paris: Chez J. B. M. Dupratz, 1797) indicates the various formulations of the calculus then available. The text is a compilation of all the different forms of the calculus written in the conviction that all the varying forms were reconcilable.

55. Lagrange defined the differential of a function by constructing its Taylor series expansion. The coefficients in the terms of the power series were the various differentials of the function. This definition assumes that every function can be expanded in a Taylor series.

56. See Grabiner, *Origins,* chap. 2.

Part III
Science and Technology

Newton Confronts the American Millwrights,
or,
Action and Reaction Are Not Always Equal

EDWIN T. LAYTON, JR.

Thanks to the researches of Robert Schofield scholars are aware of the continuing impact of Newton upon the science and technology of eighteenth-century Britain. In particular, in his study of the Lunar Society of Birmingham, Schofield shows direct links between Newton's ideas and technologists such as James Watt, Josiah Wedgwood, and other pioneers in the early Industrial Revolution.[1] In this paper I will examine Newton's impact upon American technology. As Brooke Hindle and others have demonstrated, a Newtonian tradition of science and technology existed in eighteenth-century America.[2] There are significant parallels between the American Philosophical Society of Philadelphia and the Lunar Society. The example set by American scientists and technologists affiliated with the American Philosophical Society, such as Franklin and Jefferson, no doubt encouraged American technologists to turn to science for the solution of technological problem, and a scientific style of technology had become manifest in America by the late eighteenth century. This scientific style helps to account for the rather precocious leadership that Americans demonstrated in several fields of technology. Perhaps in none was this more apparent than in water-power devices. By the twentieth century American designs had come to dominate the field of water turbines, notably the Francis "mixed flow" reaction turbine. These designs were the products of a scientific technology that drew from Newtonian mechanics incorporated within a tradition of vernacular science developed by American millwrights.

Newton's influence upon American technology was complex. The use of Newtonian science by craftsmen with little mathematical training presents a number of difficulties. Before dealing with them, it might be useful first to examine Newton's influence upon American engineers. As professional technologists, engineers might be expected to be mathematically and scientifically literate, and though professional engineers remained few in numbers until the last quarter of the nineteenth century, they constituted an important cutting edge of American technology. The first generation of American engineers (approximately 1820–50) were, for the most part, educated by self-training supplemented by on-the-job experience. Some of these, such as Uriah Boyden, had worked their way through the *Principia*.

Boyden continued to use the geometrical formalism of the *Principia* in its pristine form as late as the 1860s.[3] Others among this first generation of American engineers encountered Newton through intermediaries such as Charles Hutton. Hutton's textbook on the fluxions went through six American editions, the last in 1843.[4] The second volume of Hutton's text was a course in applied Newtonian mechanics for engineers. Hutton adapted Newton's physics and applied them to engineering problems using a somewhat simplified mathematical formalism.[5] In 1847 another self-trained American engineer, Squire Whipple, developed the first correct mathematical theory for truss bridges, which he expressed by a formalism essentially identical to that employed by Hutton.[6] Hutton's text was the first on fluxions (or the calculus) employed at West Point, where it influenced an elite of American civil engineers, since West Point engineers were encouraged to go into civilian engineering practice. They were the first American engineers to benefit from a formal college education in mathematics, science, and engineering.[7]

Though Newton's influence was direct and important for the first generation of American engineers, it is difficult to trace a continuing tradition beyond this generation. The reason is that engineers were expected to be conversant with the international literature of science and engineering. Newton provided the foundations for both, but his specific influence became intermixed with the contributions of later generations of scientists and engineers. The adoption of the Leibnitzian notation for the calculus, along with refinements of the concepts of work and energy tended to obscure Newton's influence. In the case of West Point, the "Newtonian" era was short. Texts inspired by the French mathematical and engineering tradition replaced those such as Hutton's that built directly upon Newton. A promising cadet, Dennis Hart Mahan, was sent to study engineering at French schools, and he taught civil as well as military engineering to cadets for a generation.[8] West Point was reorganized after 1816 by Sylvanus Thayer on the model of French engineering schools. Hutton's popularity gradually declined, and his text was replaced in mechanics by those of Julius Weisbach.[9] Weisbach's texts were soon established as the standard works for engineering schools such as Rensselaer Polytechnic Institute.[10] The continued printing of editions of Hutton, however, may indicate that self-trained engineers both in Britain and America found it easier to use the earlier Newtonian mechanics than the sophisticated Continental variety taught at engineering schools. In any case, American engineers at mid-century adopted an anti-theoretical ideology and self-image that tended to conceal the continued importance of mathematical theory and deductive reasoning from scientific principles, whatever their source.[11] Newton's specific influence can be detected in exceptional cases, such as those of Boyden and Whipple. Not only did they employ a Newtonian formalism, but both attempted to defend Newtonian orthodoxy against later scientific innovations.[12] But as one might expect, this was a most unusual role for an engineer to play.

The influence of Newton upon American millwrights was important for the development of a technological science for water power. But unlike engineers, the millwrights were not acquainted with the international literature of science and technology. Their training was still by an oral tradition passed from master to apprentice, supplemented by self-study. Since most millwrights worked in the hinterland far from scientific libraries, few books were available. In the typical case the millwright's scientific training was rather sketchy. An Ohio millwright, Conington Searle, deposed in 1843 that he was acquainted with hydraulics and hydrostatics, but that "my knowledge of hydraulics is a little from theory only, and what I have picked up from observations."[13]

Lacking professional organization or publications, millwrights had only limited means to communicate with each other. In particular, they had no means of establishing norms of good and bad work nor of eliminating errors once refuted. Old errors had a way of continuing from generation to generation. A handful of millwrights established a useful vernacular science for water power, but each had to struggle to get the scientific fundamentals correct. Despite the triumphs of the millwrights vernacular science, the fact that each generation had to struggle with the same fundamentals and the same errors demonstrates the value of the cognitive and institutional changes that made possible cumulative traditions such as those of modern scientists and engineers.

Given their mathematical limitations, it is perhaps surprising that craftsmen were able to construct vernacular sciences upon Newtonian foundations. The story is, indeed, rather complicated. I shall, therefore, first show how a few millwrights, notably James Rumsey and Oliver Evans, were able to create somewhat different Newtonian vernacular sciences. Secondly, I intend to show that millwrights, including the most creative ones, experienced continuing difficulties in understanding and applying Newtonian science correctly. Thirdly, I will indicate how the development of systematic testing of turbines using the Prony Dynamometer provided millwrights with an alternative approach for understanding water motors. A significant number sought to reject Newtonian science and, indeed, most theory and rest the development of the turbine almost entirely on empirical testing. Finally, I will show how this rather "Baconian" revolt against Newtonian vernacular science contributed to a highly ironic conclusion. The systematic testing tradition developed by craftsmen such as James Emerson helped lay the foundations for a great advance in technological theory that took place mainly in Germany after 1900.

Despite their mathematical and scientific limitations, American millwrights were able to develop vernacular scientific traditions that were of material benefit in the growth of a highly fruitful tradition of hydrotechnology. John. T. Desaguliers's two-volume treatise, *A Course of Experimental Philosophy*, was an important popularization of Newtonian science accessible to mill-wrights.[14] Desaguliers emphasized experiment, as the title to his book suggests, and his mathematics was quite elementary. He gave detailed rules for carrying out arithmetic calculations, particularly in his favorite field, hydraulic technology. Thus millwrights could learn from Desaguliers how to use Newton's laws of motion to reason about the motion of particles or fluid streams. This reasoning was, of necessity, mainly qualitative. For quantitative results they could translate their problems from mechanics to hydraulics where they could quantify their problems using simple arithmetical procedures provided by Desaguliers. Or they might perform experiments, also following Desaguliers's example.

James Rumsey, best known as a steamboat inventor, provides an example of Desaguliers's influence. Rumsey first gained attention for a scheme for mechanically poling a boat upstream; he attempted to drive the poling mechanism with a water wheel. The failure of this project in 1785 may have encouraged Rumsey to seek a more scientific basis for his inventions, possibly starting in the winter of 1785–86. Rumsey later claimed to have developed his idea of a reaction-powered steamboat in 1786.[15] This idea derived from the science of Benjamin Franklin and James T. Desaguliers.[16] In 1787 a rival inventor, Englehart Cruse, visited Rumsey at his home in Shepherdstown, Virginia (now West Virginia). They discussed matters relating to steam engines. In the course of their discussion, Rumsey pulled a copy of Desaguliers's book from his shelf and read excerpts to Cruse. In a later pamphlet Cruse accused Rumsey of taking credit for discoveries that were really due to Desaguliers. Cruse sarcastically suggested that "having the good fortune to meet with Monsieur Desaguliers and some Dictionaries of Arts and Sciences, Mr. Rumsey hath set up for an original genius."[17]

Disergarding the sarcasm, Cruse's statement was not wholly without foundation. As of 1787 Rumsey was still in an early phase of his scientific self-education, and his knowledge was drawn initially from Desaguliers, in good part. Rumsey's technological science and his relationship to Desaguliers can be seen in the case of his reaction mill. In 1788 Rumsey traveled to Philadelphia to seek support from the members of the American Philosophical Society for his steamboat, mill, and other inventions. His reaction mill was an improvement on Barker's mill, which Desaguliers had first described in his treatise.[18] Indeed Desaguliers's interest was probably excited by Antoine Parent's widely accepted belief in the inefficiency of existing water mills using water by impulse or action, that is, according to Newton's first law of motion. Like many other eighteenth-century scientists and engineers, Desaguliers thought a reaction mill employing the third law of motion might obviate the impediments which, according to Parent, limited the efficiency of water mills to about 15 percent. Desaguliers summarized Parent's theory, and traced through the reasoning and calculations for installing an undershot water wheel according to this theory. In developing and presenting his reaction mill in

1788, Rumsey had not gone beyond Desaguliers; he merely appealed to the authority of a widespread consensus on the probability benefits of using reaction rather than action for his steamboat and mill. Rumsey justified his reaction mill by arguing:

> When the principles of this mill are accurately examined, it will be found that there is no other method of applying the same quantity of water that can give so much motion and power in an equal time.[19]

Unfortunately for Rumsey, Parent's theory of the inefficiency of traditional water wheels was false. Parent's theory had, however, been widely accepted by both scientists and engineers. As Terry Reynolds has shown, Parent's belief that waterwheels were limited to no more than 15 percent efficiency was accepted by Leonhard Euler, Jean D'Alembert, Colin McLaurin, Willem s'Gravesande, Bernard de Belidore, Henri Pitot, and many others.[20] Indeed, Desaguliers showed more caution than some others; he accepted Parent's results only for the undershot wheel that Parent had used for purposes of analysis, though Parent explicitly claimed that his results applied to all water wheels, whether the water was applied at the top (as in the overshot wheel) or the bottom (as in the undershot).[21]

In his British patent of 24 March, 1790, Rumsey presented his own theory of his improved Barker's mill. This presentation paralleled Desaguliers's presentation of the theory of the undershot wheel according to Parent's theory, including even certain turns of phrase.

Thus Desaguliers had held that:

> The knowledge of the foregoing particulars is absolutely necessary for setting an undershot wheel to work; but the advantage to be reaped from it would still be guess-work ... if we had not an ingenious proposition of ... Mons. Parent. ...[22]

In presenting his own theory of his improved Barker's mill Rumsey held that there were four key parameters that were "absolutely necessary to be known & taken into consideration, otherwise it would be quite accidental to form the machine."[23] Perhaps most importantly, Rumsey followed Desaguliers's practice of following the presentation of the theory by a concrete calculation by way of illustration.

In Rumsey's improvement of Barker's mill, the water was admitted from below into a cylindrical tube revolving in the horizontal plane. The water was discharged from orifices on opposite sides of the ends of the rapidly rotating tube to drive the mill by reaction. Rumsey's theory explained the working of his mill in terms of centrifugal force. In using this force to greatest advantage, Rumsey sought to proportion the rotor and its orifices so that the water was discharged with the same velocity it would have in the static case for a column of water of the same height as that which drove the mill.[24] Rumsey did not justify this principle of design, but he may have been guided by the analogy that simple machines, where for a given load and input of power, an increase in speed would produce a proportionate decrease in the load moved. In any case, the theory was incorrect; mills like Rumsey's become more efficient the faster they go. Attempting to hold down the speed of his mill would only decrease its efficiency.[25]

Though Rumsey's theory was incorrect, it contained a fundamental insight. Like Desaguliers, Rumsey carried through an exemplary calculation of the application of his principle of design. Rumsey had the visual imagination to distinguish between relative and absolute motion (that is, motion of the water with respect to the rotor and the ground, respectively). Thus he realized that only the absolute velocity of the discharged water would act to drive his mill. Rumsey determined experimentally what the total centrifugal energy was and made the necessary subtraction. He concluded that the net propulsive force by reaction was only 184/630 of the total, for a theoretical limit of efficiency of only slightly more than 29 percent.[26] Clearly, this mill was not, as Rumsey had initially thought, the most efficient way of using a given fall of water.

It is not clear whether Rumsey ever became acquainted with John Smeaton's classic treatise on waterwheels, which had been published in the *Philosophical Transactions* for 1759–60. Smeaton had refuted Parent. He showed that the limit of efficiency of the undershot waterwheel was 50 percent, not 15 percent as claimed by Parent. Smeaton showed by experiment that overshot wheels were twice as efficient as undershot wheels, implying a limit of efficiency of 100 percent.[27] Whether he knew of Smeaton's work or not, Rumsey became aware that the efficiency of his mill was lower than he had initially expected. Desaguliers thought overshot wheels were from 3½ to 4 times more efficient than undershot wheels.[28] Using Parent's figure for the undershot this gave limits of efficiency for overshot wheels of between 52.5 and 60 percent, about twice the limit Rumsey had calculated for his mill.

There was no explicit recognition by Rumsey that he realized that his analysis of his mill applied equally to his steamboat (though it did). What Rumsey did was to juggle his figures to double the calculated limit in the efficiency of his mill. Rumsey did this by adding the centrifugal force of rotation to the power of the fall of water that drove his mill in calculating the total reaction. This was, of course, incorrect. The centrifugal energy was only a manifestation of the energy contained in the fall of water, not an addition to it. But by this incorrect assumption Rumsey was enabled to claim a theoretical limit of efficiency for his mill of about 58 percent. This brought the efficiency of his mill, and by implication also, his steamboat, to about the same level as the overshot according to Desaguliers. But one may question whether Rumsey really believed this mathematical juggling.[29]

Rumsey died in December 1792, before any of his major projects had come to fruition. Despite his scientific difficulties, Rumsey had done highly creative and important technological work that benefitted from his vernacular version of Newtonian mechanics. He had, just before his death, developed the prototype of the reaction wheel (a primitive turbine) that was to serve as the starting place for very fruitful American work in hydrotechnology in later years. Rumsey was not the only American to encounter difficulties with mathematical theory. William Waring, a mathematics teacher at Philadelphia's Quaker Academy, published two papers dealing with water mills and Rumsey's improvement of Barker's mill. Though probably the first American attempt to apply the calculus to technology, Waring was no more successful than Parent. Indeed, Waring's papers appear to have been modeled on those of Parent. Waring's conclusion that the theoretical efficiency of Rumsey's mill would be low was incorrect. Though the practical efficiency attainable was modest, the theoretical limit was 100 percent (approached as the velocity of the rotor approached infinity). Rumsey apparently understood this from his experiments on a model of his mill. The idea of a variable efficiency, however, was so contrary to scientific orthodoxy that Rumsey apparently preferred to publish an incorrect but orthodox theory.[30]

Rumsey's mechanical assistant and patent agent while he was in England was Joseph Barnes, his brother-in-law. While it may be doubted whether Rumsey really believed that the centrifugal force should be added to the force due to the falling water, Barnes took this doctrine as gospel truth. After Rumsey's death Barnes wrote a treatise on water mills that he submitted to the American Philosophical Society for publication. Barnes's intention was to defend Parent's theory and to refute Waring's. Unfortunately Barnes elevated to the level of a universal principle Rumsey's erroneous claim in a specific case that the propulsive force of reaction was twice that produced by action. Arguing from "intuitive" axioms, Barnes concluded that in hydraulic applications, reaction (Newton's third law) always produced twice the effect of action (Newton's first law of motion).[31] One of Oliver Evan's friends reported that Barnes's theory amounted to the claim "that two is equal to four and he can prove it."[32] Not surprisingly, the American Philosophical Society refused to publish Barnes's treatise.

A few years after Rumsey's death, Oliver Evans published his classic *The Young Mill-Wright and Miller's Guide.* This became the starting point for the vernacular science of most subsequent American inventors of hydraulic machinery. Like Desaguliers, Evans sought to place technology firmly on the foundation of Newtonian science. Nearly 40 percent of the book was taken up by a simplified presentation of mechanics and hydraulics based on several popularizations of Newtonian science. And like Desaguliers, Evans used only elementary mathematics. His readers could reason qualitatively from Newton's laws. But, except in simple cases, they had to translate their problems into hydraulics where elementary arithmetic could provide quantitive answers. But Evans differed from Desaguliers in one important respect. He attempted to found his science on the experimental tradition established by John Smeaton. Indeed, Evans reprinted the text of Smeaton's classic paper on waterwheels, so that his readers could benefit from the example of the finest experimentalist in technology of that age.[33]

Unfortunately, Evans did not present a first-hand account of Newtonian science. He attempted to improve upon or simplify Newton's laws of motion (which he did not clearly attribute to Newton). In later editions further "improvements" were introduced by Thomas P. Jones. By concealing (perhaps inadvertently) the source of the laws of motion and by "simplifying" them, Evans and his later editor, Jones, created a great deal of confusion. Jones's modification of Evans's version of Newton's second law was particularly confusing. In Newton's own phrasing this law had two clauses:

> The change of motion is proportional to the motive force impressed; and is made in the direction of the right line in which that force is impressed.[34]

But Jones's version rephrased the first clause and omitted the second. His statement of the second law was:

> The change from rest to motion, and from motion to rest, is always proportional to the force producing these changes.[35]

By suppressing the second clause of Newton's second law, Jones gave unknowing encouragement to those who, like Barnes, saw centrifugal force as an addition to the energy due to the fall of the water. Indeed, incorrect centrifugal force theories became commonplace. At the extreme these led to ideas of perpetual motion. And a number of centrifugal force schemes were promoted to bilk a credulous investing public. Only one of these need be noted. Messers Sawyer and Gwinne in 1851 presented a plan to tap centrifugal force in what they called their "New Motive Power."[36] By this time there were a number of mechanics magazines available, of which the most important for the United States was *Scientific American*. The editors heaped ridicule upon Sawyer and Gwinne's proposal to exploit centrifugal force as a new motive power. One of their readers sent a letter to the editors in which he exposed the "New Motive Power" as another perpetual motion scheme. The editors added a note to this letter:

> Our correspondent is right about the machine being *no go yet*. And according to the doctrines of Sir Isaac Newton, it never will. He says in the *Principia* that a vessel containing water and having received a circular motion, the vessel, by gradually communicating its motion to the water, will make it begin sensibly to revolve and recede by little and little from the middle and ascend to the sides of the vessel, forming itself into a concave figure, and the swifter the motion becomes the higher will the water rise, till, at last performing its revolutions in the same time with the vessel, it becomes relatively at rest to it.[37]

These arguments failed to convince the proponents of the new motive power.

Their lawyer wrote a letter of defense of Sawyer and Guinne's centrifugal force scheme; he argued that they had discovered in centrifugal force a new force in nature, one akin to gravity.[38]

Joseph Henry was among those who were disgusted by the frequent misunderstanding or conscious misrepresentation of centrifugal force by American vernacular scientists and promoters. Henry lamented

> the constant announcement in the papers of new motors, of machines moved by centrifugal force, of engines to do a large amount of work with the expenditure of an infinitesimal quantity of power. ... [39]

As improvements in Rumsey's reaction wheel made it the most important water motor in America, millwrights sought to better understand it. The columns of *Scientific American* became a significant vehicle by which millwrights communicated with one another on this and related topics. However, the long series of communications published as letters to the editors in mechanics magazines ran from the 1840s through the 1890s without a satisfactory resolution, though the problem had been successfully solved in 1846 by Zebulon Parker, a creative millwright, at the very outset of the public debate. Many millwrights had difficulty in interpreting the third law. They saw it not as a truly independent law, but as a special case of the first law, where unbalanced pressure propelled a reaction wheel. In 1848 *Scientific American* published a statement of this interpretation of reaction in a letter by one "D. T." This anonymous author reasoned as follows:

> Let us suppose an upright penstock to be filled with water. Its pressure is equal on all sides, but if we make an aperture on one side, it relieves that side of an amount of pressure according to the size of the opening ... while the pressure on the other side is the same as before.[40]

This theory was not new. James Rumsey had assumed this mode of action for his version of Barker's mill. In presenting his theory of how his reaction mill worked, Rumsey assumed that reaction acted as an unbalanced pressure on the portion of the cylinder opposite the opening. That is, he too considered the third law as a special case of the first.[41]

Though correct interpretations of the third law and of the mode by which reaction wheels worked were published from time to time, the erroneous versions continued to appear with considerable frequency until the 1890s, when the millwright as an occupational type was fast disappearing. Thus, in 1883 this issue was debated in the pages of *American Miller,* and again ten years later in the pages of *Milling.* In 1883 William Kennedy, in defending the unbalanced pressure interpretation of the third law, referred to "us direct action fellows."[42] Apparently the absorption of the third law by the first had become a recognized school of thought among millwrights. The same theory was vehemently supported by another millwright, T. W. Graham in the 1880s and 1890's. In the 1890s Graham's views were contested by an engineer,C. R. Tompkins. Graham's views are of less interest than his reasons for holding them, which were brought out in his debate with Tompkins. Graham was a staunch empiricist who thought progress in designing turbines had come from systematic testing, and he ridiculed attempts to divide the force acting in turbines into two separate forces (action and reaction).[43] Tompkins confronted Graham with the manifest absurdity of conflating the impact of a stream of water from a hose with the backward pressure exerted by the hose on the hand holding it.[44] Graham's response was to raise ideological questions. He maintained that "when such litter and rubbish is swept from the textbooks, hydraulic principles will be better understood by the masses."[45] Thus the "direct action fellows" were motivated, in part, by an egalitarian ideology that sought to keep science and technology simple enough to be understood by ordinary citizens.

It would be easy to dismiss Graham as an antiscientific crank. Undoubtedly he was. But there were at least two distinct strands in the development of the vernacular science of reaction wheels and turbines in America. One of these had much in common with Graham. Its principal spokesman was James Emerson, who did much to establish a distinctive testing tradition for American turbines. The other strand in vernacular science was the more obvious and important. It consisted of millwrights who linked vernacular scientific methods with very creative design. It was this second group that was responsible for the evolution of the two designs still most widely used, the so-called, Francis or mixed-flow reaction turbine and the Pelton impulse turbine.

I have discussed the creative role of vernacular science in the American development of the turbine in other places.[46] The most important figures were Austin and Zebulon Parker who patented a "percussion and reaction" waterwheel in 1829 and then developed appropriate theories, experimental methods, and testing procedures for turbine development. Zebulon Parker, who continued alone after his brother's death, attempted to publish his correct theory of reaction wheels in the *Journal of the Franklin Institute*. The rejection of his paper was a great loss to millwrights and inventors. Parker refuted the "unbalanced pressure" theory that conflated the first and third laws. He presented a correct theory of value to designers. But the position of the scientists associated with the Franklin Institute also deserves some sympathy, even though their decision was disastrous. Parker prefaced his paper with a lengthy comparison of the various versions of Newton's laws of motion that he found in various editions of Oliver Evans's treatise and similar sources.[47] It is a tribute to Parker's reasoning skills that he came to correct conclusions on all issues. But what must the scientists who reviewed his paper have thought of this exercise, especially when it was clear from the outset that Parker had no idea that the author of the laws of motion with which he dealt was Newton?

In rejecting Parker's naive reconstruction of Newton's laws of motion, the scientists missed other aspects of his vernacular science. Parker made several important discoveries including the use of glass-walled testing flumes with which he was able to observe, by the motion of pellets in the water, the actual movements of the water in a test wheel.[48] These and other of Parker's insights helped to fuel a highly creative tradition of vernacular science linked to design. A salient feature of this tradition was the use of Newton's laws of motion to reason about the motion of particles or streams of water moving through a mill. This had been the basis of the Parker brothers' inventions. It was employed also by John T. McCormick. McCormick developed the design that came to be called, incorrectly, the "Francis" turbine. The key to his success lay in the development of a novel and complex three-dimensional geometry for his turbine vanes. In a caveat filed with the patent office in 1875, McCormick made clear that the peculiar form he had devised for his vanes derived from an analysis of impulsive and reactive interactions between the particles of water and the rotor at various stages in the water's path.[49] Thus the Newtonian tradition of vernacular science continued to play an important role in the development of the turbine in America into the fourth quarter of the nineteenth century.

The creative innovators represented a small minority of millwrights. Many, perhaps most, millwrights agreed with Graham that scientific analysis was foolish and fruitless and that progress came through a Baconian empiricism involved in testing turbines. In America James Francis had carried the technique of testing turbines by means of a Prony Dynamometer to a level of perfection. But his tests were very expensive. James Emerson, a former sea captain and inventor, developed an improved dynamometer that reduced the costs of testing rather drastically. His testing flume at Holyoke became an important part of the turbine industry. Customers came to demand verification of manufacturer's claims to efficiency, and Emerson could provide quick and cheap test results. But Emerson's results were not scientifically elegant, and he heaped scorn upon college-educated engineers and others who sought to apply scientific theory and experiment to turbines. Thus, Emerson linked his testing to an egalitarian democratic ideology that rejected the need for esoteric scientific theory.[50]

Ironically, Emerson was one of those who provided the preconditions for a major advance in the science of the turbine. The high efficiency and low cost of McCormick's "Francis" design ensured its diffusion to Europe. But there it encountered a scientific tradition in engineering that put great emphasis upon elegant mathematical theory. While McCormick's turbine remained essentially the same in its major features, European and particularly German engineers improved the details of its design. Thus, while McCormick-type turbines could get occasional test efficiencies in excess of 90 percent, the empirical design procedures used in America led to inconsistent results, and American turbine makers would guarantee only about 80 percent efficiency for their turbines.

There were several elements in the revolution in turbine science carried out in Germany from 1890 to 1920. The "Francis" design necessitated thinking in three dimensions; it prompted the revival of Euler's three-dimensional theory, long discarded in favor of the simpler two-dimensional theory developed by Benoit Fourneyron and his successors in Europe. But the most important advance was the application of dimensional analysis and considerations of similitude to the turbine by Rudolph Camerer from 1905 to 1914. Camerer's study allowed the engineer to take scale effects into account. Only in this way could theory and experiment be united and applied to design. Changing the scale of a machine altered its properties in rather unpredictable ways. Of these the most critical was speed. Inability to predict speed behavior had been a major source of guess-work in turbine design. As Camerer investigated the problem, he found at least one of the keys to his puzzle in the American testing tradition as developed by James Emerson. In Europe turbines were individually tailor-made for each customer. In America they were built in standard sizes and sold through catalogs. Because efficiency could vary with changes in scale, it became necessary to test all of the standard sizes sold, not just one size. With his cheap methods of testing Emerson was able to test and give the main parameters for entire families of geometrically similar turbines, sometimes varying in regular increments from a diameter of nine to forty-eight inches or more. What Camerer and a number of other engineers observed was that if these values were plotted, they fell upon smooth curves, with somewhat different curves for each family of geometrically similar turbines. To Camerer these turbine "characteristics" were obvious in the American test data. It was this starting point, then, that contributed to his derivation of the correct formula for the first and most important of these characteristics, the "specific speed," which he published in 1905.[51]

There is an irony in the development of modern turbine theory. Newtonian science aided creative designers, even in the millwright's crude vernacular versions of this science. But those, like Emerson, who rejected theory of any sort and relied on a pure Baconian empiricism ended by laying the foundations for a vastly more sophisticated analytic theory of turbomachinery.

Notes

1. Robert E. Schofield, *The Lunar Society of Birmingham: A Social History of Provincial Science and Industry in Eighteenth-Century England* (Oxford: Clarendon Press, 1963).

2. Brooke Hindle, *The Pursuit of Science in Revolutionary America, 1735–1789* (Chapel Hill: University of North Carolina Press, 1965). See also Theodore Hornberger, *Scientific Thought in American Colleges, 1628–1800* (Austin: University of Texas Press, 1945), and I. Bernard Cohen, *Some Early Tools of American Science* (Cambridge: Harvard University Press, 1950).

3. See for example, Uriah A. Boyden to James B. Francis, 14 August 1867, papers of Uriah A. Boyden, National Museum of American History, Washington, D.C.

4. Charles Hutton's, *A Course of Mathematics,* with varying subtitles, went through 29 editions or reprintings, including six American printings from 1798 to 1860 (*National Union Catalog of pre-1956 Imprints,* 262: 184–86).

5. Hutton, *Course of Mathematics* . . . , (New York: n.p., 1843), vol. 2.

6. Squire Whipple, *An Elementary and Practical Treatise on Bridge Building*, 4th ed. (New York: D. Van Nostrand, 1883). The first version was published as *A Work on Bridge Building* (New York: the Author, 1847).

7. Daniel H. Calhoun, *The American Civil Engineer, Origins and Conflict* (Cambridge, Mass.: The Technology Press, 1960), 37–43.

8. George W. Cullum, "Dennis H. Mahan," *Biographical Register of the Officers and Graduates of the U.S. Military Academy at West Point* (Boston: J. F. Trow, 1891), 1: 319–25. The French influence may be seen in Dennis Hart Mahan, *An Elementary Course of Civil Engineering* (New York: Wiley and Putnam, 1937), vii, 44–53, 86–104.

9. The first English translation was by L. Gordon of an early version of Weisbach's mechanics: Julius Weisbach, *Principles of the Mechanics of Machinery and Engineering,* 2 vols. (London: H. Bailleer, 1847–48). Two American translations of a later version of Weisbach's mechanics were published in America in the 1870s; they were those of Eckley B. Coxe for Van Nostrand and Jay Du Bois for J. Wiley & Sons. I have used the Du Bois translation. See Julius Weisbach, *A Manual of the Mechanics of Engineering,* 3 vols., translated by Jay Du Bois (New York: J. Wiley and Sons, 1880).

10. "Student Note-book of Lectures at Rensselaer, 1852–1855," MS volume in Avery Library, Columbia University, New York.

11. The last American edition of Hutton's text appeared in 1843; the last London edition in 1860. See *The National Union Catalog,* 262: 186. On the antitheoretical ideology of American engineers, see Layton, "European Origins of The American Engineering Style of the Nineteenth Century," in Nathan Reingold and Marc Rothenberg, eds., *Scientific Colonialism, A Cross-Cultural Comparison* (Washington D.C.: Smithsonian Press, 1987), 151–66.

12. "Researches in Physics by Uriah Atherton Boyden," unpublished MSS, Boyden Papers, in which Boyden defended Newton's formula for the speed of sound. Squire Whipple, *An Investigation as to the Mode and Amount of Attraction Between Large and Small Bodies in Contact or at Small Distances* (Albany, N.Y.: the Author, 1877), was a defense of Newton's doctrine of central forces.

13. "Deposition of Conington W. Searle," *Parker vs. Hatfield,* U.S. Circuit Court, Ohio, Southern District, Chancery Records, ser. 1, vol. 7, Federal Archives and Records Center, number 92442R, Archives Section, Federal Records Center, Chicago, Illinois.

14. John T. Desaguliers, *A Course of Experimental Philosophy,* 2 vols. (London: Sennex, Innys, Longman et. al., 1734, 1744).

15. On Rumsey's nonscientific system for mechanically poling boats against current, see his British patent, "Applying Water and Steam Power to Machinery . . . ," no. 1738, 24 March 1790, 10–11, figs. 13, 14, 21. Rumsey dated several of his more scientific inventions to 1785–86. But his pole boat scheme kept him fully occupied until the winter of 1785. Rumsey claimed to have invented his tubular boiler in the winter of 1785, his steam pump in 1785, his reaction-powered steamboat in 1786, and his piston-engine column-of-water saw mill in 1787. Unfortunately he did not date his improvement in Barker's mill. See James Rumsey, *The Explanation and Annexed Plates of the Following Mechanical Improvements* (Philadelphia: James, 1788), 3–4, 6, 8. The steamboat is described in a pamphlet bound with the preceding, James Rumsey, *Explanation of a Steam Engine* (Philadelphia: James, 1788), second page (unnumbered). For Rumsey's life, the standard work is Ella May Turner, *James Rumsey, Pioneer in Steam Navigation* (Scottsdale Penn.: Mennonite Publishing House, 1930).

16. On Rumsey's dependence on Franklin for the reaction steamboat, see Brooke Hindle, *Emulation and Invention* (New York: New York University Press, 1981), 27, 34, 39–40.

17. Englehart Cruse, *The Projector Detected* (Baltimore: the Author, 1788), pp. 3–4.

18. Desaguliers, *Course,* 2: 459–64, plate 33.

19. Rumsey, *Explanation,* 7.

20. Terry S. Reynolds, *Stronger Than a Hundred Men, A History of the Vertical Water Wheel* (Baltimore: Johns Hopkins University Press, 1983), 207–8.

21. Ibid., 205–7; Desaguliers, *Course,* 2:424.

22. Desaguliers, *Course,* 2: 424.

23. Rumsey, British patent 1790, 5.

24. Ibid.

25. Dennis G. Shepherd, *Principles of Turbomachinery* (New York: Macmillan, 1956), 78–79, 87–88.

26. Rumsey, British patent, 1790, 6. Though Rumsey's published theory was incorrect, he got correct results from experiments that he reported to Thomas Jefferson. (See Rumsey to Jefferson, 4 October 1789, *The Papers of Thomas Jefferson*, 15: 699.)

27. John Smeaton, "An Experimental Enquiry Concerning the Natural Powers of Water and Wind to Turn Mills, and Other Machines, Depending on a Circular Motion," *Philosophical Transactions of the Royal Society*, 51 (1759–1760): 100–174.

28. Desaguliers claimed that overshot wheels did the same work as undershot mills for one tenth the water expended, but he based that on the heads typical for these two types of mills. When his figure is adjusted to equalize the head of water for each mill the advantage was about 3½–4 : 1. See Desaguliers, *Course*, 2: 531–32. Though Desaguliers's figures were incorrect, his insight was fundamental and foreshadowed the later work of Smeaton.

29. Rumsey, British patent, 1790, 5–6. Rumsey's experiments gave him the correct conclusion, that the efficiency rose with increasing speed. (See note 26, above.)

30. William Waring, "Observations on the Theory of Water Mills, etc.," and "Investigation on the Power of Dr. Barker's Mill, as Improved by James Rumsey, With a Description of the Mill," *Transactions of the American Philosophical Society* 3 (1793): 144–49, 185–93. The correct theory based on experiment was noted in Rumsey's letter to Jefferson, 4 October 1789, cited in note 26 above. The prototype reaction wheel was described in Joseph Barnes to John Vaughn, 21 September and 2 October 1792, illustration 2, Miscellaneous Manuscript Collections, American Philosophical Society, Philadelphia.

31. Joseph Barnes, "Essay on Overshot and Undershot Wheels and on Rumseys Mill," unpublished MS, American Philosophical Society, 3, 41, passim.

32. Joseph S. Sampson to Oliver Evans, 14 Sept. 1795, quoted in Greville Bathe and Dorothy Bathe, *Oliver Evans, A Chronicle of Early American Engineering* (Philadelphia: Historical Society of Pennsylvania, 1935), 53.

33. Oliver Evans, *The Young Mill-Wright and Miller's Guide* (Philadelphia: the Author, 1795). Part 1 includes both Evans's summary of mechanics and hydraulics, 1–127' and his reprinting of Smeaton's classic paper, 128–60.

34. Isaac Newton, *Mathematical Principles of Natural Philosophy on his System of the World*, translated by Andrew Mott and Florian Cajori, 2 vols. (Berkeley and Los Angeles: University of California Press, 1972), 1: 12.

35. Oliver Evans, *The Young Mill-Wright and Miller's Guide*, 9th ed., revised by Thomas P. Jones (Philadelphia: Carey Lea and Blanchard 1836), 13.

36. "There is Nothing New under the Sun," *Scientific American* 6 (14 June 1851): 309; "Motive Power—Centrifugal Force," *Scientific American* 7 (12 June 1851): 341; "The Centrifugal Force Philosophers and the New Motive Power," *Scientific American* (16 Aug. 1851): 380; "The New Motive Power—Centrifugal Force Stock," *Scientific* American (15 Nov. 1851): 69.

37. J. C. S., "The New Motive Power—Centrifugal Force," and, "Editorial Comment," *Scientific American* 6 (26 July 1851): 354.

38. Stephen P. Andrews, "To the Editor of the Scientific American." *Scientific American* 6 (2 Aug. 1851): 363.

39. Joseph Henry, "Closing Address Before the Metropolitan Mechanics Institute," in Arthur P. Molella, et al., eds., *A Scientist in American Life, Essays and Lectures of Joseph Henry* (Washington, D.C.: Smithsonian Institution Press, 1980), 61.

40. D. T., "Reaction Water Wheels," *Scientific American* 3 (29 Jan. 1848): 149.

41. James Rumsey, British patent 1790, 6.

42. William Kennedy, "Direct Action and Reaction," *American Miller* 40 (1 Feb. 1883): 60.

43. T. W. Graham, "Percussion and Reaction," *Milling* 2 (Jan. 1893): 150.

44. C. R. Tompkins, "Percussion and Reaction Wheels," *Milling* 2 (Feb. 1893): 233.

45. T. W. Graham, "The Discussion of the Turbine," *Milling* 2 (March 1893): 364.

46. Edwin T. Layton, Jr., "Millwrights and Engineers, Science, Social Roles, and the Evolution of the Turbine in America," in Wolfgang Krohn, Edwin Layton and Peter Weingart, eds., *The Dynamics of Science and Technology* (Dordrecht, Holland: D. Reidel, 1978), 61–87; and "Scientific Technology: the Hydraulic Turbine and the Origins of American Industrial Research," *Technology and Culture* 20 (Jan. 1979): 64–89.

47. Zebulon Parker, "Motion and Forces; Action and Re-action," Committee on Science and the Arts, file 490 (1846–48), Franklin Institute Archives, Philadelphia.

48. Layton, "Millwrights and Engineers," 82.

49. John B. McCormick, "Water Wheel," Caveat, U.S. Patent Office, filed 1 December 1875, National Archives, Washington, D.C.

50. James Emerson, *Treatise Relative to the Testing of Water-Wheels and Machinery,* 4th ed. (Williamsett, Mass.: the Author, 1892), 1–7, 15–46, 54–58, 120–27, passim. Emerson was not modest; he held, "I have . . . done more the past quarter century to establish a practical knowledge of milling hydrodynamics . . . than has been done by all other engineers for a century past" (p.7).

51. On the influence of American practice upon Camerer, see Rudolph Camerer, *Vorlesungen über Wasserkraftmaschinen* (Leipzig and Berlin: Verlag von Wilhelm Englemann, 1914), 295. On the discovery of specific speed, see Rudoph Camerer, "Klassifikation von Turbinen," *Zeitschrift des Vereines Deutscher Ingenenieue* 49 (4 March 1905): 308.

W. J. M. Rankine and the Scottish Roots of Engineering Science

DAVID F. CHANNELL

During the nineteenth century, the discipline of engineering underwent significant methodological and institutional changes brought about by a new interdependence between science and technology.[1] Although science and technology have become a single hyphenated concept in the twentieth century, throughout most of history they were socially and intellectually separate modes of activity. Most scientists had university training while most engineers had little or no formal education and gained knowledge of their craft through the apprenticeship system. Although there were isolated examples of the application of science to practical problems, such as Galileo's study of the strength of materials, most engineers used a conceptual framework that differed from that of science. While scientists were dealing with problems using idealized concepts such as billiard-ball atoms and Newtonian forces, engineers could find little direct practical use for those concepts. Instead they were concerned with more phenomenological ideas such as elastic fibers and pressures. By the nineteenth century, new advances in technology were becoming more and more dependent upon scientific knowledge, especially in such areas as the strength of materials, the expansive use of steam, and naval architecture; but the laws of science never seemed to fit the phenomena of the real world.[2] Engineers needed a model for a new body of knowledge—an engineering science—that would provide the missing connections between the laws of science and the actual structures, engines, and ships that they had to design.

Although both the French and the Germans contributed a great deal to the development of engineering science, one of the most influential figures in the creation and institutionalization of the field was the Scottish engineering professor William John Macquorn Rankine (1820–72). With training in both science and technology, he spent the greater part of his life trying to establish a formal connection between the two bodies of knowledge.[3] As Regius Professor of Civil Engineering and Mechanics at Glasgow University, Rankine produced a series of books including *A Manual of Applied Mechanics* (1858), *A Manual of the Steam Engine and Other Prime Movers* (1859), *A Manual of Civil Engineering* (1862), *Shipbuilding, Theoretical and Practical* (1866), and *A Manual of Machinery and Millwork* (1869), all of which went through several editions and became the standard textbooks for university-trained engineers throughout the second half of the nineteenth and well into the twentieth century. Through his position and his works, Rankine became one of the leading figures in engineering science, influencing the development of the field throughout Europe, America, and Japan.

When Rankine became Regius Professor in 1855, he realized that for the discipline of engineering to survive in the university, he could not duplicate the courses already taught in natural philosophy and chemistry, and he could not successfully compete with the established apprenticeship system. Rather, he would need to create a model or framework in which science and technology could be combined into an engineering science with its own principles, concepts, and methods. I would like to argue that Rankine derived the framework that he needed from the Common Sense philosophy of Thomas Reid and Dugald Stewart and that his concept of engineering science had its roots in Scottish thought and culture.

Through his education, Rankine was introduced to the new philosophical changes that had been brought about by the Scottish Enlightenment. His early education at Ayr Academy and Glasgow High School came at a time when the traditional grammar school approach, which emphasized the study of Latin and Greek, was being reformed to include subjects such as arithmetic, geography, natural philosophy, and chemistry, which were more useful to commerce and industry.[4] Before entering the University of Edinburgh, Rankine also received private tutoring in geometry from George Lees and studied chemistry with David Boswell Reid; both courses reflected the principles of Scottish Common Sense.[5] During his two years as a student at the University of Edinburgh, Rankine spent a great deal of his leisure time reading philosophy, especially those philosophers associated with the Scottish Enlightenment.[6] Through the writings of Dugald Stewart and the Baron Joseph DeGerando, Rankine undoubtably was introduced to the major tenets of Common Sense philosophy, especially those of Thomas Reid. His philosophical interests were reinforced in the natural philosophy courses taught by James David Forbes, who was the leading scientific exponent of Common Sense philosophy.[7] As a student in Forbes's advanced natural philosophy class, Rankine won a gold prize for an essay on "Methods of Physical Investigation."[8] Although this essay is lost, his later writings on science and engineering reflect a continued interest in the issues addressed by Common Sense philosophy.

The philosophy of Common Sense provided Rankine with two important elements that he needed to create a framework for engineering science. First, the basic assumptions of Common Sense helped him bridge the disciplinary gap that existed between science and engineering. Even in Rankine's time, most engineers were suspicious of science. For them, science was purely theoretical and abstract and therefore distinct from the practical problems of engineering. But a central tenet of Common Sense philosophy, drawn from

Francis Bacon, was that knowledge did not exist for its own sake but should be applied to some practical problem. Thomas Reid in his *Essay on the Active Powers of Man* (1788) said: "It is evidently the intention of our Maker, that man should be an active and not merely a speculative being. . . . Knowledge derives its value from this, that it enlarges our power, and directs us in the application of it."[9] Later Stewart reflected Reid in noting that "the more knowlege . . . we acquire, the better we can accommodate our plans to the established order of things."[10]

Much of Rankine's education reinforced the Baconian view that scientific knowledge was essentially practical. George Lees did not view mathematics as a purely axiomatic system, but one "intimately connected with the various branches of the Mechanical Philosophy."[11] David Boswell Reid argued that chemistry was "one of the greatest monuments of the school of Bacon."[12] For him chemistry was a subject in which the "habits of mental activity and manual dexterity go hand in hand."[13] His course did not treat chemistry as a theoretical science but as an experimental one that had practical connections to agriculture, manufacturing, and medicine.[14] Finally, James David Forbes organized his courses in natural philosophy according to the principles of Common Sense and based them on the "well known precepts of Bacon and the practice of Newton."[15] Since Forbes argued that "knowledge is good because it leads directly to useful results," his course on natural philosophy was equally divided between a study of scientific laws and their application to the arts.[16]

Rankine's justification for the existence of engineering science reflected the principles of Common Sense philosophy. In his first lecture to his engineering class at Glasgow, he observed that progress in both science and engineering had been hindered by what he called "the fallacy of a double system of natural laws"; this was the belief that theoretical, geometrical, rational laws that had been discovered by contemplation were separate and distinct from practical, mechanical, empirical laws that had been discovered by experience.[17] Although natural philosophers in the sixteenth and seventeenth centuries had overthrown this fallacy, Rankine argued that the "discrepancy between theory and practice, which in sound physical and mechanical science is a delusion, has a real existence in the minds of men; and that fallacy, though rejected by their judgments, continues to exert an influence over their acts."[18] According to Rankine, the separation of theory from practice had its most detrimental effect on engineering since "a large number of persons, possessed of an inventive turn of mind and of considerable skill in the manual operations of practical mechanics, are destitute of that knowledge of scientific principles which is requisite to prevent their being misled by their own ingenuity."[19] This is very close to Dugald Stewart's statement in *Elements of the Philosophy of the Human Mind* (1792), which said: "In the mechanical arts, it is well known, how much time and ingenuity are misapplied by those who acquire their practical skills by their own trials."[20] Just as Stewart argued that philosophical knowledge required both sense observation and an understanding of the laws that govern the mind,[21] Rankine believed that engineering science could bring about progress by connecting scientific knowlege, which emphasized the improvement of understanding and the elevation of the mind, with practical knowledge, which emphasized experience and observation.[22] Therefore, Common Sense philosophy provided a model that allowed Rankine to transcend the traditional distinctions between scientific and technological knowledge.

Common Sense philosophy provided a second element that Rankine required to create engineering science. He still needed a framework or model in which he could recast scientific theories into a form that could be used in engineering. Although science and practice depended on the same first principles, their modes of instruction might differ since practice would call "into operation a mental faculty distinct from those which are exercised by theoretical science."[23] This argument reflected the Common Sense idea that the powers of the mind could be divided into the intellectual powers of understanding and the active powers of the will.[24]

For Rankine the difference in the modes of instruction in science and practice seemed to center on the use of hypotheses. Most of his work in science, especially his studies of the laws of heat and light, was based on a new theory of matter that he labeled the hypothesis of molecular vortices.[25] According to this hypothesis, the atoms of matter consisted of a nucleus enveloped by an elastic atmosphere. He also postulated that heat was the *vis viva* of the revolutions of the atmospheres and that light was transmitted by the oscillations of the nuclei. A key element of the molecular vortex was that it could be described mechanically. The hypothesis of molecular vortices provided Rankine with a powerful model for solving scientific problems. First optical problems, such as double refraction, could be explained by the atmosphere's damping the vibrations of the nuclei in certain directions.[26]

The model could also explain the relationships between heat and light as the transfer of motion from the atmospheres to the nuclei and vice versa. But the most successful application of the hypothesis came in the study of the theory of thermodynamics. In a series of papers, Rankine used his hypothesis to derive equations for the specific heat of gases. He then derived a general theory of the mechnical action of heat based on the idea that heat could be related to work because the rotating atmospheres tended to expand their volumes when they absorbed heat. By showing that a certain quantity of heat disappeared during the expansion of the atmospheres, Rankine was able to derive a version of the second law of thermodynamics that placed him with Rudolf Clausius and William Thomson as a discoverer of the law.[27]

Although Rankine's scientific research was based on the molecular vortex, the hypothesis played no direct role in his technological works. For example, in his *Manual of the Steam Engine,* the engineering work most closely associated with his scientific research, Rankine stated that "it is possible to express the laws of thermodynamics in the form of independent principles, deduced by induction from facts of observation and experiment, without reference to any hypothesis as to the occult operations with which the sensible phenomena may be conceived to be connected; and that course will be followed in the body of the present treatise."[28]

During the nineteenth century, the role of hypotheses in scientific theory was the subject of much debate within Common Sense philosophy. Thomas Reid had argued strongly against the use of hypotheses and argued instead in favor of the Baconian inductive method: " ... no regard is due to the conjectures or hypotheses of philosophers, however ancient, however generally received. ... What can fairly be deduced from facts duly observed or sufficiently attested, is genuine and pure."[29] But Dugald Stewart realized that a knowledge of nature based on observed facts could be problematic. He said: "The knowledge which is acquired of the course of Nature by mere observation, is extremely limited, and extends only to cases in which the uniformity of the observed phenomena is apparent to our sense."[30] This uniformity would occur only if a single law of nature were operating, but in most cases several different laws will be combined and "although each law may take place with the most complete uniformity, it is likely that nothing but confusion will strike the mere observer."[31] Stewart argued that before the order of nature can be discovered, "we must employ our reasoning powers in comparing a variety of instances together, in order to discover ... the simple laws which are concerned in the phenomenon under consideration."[32]

But unlike Reid, Stewart was willing to employ hypotheses as part of the reasoning powers used to discover the laws which described natural phenomena. He said: "although a knowledge of facts must be prior to the formation of a legitimate theory; yet a hypothetical theory is generally the best guide to the knowledge of connected and useful facts."[33] As long as hypotheses were properly seen as principles based on probable facts and not mistaken for true principles, Stewart believed they could contribute to the creation of an inductive theory. He quoted Dr. John Gregory, who argued that hypotheses "are the first rudiments or anticipations of Principles."[34] But according to Stewart, the principles that were anticipated by hypotheses could not be considered the truth until they were empirically tested.[35] That is, although hypotheses could be a preliminary step toward a complete theory, one had to return to the methodology of Reid to recast the principles gained from hypotheses into an inductive theory.

Stewart's philosophical framework provided Rankine with the model that he needed to transform science into engineering science. As indicated in his *Manual of the Steam Engine,* Rankine realized that his scientific-hypothetical theories could not be simply applied to engineering problems that required a more inductive approach. His most explicit explanation of the relationship between a hypothetical theory and an inductive theory was put forward in his essay "Outlines of the Science of Energetics," published in 1855, the same year that he became Regius Professor at Glasgow.[36] Although many scholars have treated this essay as a major contribution to the creation of a postivisitic theory of thermodynamics, most of the examples he used in the work were drawn from engineering, not science. The terminology that he developed in the essay can be found in his engineering works, while he continued to use his vortex hypothesis in his science even after the publication of the essay.[37] I would argue that he derived the essay from his interest in Stewart's Common Sense philosophy and that it provides the key to understanding the relationship between Rankine's science and technology, which is the basis of his engineering science. As such, the essay shows how Scottish Common Sense philosophy provided a methodological framework for modern engineering.

According to Rankine, the development of a physical theory takes place in two stages: first, through either natural observation or experimentation, relationships between phenomena come to be expressed as formal laws; second, the formal laws of an entire class of phenomena are reduced to a simple system of principles from which all formal laws can be deduced.[38] This program reflects Stewart's statement: "the highest, or rather the only proper object of Physics, is to ascertain those established conjunctions of successive events, which constitute the order of the Universe; to record the phenomena which it exhibits to our observations or which it discloses to our experiments; and to refer these phenomena to their general laws."[39] This still leaves open the problems of how the laws expressing relationships between observed phenomena were to be reduced to a simple system of principles.

Rankine argued that two methods could be used to formulate a physical theory. At first, one could use the hypothetical method, where conjectures not apparent to the senses are used to deduce laws in a new class of phenomena based on modifications of some previously known laws. Since the laws of mechanics were well known, this led to the adoption of mechanical hypotheses such as the wave theory of light or the molecular theory of heat. Reflecting Stewart, Rankine argued that hypothetical theories had been quite successful in reducing the laws of complicated phenomena "to a few simple principles, and by anticipating laws afterwards verified by observation."[40] But there were also disadvantages to the hypothetical method. Rankine noted that the success of such hypotheses led to a tendency "to explain away, or set aside, facts inconsistent with these hypotheses, which facts, rightly appreciated would have formed the basis of true theories."[41] It is just such "inconsistent facts" that are most important to a practical theory of technology.

There was also a second way to formulate a physical theory, one that would overcome the disadvantages of the hypothetical method. Rather than supposing "physical phenomena to be constituted, in an occult way," of mechanical processes, the abstractive method assigned a name or symbol to the properties that a group of phenomena had in common "as perceived by the senses."[42] Through a process of induction from the observed facts, a series of more and more general laws could be generated. But most importantly, the abstractive method would not eliminate the need for hypotheses since Rankine, like Stewart, believed that "a hypothetical theory is necessary, as a preliminary step, to reduce the expression of the phenomena to simplicity and order before it is possible to make any progress in framing an abstractive theory."[43]

An example of the application of the abstractive method was the creation of the science of energetics, which used the concept of energy to reduce physical phenomena to a set of general laws. In his science of energetics, Rankine attempted to show that the relationships between observed phenomena could be reduced to specific instances of the transfer and transformation of actual (kinetic) and potential energy.[44] Rather than hypothesizing some nonobservable process, the concept of energy was used to label the general "capacity to effect changes," which was a common property of all matter.[45] Using the term "accident" to refer to any variable state of a substance, and "effort" to refer to any cause that varied an accident, Rankine defined "work" as any variation of an accident by an effort.[46] Therefore, since energy was the capacity to do work (or effect changes), physical laws could be reduced to relationships between efforts (which cause change) and states of substances, or accidents (which were capable of change).

This formulation of energetics is remarkably similar to the model of engineering science that Rankine put forward in his second Glasgow University lecture, entitled "On the Science of Engineering."[47] In this lecture, he argued that scientific progress in engineering took place "when the results of experience and observation on the properties of materials which are used in a class of structures, and the laws of the actions which take place in a class of machines have been reduced to a science."[48] For example, in his theory of machines, Rankine began by treating a machine as a set of moving points so that its actions could be reduced to the science of kinematics, which described the transmission and modification of motion.[49] When he then treated the machine as a material body, its motions depended upon the properties of materials, such as the coefficient of friction, which described the transformation and modification of force.

Just as Rankine's energetics had reduced physical phenomena to things that could cause changes (efforts) and substances that could undergo changes (accidents), his engineering science reduced technology to the laws of actions and the properties of materials. Rankine's engineering science can therefore be seen as an attempt to recast his hypothetical model of science into an abstractive inductive theory. For example, in his essay on energetics, he used examples drawn from engineering, such as stress and strain and the efficiency of engines, to explain the abstractive method while he used expamples drawn from science, such as the wave theory of light and the molecular theory of heat, to explain the hypothetical method.[50] Also, the terminology developed in his abstractive method, such as "effort" and "accident" were not used in

Rankine's scientific works but were used in his engineering texts such as his *Manual of Applied Mechanics* and *Manual of the Steam Engine*.[51] Rankine's engineering science reflected both Stewart's argument for the use of hypotheses and Reid's support for a purely inductive method. For Rankine technology was not simply applied science. Although scientific hypotheses were useful in reducing physical phenomena to simplicity and order, the resulting hypothetical model would have to be transformed into an abstractive inductive theory before the hypotheses could be incorporated into technology.

Some conclusions can be drawn from this study of Rankine's engineering science. The Common Sense philosophy of Reid and Stewart provided Rankine with the conceptual and methodological frameworks that he needed to bring together science and technology into engineering science. Several scholars have studied the relationship between Common Sense philosophy and science, but the emphasis has usually been on the changes brought about by Common Sense on the content and methodology within a particular science, such as the creation of an experimental tradition within physics or chemistry.[52] But, as a study of Rankine has shown, Common Sense philosophy may have played an equally important role in the transformation of science into engineering science. One of the most long-lasting influences of the Scottish Enlightenment may be its contribution to the philosophical framework of modern engineering.

Notes

This study is based on work supported by the National Science Foundation under Grant No. SES-8015514. An earlier version of this paper was presented at the International Conference on the Philosophy of Thomas Reid, at King's College, Aberdeen, Scotland, on 2–4 September 1985.

1. In this paper the term technology refers to the body of knowlege, methods, and materials that form the basis of the discipline of engineering.

2. See Nathan Reingold and Arthur Mollela, eds., "The Interaction of Science and Technology in the Industrial Age," *Technology and Culture* 17 (1976): 621–742.

3. For biographical information on Rankine, see David F. Channell, *Scottish Men of Science—William John Macquorn Rankine, F.R.S.E., F.R.S.* (Edinburgh: Scotland's Cultural Heritage, 1986); P. G. Tait; "Biographical Memoir," in W. J. M. Rankine, *Miscellaneous Scientific Papers* edited by W. J. Millar (London: Charles Griffin, 1881), ix–xxxvi; Lewis D. B. Gordon, "Obituary Notice of Professor Rankine," *Proceedings of the Royal Society of Edinburgh* 8 (1872–75): 296–306; and E.M. Parkinson, "Rankine, William John Macquorn," in C.C.Gillispie, ed., *Dictionary of Scientific Biography,* (New York: Scribners, 1971), 11: 291–95.

4. See James Cleland Burns, ed., *The History of the High School of Glasgow* (Glasgow: David Bryce, 1878); and *Air Academy and Burgh Schule, 1233–1895* (Ayr: Ayrshire Post, 1895).

5. See George Lees, *Elements of Arithmetic, Algebra, and Geometry, for Use of the Students in the Edinburgh School of Arts* (Edinburgh: Adam Black & William Tait, 1826); and David Boswell Reid, *Elements of Practical Chemistry* (Edinburgh: MacLachlan and Stewart, 1830).

6. Tait, "Biographical Memoir," xxi.

7. See George Elder Davie, *The Democratic Intellect* (Edinburgh: University of Edinburgh Press, 1961). 169–200; and Richard Olson, *Scottish Philosophy and British Physics 1750–1880* (Princeton: Princeton University Press, 1975), 225–26.

8. Tait, "Biographical Memoir," xxi.

9. Thomas Reid, *The Works of Thomas Reid, D.D.*, 8th ed. edited by William Hamilton, (Edinburgh: James Thin, 1895), 2: 511.

10. Dugald Stewart, *Outlines of Moral Philosophy: For the Use of Students in the University of Edinburgh* (Edinburgh: William Creech, 1793), 3.

11. Lees, *Elements of Arithmetic,* 3.

12. David Boswell Reid, *Remarks on the Present State of Practical Chemistry and Pharmacy with Suggestions as to the Importance of an Extended Practical Course* (Edinburgh: Neill and Co., 1838), 3–4.

13. Reid, *Elements of Chemistry,* vii.

14. Reid, *Remarks,* 3.

15. James David Forbes, *The Dangers of Superficial Knowledge: An Introductory Lecture to the Course of Natural Philosophy in the University of Edinburgh* (London: John Parker, 1849), iii.

16. Ibid., 7. Although there are no lecture notes from Rankine's years, Scottish professors repeated their courses *verbatim* year after year. Therefore, I have used the lecture notes taken by Balflour Stewart, "Abridgement of Prof. Forbes Lectures on Natural Philosophy," 7 vols., 1845–46, University of Edinburgh Archives, Dc. 7.101–7.

17. W. J. M. Rankine, *Introductory Lecture on the Harmony of Theory and Practice in Mechanics* (London: Richard Griffin, 1856), 4–5.

18. Ibid., 9–10.

19. Ibid., 13–14.

20. Dugald Stewart, *Elements of the Philosophy of the Human Mind* (London: Thomas Tegg, 1843), 27.

21. Dugald Stewart, *Outlines of Moral Philosophy: For the Use of Students in the University of Edinburgh* (Edinburgh: William Creech, 1793), 4–5.

22. Rankine, *Harmony of Theory and Practice,* 17–19.

23. Ibid., 20.

24. Thomas Reid, *Works,* 1: 242.

25. W. J. M. Rankine, "On the Mechanical Action of Heat Especially in Gases and Vapours," in *Miscellaneous Scientific Papers,* 234.

26. See W. J. M. Rankine, "On the Vibrations of Plane Polarized Light," *Philosophical Magazine,* 4th ser., vol. 1 (1851): 441–46; and W. J. M. Rankine, "On the Axes of Elasticity and Crystalline Forms," *Philosophical Transactions of the Royal Society* 146 (1856): 361–86.

27. See Rankine, "Mechanical Action of Heat," and Rankine, "On the Second Law of Thermodynamics," *Philosophical Magazine,* 4th ser., vol. 30 (1865): 241–45.

28. W. J. M. Rankine, *Manual of the Steam Engine and Other Prime Movers* (London: Richard Griffin, 1859), xviii.

29. Thomas Reid, *Works,* 1: 236.

30. Stewart, *Moral Philosophy,* 4.

31. Dugald Stewart, *Elements of the Philosophy of the Human Mind,* 2 vols. (Boston: Wells & Lilly, 1814), 1: 271.

32. Ibid.

33. Ibid., 2: 332.

34. Ibid., 2: 337.

35. Ibid., 1: 371–73.

36. W. J. M. Rankine, "Outlines of the Science of Energetics," in *Miscellaneous Scientific Papers,* 209–28.

37. For example, see Olson, *Scottish Philosophy,* 271–87.

38. Rankine, "Science of Energetics," 209.

39. Stewart, *Philosophy of Mind,* 2: 261.

40. Rankine, "Science of Energetics," 212. Also see W. J. M. Rankine, "On the Use of Mechanical Hypotheses in Science, Especially in the Theory of Heat," *Proceedings of the Royal Philosophical Society of Glasgow* 5 (1864): 132.

41. Rankine, "Science of Energetics," 212.

42. Ibid., 210, 213.

43. Ibid., 213.

44. For a description of Rankine's concept of energy, see David F. Channell, "Rankine, Aristo-

tle, and Potential Energy," *The Philosophical Journal* (Royal Philosophical Society of Glasgow) 14 (1977): 111–14.

45. Rankine, "Science of Energetics," 213.

46. Ibid., 214–16.

47. W. J. M. Rankine, *Introductory Lecture on the Science of Engineering* (London: Richard Griffin, 1857).

48. Ibid., 6.

49. W. J. M. Rankine, *Manual of Applied Mechanics* (London: Richard Griffin, 1858), 421–22.

50. Rankine, "Science of Energetics," 212, 215, 226.

51. See Rankine, *Applied Mechanics,* 476; and Rankine, *Steam Engine,* 30.

52. For example, see Arthur Donovan, "William Cullen and the Research Tradition of Eighteenth Century Chemistry," in R. H. Campbell and Andrew Skinner, eds., *The Origins and Nature of the Scottish Enlightenment* (Edinburgh: John Donald, 1982); and Olson, *Scottish Philosophy.*

The "Rediscovery"
of Kirchhoff's Circuitry Laws

ANDREW J. BUTRICA

Probably every introductory electrical engineering textbook[1] contains the following two laws stated in one form or another:

1. The algebraic sum of the currents entering any junction (node) in a network of conductors is always zero ($\Sigma I = 0$).
2. The algebraic sum of the potential drops around any closed loop in a network of conductors is always zero ($\Sigma E = 0$).

Commonly called Kirchhoff's Laws or Formulas, they are as fundamental as Ohm's Law to our understanding of electrical circuitry. Yet, strangely enough, over twenty years passed from their formulation by the German scientist Gustav Kirchhoff (1824–87)[2] in 1845 to their first use by telegraph engineers in France. The story of the French telegraph engineers's "rediscovery" and naming of these laws illustrates how a scientific discovery takes on a new life in engineering practice.[3]

An article published in the *Journal de physique* of 1872 pointed out that electricians had ignored Kirchhoff's Formulas and that there was a need to fill in this gap.[4] The failure to use Kirchhoff's equations prior to 1872 was not unique to France. For example, the article on electricity in the 1860 edition of the *Encyclopedia Britannica* failed to mention them, but by the appearance of the next edition in 1879, Kirchhoff's Formulas had become sufficiently commonplace that his laws of "current distribution in a network of linear circuits" were associated with Kirchhoff's name and explained in the article on electricity.[5] How, then, after more than twenty years of being unsung and nameless, did the name of an eminent mathematical physicist and his equations for electrical circuits become part of the vocabulary of practicing electricians, especially telegraph engineers, in France?

1. Kirchhoff's Laws

Gustav Kirchhoff had been interested in electrical theory from his earliest work as a student under Professor Franz Neumann (1798–1895)[6] in 1845 until his death on 17 October 1887.[7] In articles published between 1845 and 1848 in the *Annalen der Physik,* Kirchhoff extended Ohm's law for linear conduction to the case of electrical conduction in three dimensions (through a wire instead of along a straight line).[8] To do this, he employed an analogy between the flow of heat and electricity based upon the concept of potential and derived his formulas through the mathematical language of the calculus. Although he suggested statements of his laws as early as 1845, Kirchhoff did not fully prove them until his 1848 article, "On the Application of Formulas for the Intensity of Galvanic Force," in which he gave the following statement of his two laws:[9]

Let the system consist of *n* wires, in which the resistances are w_1, $w_2 \ldots w_n$, and in which the intensities of the currents are J_1, $J_2 \ldots J_n$, $\ldots$ then whenever wires 1, 2, $\ldots r$ form a closed figure [a loop], $w_1 J_1 + w_2 J_2 + \ldots + w J_r$ equals the sum of all the potential differences $\ldots$ [and] when wires 1, 2, $\ldots$ meet in a point: $J_1 + J_2 + \ldots J_p = 0$.

Kirchhoff's Laws thus initially appeared in a German scientific journal expressed algebraically. Their explanation was couched in the mathematical language of the calculus. Neither the calculus nor the German language posed a barrier to their discovery by French telegraph engineers. Both the original 1845 and the later 1848 articles were translated and appeared in the French scientific journal the *Annales de chimie et de physique,* though not until 1854.[10] Moreover, while Kirchhoff developed his equations through the calculus, their utilization did not require an understanding of calculus since Kirchhoff expressed them algebraically. Nonetheless, Kirchhoff's Laws remained outside the set of mathematical tools taught to French telegraph engineers until their "rediscovery" around 1870.

2. Telegraph Instruction

Beginning in 1844, the head of the French telegraph network began recruiting engineers exclusively from among the science-trained graduates of the École Polytechnique.[11] Organized in 1794, the École Polytechnique taught science (mostly physics and chemistry) and other courses to aspirants of positions in various military and civilian state services such as the Corps du Génie (military engineering), the Corps des Mines, and the Corps des Ponts et Chaussées.[12] By 1858, the rush to hire *polytechniciens* had left its mark on the Telegraph Administration's upper ranks, where they made up two-thirds of all engineers.[13] After 1858, however, the Telegraph Administration hired fewer and fewer *polytechniciens* in order to provide greater opportunities for internal promotion. Between 1865 and the establishment of new recruitment guidelines in 1876, graduates of the school did not enter the telegraph service at all.[14]

But taking classes at the École Polytechnique did not ensure that a telegraph engineer or any other student learned Kirchhoff's formulas. The textbooks used to teach physics between 1845 and 1870, for example, did not include a discussion of Kirchhoff's Laws.[15] Furthermore, the practical and theoretical instruction provided student engineers at the Telegraph Administration's school in Paris similarly omitted Kirchhoff's circuitry equations.[16] But the absence of these formulas in the École Polytechnique and Telegraph Administration courses taken by engineer trainees did not mean that the *polytechniciens* who taught at the Telegraph Administration's school were totally unaware of the articles in which Kirchhoff set forth and proved the equations that bear his name.

Édouard Ernest Blavier (1826–87) and Eugène Gounelle (1821–64), both telegraph engineers and graduates of the École Polytechnique, taught neophyte telegraph engineers until 1862, when Claude Marie Guillemin (1822–74), physics and chemistry teacher at the École Militaire de Saint-Cyr, and Joseph Lagarde (1829–97), a telegraph engineer and graduate of the École Polytechnique, assumed their duties.[17] However, the textbook used to teach student engineers remained the same, that written by Blavier and Gounelle and first published in 1857. A second edition, written by Blavier alone, appeared between 1865 and 1867.[18]

The two engineers undertook an extensive survey of the theory of electrical conduction through variously-shaped media, such as planes and cylinders, that appeared as a series of articles in the Telegraph Administration's own *Annales télégraphiques,* edited by Blavier, in 1859 and 1860. Among the various studies cited were those of Kirchhoff, particularly his 1845 and 1848 articles that set forth and proved his circuit analysis equations. Blavier and Gounelle even supplied footnote references to Kirchhoff's original German–language articles. Yet they failed to mention his formulas.[19] The real obstacle to the integration of Kirchhoff's Laws into the vocabulary of French telegraph engineers was neither the German language nor the calculus of Kirchhoff's articles but the demands of telegraph practice.

3. The Demands of Telegraph Practice

Every day telegraph engineers resolved technical problems arising within the network. These problems often consisted of broken wires, the intertwining of two or more adjacent wires, and the breakdown of insulation. Kirchhoff's Laws provide general equations for the mathematical analysis of such circuits and the location of the problem, called a "fault." Nevertheless, as long as telegraph lines remained in the air, Ohm's Law and equations derived from it provided a sufficient mathematical and practical means for analyzing telegraph circuits.

The mathematical method for ascertaining the point at which current deviated from a telegraph line, as described in the textbooks used to teach engineers before the "rediscovery" of Kirchhoff's Laws,[20] depended upon certain assumptions about the nature of the telegraph circuit and the application of Ohm's Law. The same solution applied regardless of the cause of the fault. Telegraph engineers assumed that the resistance from the point of current deviation to the ground was expressible in the same units of resistance as the line wire, that is, the standard resistance unit adopted by the Telegraph Administration. They also assumed that the resistance of the earth in the circuit was null and that the emf of the battery was equal to one. In the case of longer lines, the internal resistance of the battery was assumed to be zero. Before the advent of the problem, the amount of current on the line, following Ohm, was $I = 1/a$, where I was the current, 1 the assumed emf of the battery, and a the resistance of the line including the electromagnet of the receiving instrument. The circuit from the fault to ground and thence to the next station being considered as a single wire, the telegraph line therefore became a branched circuit capable of being treated as a special application of Ohm's Law.

With this extension of Ohm's Law to the location of telegraph line faults, French telegraph engineers neither needed nor used Kirchhoff's Laws for the approximate quarter of a century that followed their publication in Germany. After 1867, however, knowledge of his laws and their identification with Kirchhoff's name spread quickly and became a part of the engineers' vocabulary and set of mathematical tools. The reasons for this change were multiple: the advent of underwater telegraph cables; the French government's commitment to laying, owning, and maintaining its own cables; the numerous ruptures to which these cables were susceptible; and the investigations of a graduate of the École Polytechnique.

4. Underwater Telegraph Cables

The first underwater telegraph cable of any considerable length anywhere was that laid between Great Britain (Dover) and France (Calais) in 1851.[21] From that beginning, the construction and ownership of international underwater telegraph cables rapidly became a British monopoly.[22] In France, the construction and immersion of underwater telegraph cables took place on a far more modest scale, initially either crossing rivers or connecting the land network to offshore islands. In 1859 the Telegraph Administration laid twenty-eight such cables totaling 180 kilometers, and by 1889, these cables totaled 265 kilometers.[23] Nonetheless, more significant for the telegraphic application of Kirchhoff's Laws and the development of French underwater telegraphy were those cables that connected France to North Africa.

In 1853 the French government signed an agreement with the British entrepreneur of the Dover-Calais cable for the construction and laying of an underwater telegraph cable between France and Algeria. The cable was to cross the Mediterranean Sea indirectly by way of Corsica and Sardinia. After one cable rupture too many and the entrepreneur's failure to complete the project in a timely fashion, the French contracted with another British firm,

Glass, Elliot & Co., on 13 April 1860. Glass, Elliot & Co. had built the first, but unsuccessful, Atlantic telegraph cable and planned to run the French Mediterranean cable across the island of Menorca. It began in April 1860, hoping to finish by the end of August, but a ship accident postponed work until January 1861, when the cable broke. The French signed another contract with Glass, Elliot & Co. for a "direct" cable to Algeria. The company succeeded in connecting France and Algeria through Menorca, but a storm interrupted the cable between Menorca and Algeria on 25 September 1862.[24]

The French immediately planned another underwater telegraph cable to connect with North Africa. In July 1863, an expedition took soundings for a cable to run from Oran, Algeria, to Cartagena, Spain, the purpose of landing in Spain being to shorten the route and thus lessen the chances of failure. Charged with manufacturing and immersing the Oran-Cartagena cable was the Berlin firm of Siemens & Halske.[25] For the first time, the French took an active part in the laying of an underwater telegraph cable. They purchased a British steamer, refitted it in London with cable-laying equipment, and rechristened it the *Dix-décembre* (the *Ampère* after 1870). The *Dix-décembre* began laying the Oran-Cartagena cable on 12 January 1864. After a few hours, the cable broke. The expedition began again on 28 January 1864 and, after twelve hours, the cable broke once more.[26]

The French commitment to the development of a French underwater telegraph cable technology and industry also led to the creation of a center for the laying and repair of cables in Toulon. Overseeing the Toulon works was François Ailhaud (1824–79), the telegraph engineer who later success-fully duplexed the Hughes telegraph and pioneered in French underwater telegraphy.[27] Among those working under Ailhaud was a recent graduate of the École Polytechnique, Jules Raynaud (1843–88).[28]

5. Jules Raynaud

At the Toulon cable center, Raynaud was responsible for electrical measurements,[29] work that involved both the analysis of telegraph circuits and electrical units. Electrical measurement, particularly of resistance, grew in importance with underwater telegraphy. The frequent ruptures of cables led the Telegraph Administration to carry out a number of experiments regarding the determination of the breaking point of underwater conductors.[30] In testing cables, telegraph engineers charged with overseeing the manufacture and immersion of underwater cables measured (1) the resistance of the conductor(s), (2) the resistance of the insulating covering, and (3) the cap-acitance of the cable. Once a cable was submerged, however, the location of a fault required an unprecedented level of precision that was not possible with the standard resistance unit used on land lines.[31] A redefinition of that unit therefore became necessary.[32]

The resistance unit adopted by the Telegraph Administration, one kilometer of iron wire four millimeters in diameter, varied widely with temperature and the kind of iron used. Resistance units exhibited by the French telegraph apparatus manufacturing firms of Breguet and Digney at the 1862 London International Exposition, for example, measured 9.266 and 10.420 Siemens resistance units, respectively.[33] The Siemens (Siemens & Halske) unit was the resistance of a column of mercury one meter long and one square millimeter in area at 0°C. It was less susceptible to deviations caused by impurities in the resisting medium and variations in ambient temperature and hence a far more reliable unit of resistance than that used by the Telegraph Administration for underwater telegraph cables.[34] Undoubtedly, this resulted from the experience of Siemens & Halske in underwater telegraphy.

In addition to requiring a redefinition of the standard unit of resistance, the location of faults along submerged cables posed other problems. Because the resistance of the path followed by such currents was extremely small in relation to the resistance of the entire cable, use of the usual electrical measurement techniques was difficult and insufficiently reliable.[35] Like those for finding current deviations on overhead lines, the methods for locating cable faults relied upon the application of Ohm's Law.[36] The most important of these methods was that developed by Blavier and used by English and German engineers.[37]

When Raynaud began working on electrical measurements at the Toulon cable center, he undertook a long series of experiments and an extensive study of the literature regarding the applications of Ohm's Law to telegraphy.[38] Simultaneously he wrote a thesis on the application of Ohm's Law to underwater telegraph cables. Published in 1870, the thesis earned him the degree of *licencié ès sciences physiques et ès sciences mathématiques* from the University of Paris.[39] Raynaud's coursework at the École Polytechnique equipped him with the mathematical language and physics theory of electrical potential needed to understand and take advantage of Kirchhoff's electrical articles. He could have discovered these articles as part of his investigation of Ohm's Law, either on his own or through Henri Sainte-Claire Deville (1818–81), professor of chemistry at the École Normale and the Faculty of Science in Paris,[40] who was one of the examiners on Raynaud's thesis.[41] At the École Normale chemistry laboratory, Sainte-Claire Deville had supervised the work of Lucien de la Rive,[42] who in 1863, a year after Raynaud's graduation from the École Polytechnique, published an article employing Kirchhoff's Formulas.[43] What is more likely the case, however, is that the key was the team of Édouard Ernest Blavier and Eugène Gounelle.

6. Blavier and Gounelle

In May 1853 the Telegraph Administration charged Blavier with installing French telegraph equipment in London and conducting tests along the Paris-London direct line, a line that contained overhead, underground, and underwater (the Channel cable) sections. From the beginning, Blavier observed the difficulty of obtaining normal or rapid transmission speeds.[44] The slowness of underwater telegraphy formed the focus of a report he prepared with his colleague Eugène Gounelle on the 1858 transatlantic cable.[45] Blavier and Gounelle were thus led into a study of the nature of electrical conduction in media of varying shapes and undertook an extended review of the literature, publishing relevant articles in the *Annales télégraphiques* between 1858 and 1860. In addition to those by Kirchhoff, they presented works by Georg Simon Ohm (1789–1854), Willem Smaasen (1820–50), and William Thomson, Baron Kelvin of Largs (1824–1907).[46] As they stated at the outset, the exericse was directly linked to the development of underwater telegraphy:[47]

> The studies about which we are going to speak are already a bit old, but the planned construction of a large number of underwater lines whose success may be greatly influenced by the speed with which electricity is conducted requires that we place before the reader's eyes a résumé of the experiments already undertaken....

Beginning in 1860, their interest in the difficulties raised by cable telegraphy drove the two telegraph engineers to follow, and reproduce in the pages of the *Annales télégraphiques,* the work of the British Association for the Advancement of Science leading to the creation of the 1862 electrical units, including the ohm as a standard of resistance.[48] Blavier and, until has death in 1864, Gounelle thus closely pursued the interrelated subjects of electrical propagation theory, electrical units, and underwater telegraphy.[49] If anyone, then, was to have directed the attention of Jules Raynaud, a telegraph engineer at the Toulon cable center, in the direction of Kirchhoff's work, it would have been Blavier and Gounelle.

7. The "Rediscovery" of Kirchhoff

Raynaud first brought Kirchhoff's Laws to light in 1867 in an article published in the French Academy of Science's *Comptes rendus* and titled "On a Practical Means of Determining the Voltaic Constants of a given Pile." There he showed the practicality of using "the laws of Ohm and Kirchhoff" to find an unknown resistance in a fashion similar to that employed in the Wheatstone bridge.[50] In addition to drawing attention to the equations and his own work in the prestigious and widely circulated Academy journal, Raynaud spotlighted that article (and indirectly Kirchhoff's Laws) with a letter in 1872 to the Royal Society of London. Published in the Society's *Proceedings,* the letter pointed out both the existence of the 1867 article and the usefulness of Kirchhoff's equations.[51]

Raynaud demonstrated the utility of Kirchhoff's formulas to a far greater extent, however, in his thesis, "Experimental Research on Ohm's Laws and their Application to Electrical Tests on Underwater Cables," published in 1870. There he stated the laws as follows:[52]

1) If several conductors meet in a point, the sum of the intensities of the currents which cross it is equal to zero, considering as positive the currents which flow toward the junction point and as negative those that flow from it. . . .
2) In a series of conductors forming a closed figure [a loop], the sum of the products of the intensities [currents] and the resistances is equal to the sum of the electromotive forces in the same circuit.

After introducing these two principles for the analysis of any given network of conductors, Raynaud pointed out:[53]

These two laws furnish a certain number of equations that permit the resolution of the most complicated problems of networks, among which the following is very remarkable, in that it provides a precise means of measuring the resistance of a circuit.

What followed was a demonstration of the applicability of Kirchhoff's Laws to the simplification of the calculations of Wheatstone's bridge (with applications for underwater telegraphy), the determination of the difference in electromotive forces of two voltaic piles, the location of a break in an underwater cable, and the reckoning of the quantity of charge in an underwater telegraph cable, among other uses.[54] Kirchhoff's equations could now be brought to play in underwater telegraphy.

Raynaud further diffused knowledge of Kirchhoff's equations in 1872 and 1873, when he published three articles in the *Journal de physique*,[55] a publication of the Société Française de Physique. As in his thesis, upon which he based these articles, Raynaud gave the reader a verbal, rule-of-thumb formulation of the two laws, as well as a mathematical shorthand version: "$\Sigma i = 0$" and "$\Sigma (ir - e) = 0$".[56] The formulas had emerged from their obscurity but had yet to find their place (and their name) in the routine instruction of telegraph engineers.

8. The École Supérieure de Télégraphie

The opening of the École Supérieure de Télégraphie in 1878 furnished the means for uniformly training telegraph engineers in the nature and use of Kirchhoff's Laws. As early as 1845 the head of the Telegraph Administration had attempted to establish a similar school for the advanced theoretical and practical training of telegraph engineers but had failed to receive the necessary funding from the legislature.[57] France's defeat in the 1870 war against Prussia and the emergence of the debt-burdened Third Republic meant budget cuts for the Telegraph Administration[58] and the unification of the postal and telegraph bureaucracies, an old notion that had been surfacing and resurfacing in the French legislature since 1828.[59] The lure of reduced bureaucratic costs and the apparent success of other countries like Britain encouraged the National Assembly to vote the joining of the postal and telegraph authorities into law on 6 December 1873.[60]

The two administrations did not actually merge until the presidential decrees of 27 February and 20 March 1878 transferred the telegraphs from the minister of the interior to the minister of finances and organized the administrative council of the posts and telegraphs. Adolphe Cochery (1828–1900) became first the undersecretary of state to the minister of finances then, after 30 January 1879, the first minister of posts and telegraphs, a position he held during several cabinet changes until 7 January 1886.[61] The political importance of his position, particularly as a minister, and his longevity as head of the French telegraphs permitted Cochery to institute and oversee important and vast changes in the telegraph service, among them the establishment of a school for advanced training in telegraph engineering.

Cochery believed that the telegraph network needed "the special science of an engineer." To train telegraph engineers who would be "not only knowledgeable of current science, but even prepared to hasten its progress," Cochery created the École Supérieure de Télégraphie on 12 July 1878.[62] The main purpose of the school was to train engineers for the telegraph service. After successful completion of the school's two–year program, graduates received a job and the title subengineer ("*sous-ingénieur*") in the technical division of the telegraphs. Nevertheless, the school was also open to anyone, Frenchman or foreigner, who passed the demanding entrance exams.[63]

The entrance examinations were given in four parts. The first covered French composition and served to sift out prospective applicants; those failing did not proceed through the remaining exams. After French composition came a written physics and chemistry examination, followed by a mechanical drawing test. The last hurdle was an oral exam covering a wide range of topics, including integral and differential calculus, mechanics (kinematics, statics, and dynamics), physics (heat, dynamic electricity, acoustics, and optics), chemistry (metals and organic chemistry), mechanical drawing, a foreign language (German or English), and the subjects taught in the class of special mathematics of the lycées. The contents of the special mathematics class comprised various mathematical topics such as trigonometry, descriptive geometry, algebra, and "pre-calculus" (the idea of limits, for example) as well as elementary notions of physics (basic mechanics, static electricity, and magnetism) and general chemistry.[64]

Once admitted, students took a number of courses and seminars dealing with various theoretical and practical aspects of telegraphy and other electrical technologies such as telephony and electric lighting. The faculty of the École Supérieure de Télégraphie consisted entirely of telegraph engineers who had graduated from the École Polytechnique and the school's director was E. E. Blavier.[65]

In the first year, students took two courses in telegraph and postal operations taught by Jean Joseph Richard (1826–88)[66] and a course that dealt exclusively with telegraph apparatus given by Charles Bontemps (1839–84).[67] Rounding out the first year's coursework was a special physics class, called "*physique appliquée à la télégraphie*," taught by Ernest Mercadier

(1836–1911),[68] who also held the most influential post at the École Polytechnique, that of *"directeur des études."*[69] The course concentrated exclusively on electricity, magnetism, and electromagnetism and interwove practical applications, such as the construction and operation of specific varieties of relays and other telegraph and electrical apparatus, into discussions of general theoretical matters. Although the intent of the course, as suggested by its title, was to provide an understanding of physics as it related to telegraphy, the actual lessons often went beyond telegraphy into telephony, electrical lighting, and dynamos.

In the second year, students took Richard's course on the construction of telegraph lines—which included underwater and underground cables as well as the usual aerial lines—as well as Mercadier's course in *"chimie appliquée à la télégraphie,"* which treated conductors, resistors, insulators, and, for the most part, battery theory. Two other second-year courses dealt with electrical measurement, one of which involved practical laboratory exercises. The teacher of these classes was Jules Raynaud, who included a discussion of Kirchhoff's Laws and their applicability to telegraphy in them.[70]

9. Conclusion

Through Raynaud, then—as a graduate of the École polytechnique and the University of Paris, as a telegraph engineer challenged by the new difficulties posed by underwater telegraph cables, and, finally, as a teacher of telegraph engineers—the name of Kirchhoff ceased to be associated solely with mathematical physics and became a part of the telegraph practitioner's vocabulary as Kirchhoff's Laws in France. The transference of Kirchhoff's equations from mathematical physics to the day-to-day work of telegraph engineers demonstrates the complex of factors involved when scientific discoveries acquire a new life as engineering conventions.

Individuals and institutions clearly were significant factors. Raynaud "rediscovered" Kirchhoff's equations, and educational institutions such as the University of Paris and the École Polytechnique played a role by intellectually preparing him. Raynaud was assisted in the "rediscovery" and diffusion of Kirchhoff's Laws by scientific and technical journals, like the *Annales télégraphiques,* the *Journal de physique,* and the *Proceedings of the Royal Society of London,* journals which were organs of scientific and technological institutions. Raynaud was the individual and the École Supérieure de Télégraphie the institutional vehicle which assured the routine incorporation of Kirchhoff's equations into the general body of telegraph engineering knowledge. In turn, both the school and Raynaud's personal research were shaped by contemporary political and bureaucratic circumstances.

Certainly, the Third Republic's desire to cut expenses, partly as a consequence of the war against Prussia, assisted those who wished to unify the postal and telegraph bureaucracies. The merger of the two administrations then furnished an opportunity for the new minister of posts and telegraphs to reorganize the telegraphs in a fashion to his liking, the partial result of which was the creation of the École Supérieure de Télégraphie. Earlier, the government's commitment to laying and maintaining French-controlled underwater telegraph lines led to the purchase of the *Dix-décembre* and the establishment of the Toulon cable works and furnished the impetus and the institutional context that brought Raynaud to the problems of underwater telegraphy.

If it had not been for those problems, however, Raynaud may not have interested himself in Kirchhoff's Laws. Ohm's Law furnished a sufficient mathematical means for locating faults on land lines. Underwater cables were different. The unit of resistance that served land lines was unreliable and insufficiently precise for submerged cables. Furthermore, the great difference between the resistance of the entire cable and that of the path of the fault current made the normal techniques of electrical measurement difficult and untrustworthy. The day-to-day problems of telegraph engineers had changed with the coming of cable telegraphy. Thus new technological demands set Raynaud on a course that eventually brought him to Kirchhoff's equations.

The advent of a scientific discovery did not mean its immediate adoption by engineers. Blavier and Gounelle were acquainted with Kirchhoff's work yet found no reason to integrate his formulas into the instruction of telegraph engineers. They were primarily interested in Kirchhoff's concepts of electrical conduction. The pattern for the transferral of knowledge from the scientific to the engineering community that emerges from this study is not one of a "supply-push" from science into engineering, but that of a "demand-pull" created by the day-to-day technical problems posed by cable telegraphy. Without the need to solve a common technological difficulty, Raynaud would not have "rediscovered" Kirchhoff's Laws. The exact circumstances under which this "demand-pull" operated, however, were shaped by individuals, institutions, journals, and the political climate.

Notes

1. The following are electrical engineering textbooks presently used in teaching college-level electrical engineering, with page numbers indicating the introduction of Kirchhoff's Laws into the course matter: William H. Hayt, Jr., and Jack E. Kemmerly, *Engineering Circuit Analysis,* 3d ed. (New York: McGraw-Hill Book Co., 1978), 34–37; Jacob Millman, *Microelectronics: Digital and Analog Circuits and Systems* (New York: McGraw-Hill Book Co., 1979), 708–9; Robert Grover Brown, Robert A. Sharpe, William Lewis Hughes, and Robert E. Post, *Lines, Waves, and Antennas: The Transmission of Electrical Energy,* 2d ed. (New York: John Wiley & Sons, 1973), 13; Eugene C. Lister, *Electric Circuits and Machines,* 5th ed. (New York: McGraw-Hill Book Co., 1975), 31–32; and, A. E. Fitzgerald, David E. Higginbotham, and Arvin Grabel, *Basic Electrical Engineering,* 5th ed. (New York: McGraw-Hill Book Co., 1981), 24–28.

2. For biographical information on Kirchhoff other than that provided in L. Rosenfeld's article in Charles Coulton Gillispie, ed., *Dictionary of Scientific Biography* (New York: Charles Scribner's Sons, 1973), 7: 379–83, see Robert von Helmholtz, "A Memoir of Gustav Robert Kirchhoff," translated by Joseph de Perott, in *Annual Report of the Board of Regents of the Smithsonian Institution to July, 1889* (Washington, D. C.: Government Printing Office, 1890), 527–40, and A. S. Everest, "Kirchhoff, Gustav Robert, 1824–1887," *Physics Education* 4 (1969): 341–43.

3. For a discussion of the transferral of knowledge from scientific to the engineering communities, see Edwin Layton, "Mirror-Image Twins: The Communities of Science and Technology in 19th-Century America," *Technology and Culture* 12 (1971): 562–80, and "American Ideologies of Science and Technology," *Technology and Culture* 17 (1976): 688–700. Hugh Aitken, *Syntony and Spark: The Origins of Radio* (Princeton: Princeton University Press, 1985), chapters one and six, also discusses the subject and includes references to the role of economic factors.

4. Jules Raynaud, "Des lois de la propagation de l'électricité dans l'état permanent," *Journal de physique* 12 (1872): 305; and Raynaud, "Notice sur la carrière administrative et les travaux scientifiques de E. E. Blavier," *Annales télégraphiques* 3d. ser., 14 (1887): 28.

5. James D. Forbes, "Dissertation Sixth: Exhibiting a General View of the Progress of Mathematical and Physical Science, Principally from 1775 to 1850," in *The Encyclopedia Britannica or Dictionary of Arts, Sciences and General Literature*, 8th ed. (Edinburgh: Adam and Charles Black, 1860) 1: 982–87; and *The Encyclopedia Britannica: A Dictionary of Arts, Sciences and General Literature*, 9th ed. (Edinburgh: Adam and Charles Black, 1879), 8: 42–43.

6. For Franz Neumann, see Luise Neumann, *Franz Neumann: Erinnerungsblätter von seiner tochter* (Tubingen & Leipzig: J. C. B. Mohr, 1904); W. Voigt, "Gedächtnissrede auf Franz Neumann," in *Franz Neumanns Gesammelte Werke* (Leipzig: B. G. Teubner, 1906) 1: 1–19; and Paul Volkmann, *Franz Neumann: Ein beitrag zur Geschichte deutscher Wissenschaft* (Lepizig: B. G. Teubner, 1896).

7. During the 1850s, Kirchhoff published electrical works in the *Annalen der Physik*. As late as the 1870s and 1880s, he continued to publish works on electrical theory, including five articles in the *Monatsberichte der Königlich Preussichen Akademie der Wissenschaften zu Berlin*. He also took an active part in electrical institutions and attended international electrical congresses, such as that at Paris in 1881 (Ministère des postes et des Télégraphes, *Congrès international des Electriciens, Paris 1881* [Paris: G. Masson, 1882], 17) and joined organizations such as the Société Internationale des Electriciens, of which he was a founding member. See "Liste des Membres Fondateurs (15 Janvier 1884)," *Bulletin de la Société internationale des Electriciens* 1 (1884): 49.

8. Kirchhoff, "Ueber den Durchgang eines elektrischen Stromes durch eine Ebene, insbesondere durch ene kreisförmige," *Annalen der Physik* 64 (1845): 497–514; Kirchhoff, "Nachtrag zu dem Aufsatze: Ueber den Durchgang eines elektrischen Stromes durch eine Ebene, insbesondere durch eine kreisförmige," *Annalen der Physik* 67 (1846): 344–49; Kirchhoff, "Ueber die Auflösung der Gleichungen, auf welche man bei der Untersuchung der linearen Vertheilung galvanischer Ströme geführt wird," *Annalen der Physik* 72 (1847): 497–508; and, Kirchhoff, "Ueber die Anwendbarkeit der Formeln für die Intensitäten der galvanischen Ströme in einem Systeme linearer Leiter auf Systeme, sie zum Theil aus nicht linearen Leitern bestehen," *Annalen der Physik* 75 (1848): 189–205.

9. Kirchhoff, "Ueber die Anwendbarkeit der Formeln," 205: "Besteht das System aus n Drähten, deren Widerstände w_1, w_2 ... w_n sind, und deren Ströme die Intensitäten J_1, J_2, ... J_n haben ... dass immer, wenn die Drähte 1, 2, ... reine geschlossene Figur bilden, $w_1 J_1 + w_2 J_2 + \ldots + w J_r$ gleich der Summe aller Spannungsdifferenzen ist ... wenn die Drähte 1, 2, ... p in einem Punkte zusammenstofsen: $J_1 + J_2 + \ldots J_p = 0$."

10. *Annales de chimie et de physique*, 40 (1854): 115–27 and *Ibid.*: 327–33.

11. C. P. Marielle, *Répertoire de l'École Impériale Polytechnique* (Paris: Mallet–Bachelier, 1855), 206–7; P. Leprieur, *Répertoire de l'École Impériale Polytechnique* (Paris: Gauthier–Villars, 1867), 63; and "Ordonnance du roi portant règlement sur le service de la télégraphie," 24 August 1833, in Ministère de l'intérieur, Direction générale des lignes télégraphiques, ed., *Lois et règlements* (Paris: Imprimerie Impériale, 1859), 1. This work is a compilation of laws, decrees, and other administrative notices dealing with French telegraphy for the period 1833 through 1854. Subsequent volumes cover later years. Only the first volume has pages numbered sequentially. Since items in the entire series are arranged chronologically within each volume, an entry's date indicates the volume and the location within that volume. Therefore, future references to the work give the title of the entry, its date, and the title *Lois et règlements*.

12. For the most recent book-length works on the École polytechnique, see Jean Pierre Callot, *Histoire de l'École polytechnique* (Paris & Limoges: Charles Lavauzelle, 1982), and Terry Shinn, *Savoir scientifique et pouvoir social: L'École polytechninque, 1794–1941* (Paris: Presses de la fondation nationale des sciences politiques, 1980), 225–43 contain an extensive bibliography.

13. Based upon a comparison of the names found in Marielle, 205–07 and M. Vallée, *Annuaire des lignes télégraphiques suivi des décrets et arrêtés concernant les fonctionnaires et agents* (Paris: Paul Dupont, 1858), 9–13.

14. Société amicale de secours des anciens élèves et société des amis de l'École Polytechnique, *Annuaire des anciens élèves de l'École polytechnique* (Paris: Gauthier–Villars, 1952), page P.1 and Ernest Mercadier, "Les télégraphes," in École Polytechnique, *Livre du centenaire, 1794–1894* (Paris: Gauthier–Villars et fils, 1897), 3: 283–84.

15. Auguste Bravais, *École polytechnique, 2ème division, Cours de physique, 1845–1846* (Paris: n.p., 1845–46); Jules Jamin, *Cours de physique de l'École polytechnique* (Paris: Mallet–Bachelier, 1858–1866); Jules Jamin, *Cours de physique de l'École polytechnique,* 2d ed., 3 vols. (Paris: Mallet–Bachelier, 1863–69); Jamin, *Cours de physique de l'École polytechnique,* 3d ed., 3 vols. (Paris: Mallet–Bachelier, 1868–71). See also Joseph Jacquet, "L'Enseignement de l'électricité à l'École Polytechnique: Premiers résultats d'un sondage dans les cours de l'École depuis sa création," 359–70 in Fabienne Cardot, ed., *La France des électriciens, 1880–1980* (Paris: Presses Universitaires de France, 1986).

16. E. E. Blavier and E. G. Gounelle, *Résumé des cours faits à l'Administration des lignes télégraphiques, 1858–1859,* 2 vols. (Paris: Napoléon Chaix et Cie., 1858–59) and Blavier, *Nouveau traité de télégraphie électrique,* 2 vols. (Paris: Eugène La croix, 1865–67).

17. Blavier and Gounelle were the instructors of telegraph engineer trainees according to Circulaire no. 227, 11 March 1859, *Lois et règlements.* Marielle, *Répertoire,* 20, 101, indicates their graduation from the École Polytechnique. Gounelle's obituary, E. E. Blavier, "Eugène Gounelle," *Annales télégraphiques* 2d. ser. 7 (1864): 92–96, provides biographical data. A large number of biographical articles on Blavier exist, the lengthiest and most useful being that in the *Annales télégraphiques* 3, 13 (1886): 566–74; 3d ser., 14 (1887): 5–44, 369–401; 3d ser., 16 (1889): 97–114, 193–218. "Rapport sur les appareils télégraphiques de l'Exposition d'électricité de 1881," Archives, Academy of Science, Paris, states that Blavier taught telegraphy from 1858 through 1862. Jules Bourdin, "Blavier et son oeuvre," *La lumière électrique* 23 (1887): 645 states that Guillemin and Emile Burnouf were the instructors who replaced Blavier and Gounelle. I doubt that Burnouf actually taught telegraphy in Paris for two reasons: (1) the "Tableau des membres composant l'Académie de Stanislas suivant l'ordre de réception," for 1862 through 1865 lists Burnouf as a *"membre titulaire"* and as "Professeur de Littérature ancienne à la Faculté des Lettres," at Nancy. For him to have been a *"membre titulaire"* of the Académie, Burnouf would have had to reside in Nancy, where he held a teaching position. *Mémoires de l'Académie de Stanislas* (1862): 473; (1863): 669; (1864): 421; and (1865): 463. (2) An unidentifiable form dated 8 November 1862 in his personnel folder, "Lagarde," F(90) 20,539, Archives Nationales, Paris, states that Joseph Lagarde, a graduate of the École Polytechnique, taught the theoretical course at that time. Lagarde probably replaced Gounelle, and Guillemin followed Blavier. Also, there appears to be a problem with Guillemin's first name. Bourdin and "M. Alexandre Guillemin," *Annales télégraphiques* 3d ser., 1 (1874): 114 calls him Alexandre, and B. W. Feddersen and A. J. von Oettingen, eds., *J. C. Poggendorff's Biographisch-Literarisches Handwörterbuch zur Geschichte der Exacten Wissenschaften* (Leipzig: Johann Ambrosius Barth, 1898), 3: 563, gives his name as Claude Marie. Nonetheless, the same biographical information appears in both sources.

18. Circulaire no. 227; Blavier and Gounelle *Cours théorique et pratique;* and Blavier, *Nouveau traité.*

19. Blavier and Gounelle, "Théorie de la propagation de l'électricité," *Annales télégraphiques* 2d ser., 2 (1859): 218–44; 381–404; 2d ser., 3 (1860): 26–45, 135–56; 2d ser., 2 (1859): 241–43.

20. Blavier and Gounelle, *Cours théorique et pratique,* 1: 153–59 and Blavier, *Nouveau traité,* I: 91–94 and 338–47.

21. John Watkins Brett, *Origin and Progress of the Oceanic Electric Telegraph* (London: Johnson, 1858), provides a history of the Calais-Dover cable by the entrepreneur responsible for the manufacture, laying, and maintenance of the cable.

22. A. L. Ternant, "Les câbles sous-marins," *Annales industrielles* 2 (1870–71): 1141. "Nomenclature des câbles sous-marins du globe," *Journal télégraphique* 13 (1889): 214–43 provides an overview of world underwater telegraph cables that quantitatively demonstrates the British control of cables.

23. *Annales télégraphiques* 2d ser., 2 (1859): 444 and "Nomenclature des câbles sous-marins du globe," 218.

24. Eugène Wünschendorff, *Traité de télégraphie sous-marine* (Paris: Baudry et Cie., 1888), 9–14, 17, 28–30, and Maurice du Colombier, "Notice sur le câble d'Alger," *Annales télégraphiques,* 2d ser., 5 (1862): 105–40.

25. For an English–language history of Siemens & Halske, see Sigfrid von Weiher and Herbert Goetzeler, *The Siemens Company: Its Historical Role in the Progress of Electrical Engineering,* translated by G. N. J. Beck (Berlin & Munich: Siemens Aktiengesellschaft, 1977).

26. Wünschendorff, *Traité,* 30–35; "Rupture du câble d'Algérie," *Annales télégraphiques* 2d ser., 6 (1863): 107–9; Henri Blerzy, "Revue de télégraphie sous-marine," *Annales télégraphiques* 407–8, 584–601; *Annales télégraphiques* (1864): 174–90 and 760–61; and Session de 1870, Budget de l'exercise 1871, "Projet de loi pour la fixation des recettes et des dépenses ordinaires et extraordinaires de l'exercise 1871," AD XVIII(F) 869, Archives Nationales, Paris.

27. *Journal télégraphique* 4 (1879): 497 and *Annales télégraphiques* 3d ser., 6 (1879): 493. Biographical sketches on Ailhaud appeared in the *Annales télégraphiques* 3d ser., 6 (1879): 493–96 and *Journal télégraphique* 4 (1879): 497.

28. "M. F. E. J. Raynaud," *Journal télégraphique* 12 (1888): 18.

29. "M. F. E. J. Raynaud," 19.

30. Jules Raynaud, *Recherches experimentales sur les lois de Ohm et leurs applications aux essais électriques des câbles sous-marins* (Paris: A. Parent, 1870), 7.

31. H. Blerzy, "Essais électriques des câbles sous-marins," *Annales télégraphiques* 2d ser., 5 (1862): 332–35.

32. Blerzy, "Essais électriques," 336.

33. A. L. Ternant, *Manuel pratique de télégraphie sous-marine* (Paris: E. Lacroix, 1869), 87.

34. Blerzy, "Essais electriques," 337.

35. Blavier, *Nouveau traité,* 1: 355.

36. Blerzy, "Essais électriques," 340–61.

37. Raynaud, "Notice sur la carrière administrative et les travaux scientifiques de E. E. Blavier," 18–19.

38. *Journal télégraphique* 12 (1888): 19, and Raynaud, *Recherches experimentales,* 7.

39. Jules Raynaud, *Recherches experimentales.* According to Albert Maire, *Catalogue des thèses de sciences soutenues en France de 1810 à 1890 inclusivement* (Paris: H. Welter, 1892), 67, Raynaud did his thesis work at the University in Paris. The title page of his thesis states that Raynaud was a graduate of the École Polytechnique and *licencié ès sciences physiques et ès sciences mathématiques.*

40. Poggendorff, *Biographisch Handwörterbuch,* 2: 737–38, 4: 1162–63, and *Revue scientifique* 3 (1882): 1–8.

41. Raynaud, *Recherches experimentales,* title page, states that Raynaud's two examiners were Jules Jamin (Raynaud's physics professor at the École Polytechnique) and Henri Sainte–Claire Deville.

42. Lucien de la Rive, "Méthode de M. W. Thomson pour la mésure de la conductibilité électrique: Application aux métaux fondus," *Comptes rendus hebdomadaire de l'Académie des Sciences, Paris* 57 (1863): 700, states that de la Rive worked at the École Normale chemistry laboratory under Henri Sainte-Claire Deville. I have found no biographical information on Lucien de la Rive.

43. Lucien de la Rive, "Sur le nombre d'équations indépendantes dans la solution d'un système de courants linéaires," *Archives des sciences physiques et naturelles* 17 (1863): 105–12.

44. Blavier, "Expériences sur la ligne souterraine de Calais à Londres en 1853," *Bulletin des séances de la Société météorologique* (1854): 5, offprint found in Blavier's dossier, Archives, Academy of Science, Paris.

45. Blavier and Gounelle, "Rapport sur les documents fournis au sujet de la pose du câble transatlantique," F(90) 20, 847, Archives Nationales, Paris.

46. The series began as Gounelle, "Résumé des travaux faits pour déterminer la vitesse de propagation de l'électricité dans les conducteurs aériens et sous-marins," *Annales télégraphiques* 2d ser., 1 (1858): 239–73, and continued as Blavier and Gounelle, "Théorie de la propagation de l'électricité," *Annales télégraphiques* 2d ser., 2 (1859): 218–44; 381–404; 2d ser., 3 (1860): 27–45; 135–56. The fullest biography of Ohm is Heinrich von Füchtbauer, *Georg Simon Ohm: ein Forscher wächst aus seiner Väter Art* (Berlin: VDI-verlag gmbh, 1939), 2d ed. (Berlin: F. Dünmlers, 1947). Very little is available on Smaasen aside from that provided in Poggendorff, *Biographisch Handwörterbuch* 2: 943. On the other hand, abundant biographical information exists for Thomson. The most complete remains that of S. P. Thompson, *The Life of William Thomson, Baron Kelvin of Largs,* 2 vols. (London: Macmillan, 1910).

47. Blavier and Gounelle, "Théorie de la propagation de l'électricité," 239: "Les travaux dont

nous allons parler sont déjà un peu anciens, mais la construction projetée d'un grand nombre de lignes sous-marines, sur la réussite desquelles la vitesse de propagation de l'électricité peut avoir une grande influence, nous engage à mettre sous les yeux de nos lecteurs le résumé des expériences qui ont faites à diverses reprises."

48. "Rapport du Comité anglais chargé, par le Conseil privé du commerce et la Compagnie du télégraphe transatlantique, de l'enquête sur la construction des câbles télégraphiques sous-marins," *Annales télégraphiques,* 2d ser., 5 (1862): 233–56, 297–331, 417–55. The work of the B.A.A.S. Committee on Electrical Standards was summarized by its reporter in Henry Charles Fleeming Jenkin, ed., *Reports of the Committee on Electrical Standards* (London and New York: E. & F. N. Spon, 1873). For a more contemporary although less detailed and authoritative history of the ohm, see L. G. A. Sims, "The History of the Determination of the Ohm," *Bulletin of the British Society for the History of Science* 2 (1957): 57–61.

49. Butrica, "Toward the 'Pre-history' of Standardization: Reflections on Electrical Units in France before 1881," paper read at the conference "Internationalisation et standardisation en électricité, 1850–1914," July 1988, C.R.H.S.T., La Villette, Paris, discussed this as well as the role of Blavier and Raynaud in importing the physics of James Clerk Maxwell into French electrical engineering in the 1870s, a subject that deserves far more attention.

50. Raynaud, "Sur un moyen de déterminer les constantes voltaïques d'une pile quelconque," *Comptes rendus* 65 (1867): 170–72.

51. "On a mode of Measuring the Internal Resistance of a Multiple Battery by Adjusting the Galvanometer to Zero," *Proceedings, Royal Society of London* 20 (1872): 159, states that Raynaud's letter to Secretary George Gabriel Stokes was received 11 January 1872.

52. Raynaud, *Recherches experimentales,* 16: "1) Si plusieurs conducteurs concourent en un même point, la somme des intensités des courants qui les traversent est égale à zero, en considerant comme positifs les courants qui se dirigent vers le point de jonction, et comme negatifs ceux qui s'éloignent.... 2) Dans une série de conducteurs formant une figure fermée, la somme des produits des intensités par les resistances est égale à la somme des forces électro–motrices dans le même circuit."

53. Ibid., 17: "Ces deux lois fournissent un certain nombre d'équations qui permettent de résoudre les plus compliqués des dérivations, parmi lesquels le suivant est très-remarquable, en ce qu'il donne un moyen précieux de mésurer la resistance d'un circuit."

54. Ibid., 17, 30, 56, 112.

55. Raynaud, "Des lois de la propagation," 305–21; "Courants dérivés; Lois de Kirchhoff," *Journal de physique* 2 (1873): 86–98; and "Résolution des équations fournies par les lois de Kirchhoff, pour la distribution des courants électriques dans un système quelconque de conducteurs linéaires," Ibid., 161–71.

56. Raynaud, "Courants dérivés," 92.

57. Alphonse Foy, "Rapport à M. le Ministre de l'intérieur," dated 8 November 1845, F(90) 1454*, Archives Nationales, Paris, and *Moniteur universel,* 26 May 1845, supplement, ix, and 12 June 1845, 1666.

58. For example, Chambre de Députés, Session de 1875, Budget de l'exercise 1876, "Projet de loi pour la fixation des recettes et des dépenses de l'exercise 1876," AD XVIII(F) 918, Archives Nationales, Paris, regarding the telegraph budget, states: "L'Administration, préoccupée de réduire ses dépenses au strict nécessaire, n'a demandé pour les exercises 1874 et 1875 aucune augmentation sur le crédit de 8,058,600 francs, qui avait été ouvert sur le budget de 1873 pour les traitements de ses fonctionnaires et agents."

59. Emile Delfieu, *Le Monopole télégraphique et téléphonique* (Nimes: P. Gellion et Bandini, 1918), 14–17, and Charles Rolland, "Rapport fait au nom de la Commission des services administratifs (postes et télégraphes)," *Journal officiel de la République française,* 10 July 1872, 4700–4701. A useful historical overview of the postal and telegraph fusion can be found in Adolphe Cochery, "Rapport adressé au Président de la République sur l'organisation des services des postes et des télégraphes avant et depuis l'année 1878," *Journal officiel de la République française,* 19 June 1884, 3153–61.

60. Delfieu, *Le Monopole télégraphique,* 17.

61. "Bulletin administratif," *Annales télégraphiques,* 3d ser., 5 (1878): 227–28, and Michel de Cheveigné and Pierre Lajarrige, "Le téléphone de 1881–1889," in Centre national d'Études des Télécommunications, ed., *Chroniques téléphoniques et télégraphiques* (Paris: Centre national d'Etudes des Télécommunications, 1982), 91–93.

62. "Circulaire no. 17," dated 12 July 1878, *Bulletin mensuel des postes et télégraphes* 1 (June 1878): 143–44, and "L'École Supérieure de Télégraphie," *Annales télégraphiques* 3d ser., 5 (1878): 558. On the École Supérieure de Télégraphie, see Butrica, "The École Supérieure de Télégraphie and the Beginnings of French Electrical Engineering Education," I.E.E.E. *Transactions on Education* E-30 (August 1987): 121–29. For the development of French electrical engineering education in general, see Butrica, "Telegraphy and the Genesis of Electrical Engineering Institutions in France, 1845–1895," *History and Technology* 3 (1987): 365–80, and Butrica, "La formation des ingénieurs-électriciens en France, 1845–1895: Du secteur public au secteur privé," *Bulletin d'histoire de l'életricité* no. 11 (June 1988): 79–85.

63. "Circulaire no. 17," 145–46 and Cochery, "Rapport," 3164.

64. "Programme des connaissances exigées pour l'entrée à l'École supèrieure de Télégraphie," *Annales télégraphiques* 3d ser., 5 (1878): 561–66; "Circulaire no. 17," 146; and "Programme des connaissances exigées pour l'admission à l'École polytechnique, 1872," *Journal officiel de la République française,* 5 February 1872, 838–43.

65. Based upon a comparison of the names of the school's teachers from "Personnel de l'École supérieure de Télégraphie pour l'année scolaire 1878–1879," *Annales télégraphiques* 3d ser., 5 (1878): 567, with the graduates of the École Polytechnique listed in C. Marielle, *Répertoire,* 205–7. The discussion of the school's curriculum that follows is based upon École Supérieure de Télégraphie, *Sommaire des cours, 1880–1881* (Paris: Ministère des Postes et des Télégraphes, 1881), from the Central Library of the Ministry of P.T.T., Paris.

66. For Jean Richard, see his obituary in *Journal télégraphique* 12 (1888): 72–73.

67. Necrological notices on Bontemps appeared in *Annales télégraphiques* 3d ser., 11 (1884): 183–84, and *Journal télégraphique* 8 (1884): 148.

68. Ernest Mercadier's personnel folder is extant, F(90) 20, 542, Archives Nationales, Paris. An obituary appeared in *Journal télégraphique* 35 (1911): 202.

69. Shinn, *Savoir scientifique,* 42, discusses the duties of the *"directeur des études"* at the École Polytechnique.

70. "Personnel de l'École supérieure de Télégraphie pour l'année scolaire 1878–1879," 567, and École Supérieure de Télégraphie, *Sommaire des cours, 1880–1881.*

Warner and Swasey at the Naval Observatory:
A View of the
Science-Technology Relationship

EDWARD JAY PERSHEY

Machine tools and telescopes are two kinds of machines. One can be used to build the other, and each has a specialized role in a society, such as America, that has come to value science and technology. At the end of the last century the Cleveland machine-tool company of Warner & Swasey brought those two machines together in a way that illuminates for us the nature of the development of the science-technology relationship in America. Particularly interesting is the case of the government contract to rebuild the large astronomical instrument for the U.S. Naval Observatory at Washington, D.C., in 1891 and how fervently the two Cleveland machinists strove to obtain the work. The actions of Worcester Warner and Ambrose Swasey in this area clearly show that science and technology are related in social and political ways more than in any other.

Much has been written by historians of science and even more by historians of technology about the changing nature of the science-technology relationship at the end of the nineteenth century, especially in America, and the way that science and technology have become at times indistinguishable.[1] In analyzing the relationship between the two, very often the issue of instrumentation for scientific research is cited as one of the obvious connections between the cognitive structure of technology and that of science. At least one investigator, an economic historian, has extended this idea to suggest that it is technology that defines the research arena for science by setting the parameters for funding through the design of instrumentation.[2] In sociological studies of the science-technology relationship, there is a basic underlying assumption that the kind of knowledge involved in both is so similar as to suggest a shared cognitive base. Although science and technology may seem to be doing very similar things and learning the same kinds of things (they may *now,* in fact, share a methodology), recent historical studies have shown that they are quite different.[3] The differences, I would suggest, are more telling than any similarities and have largely to do with the end product of technology: things.

While the complexity of the current twentieth-century society finds *scientists* busily researching improved nonmetallic building materials and *engineers* doing high-level theoretical engineering research on the strength of materials, one of the ways to distinguish between the two worlds is still to look at their ultimate goals. For engineering and technology in general, the test of the knowledge-base lies ultimately in working machines and systems that function in some predictable and usable manner in the real world. In the crude dictum of a curator, you ought to be able to paint a number on it and catalog it. In contrast, the test of the scientific knowledge-base lies in the relationship of that knowledge to the theories involved in the cognitive constructions used to explain the nature of the real world. This rather simplistic differentiation, however, works in subtle ways to help us define science and technology.

When one looks at technologists providing instrumentation for science, as with the machine-tool firm of Warner & Swasey, the relationship between science and technology is certainly less cognitive than it is socioeconomic and political. More often than not, these two human endeavors are brought together for reasons having little to do with technical or scientific ideas, or even any shared methodology. Interesting connections between the two occur not in the knowledge base, but in the surrounding culture and society.

At the end of the nineteenth century the machine-tool company of Warner & Swasey built the largest astronomical telescopes in the world. The 40-inch Yerkes refractor, the 36-inch Lick refractor, and the 26-inch Naval Observatory refractor were all the products of this Cleveland machine-tool shop. In the twentieth century Warner & Swasey went on to build several large reflecting telescopes as well: the 72-inch for the Dominion Astrophysical in Canada, and the 82-inch MacDonald reflector at the University of Texas. The company used the impressive images of these behemoths of science to promote its line of machine tools. The only study of the Warner & Swasey telescopes has analyzed them as indicators of the professionalizaton and maturation of American engineering within the intellectual framework of an engineering discipline that mirrors science.[4] However, if one looks at why Warner & Swasey chose to build telescopes to sell machine tools and at how aggressively the company sought the various contracts, especially that for the Naval Observatory in the 1890s, a different picture of the relationship between science and technology emerges.

When the activity of Worcester Warner and Ambrose Swasey is studied, it can be seen that although both were deeply interested in astronomy and wanted to acquire the social trappings of science through their telescope building, neither had much insight into what astronomers did not nor were either interested in the ends of their work—the knowledge they produced. Their interest and understanding of the science was at a popular level and their sensitivity to science as a body of knowledge was no greater than that of the nonscientist or nonengineer. Trying to understand the science-technology relationship through cognitive content or even methodology in this case leads nowhere. However, if one looks at the economic relationships between building telescopes and the political leverage gained by building such instruments for the government, then the relationship between the technology of Warner & Swasey and the science of astronomy becomes more interesting and understandable.

Technology in the nineteenth century became increasingly complex, and technologists became increasingly removed from the norms of everyday life. While millwrights, architects, and engineers were always unusually esoteric members of society, during the nineteenth century they also joined the even smaller ranks of the college-educated, making themselves doubly removed from the general populace. While neither Warner nor Swasey received degrees in engineering, they supported and encouraged this movement in their chosen profession and became directly involved in engineering education at the Case School of Applied Science in Cleveland. While scientists were not very high on the social ladder, they nevertheless held professorial appointments at institutions of higher learning. Engineers coupled this unique social status with the financial opportunity afforded to them through ownership of a technologically-based business. This combination of intellectual and financial prestige promised to gain for the engineers the social status that they sought.

Typical of nineteenth-century mechanics, Worcester Warner and Ambrose Swasey had served apprenticeships in a New-Hampshire machine shop in the 1860s. Their first jobs were with the machine-tool company of Pratt & Whitney in Hartford, Connecticut, in the 1870s. In 1880, with money saved from various in-house contract work at Pratt & Whitney, Warner and Swasey formed their own company to produce a line of small machine tools, mainly for the brass-working trade. Personal, hobby-level interest in astronomy suggested to them that instrument-making would be a showcase for their abilities as machine designers and fabricators. Very quickly after building only a few small astronomical telescopes in the early 1880s, the two men successfully bid on and built the mounting for the Lick Observatory's great telescope, then the largest such instrument in the world. Their success in the mechanical design and construction of the Lick telescope made their corporate identity known throughout the world. The market for their line of turret lathes and other machine tools expanded dramatically into the 1890s. It was the promise of this expanded market that turned the two machinists into instrument-makers, rather than any feel for astronomy as a science or their personal desire to be known as scientists.[5]

The growing *popularity* of science suggested to Warner and Swasey and other contemporaries (Edison, for example) various advantages of associating their work with science. The general public had, surprisingly, provided a broad base of financial support for science, particularly astronomy, since the great comet of 1843.[6] A long-lasting, though ultimately small, amateur astronomical community can trace its roots in America to the mid-nineteenth century. The dark skies of pre-electric-light America afforded an increasingly educated populace impressive celestial views that have not survived into today's urbanized society. In this respect alone, the nineteenth century may represent an unusual and perhaps unique combination of urbanized intelligentsia and viewable skies.

After their success with the huge Lick mount, Warner & Swasey actively sought contracts to build more of these impressive advertisements to engineering skill.[7] Its timing was right, for in the early 1890s, the U.S. Naval Observatory, in Washington, D.C., was being moved out to Georgetown Heights and the main instrument, an 1872 26-inch refractor originally built totally by the well-known optical firm of Alvan Clark & Sons, needed an improved mechanical mounting. While the Clarks felt it only proper that they redo the work of the firm, since it had been the elder Alvan Clark and founder of the firm who had designed the original mounting, Warner & Swasey bid on the work and brought various political pressures to bear on the navy to win the contract.

While these early contracts to build large astronomical telescopes were quite lucrative for Warner & Swasey, they did not represent a large proportion of the total earnings of the company.[8] However, it should be noted that the Naval Observatory and Yerkes Observatory contracts were obtained in the 1890s, during a severe economic downturn. There is no doubt that this esoteric instrument-making by Warner & Swasey allowed it to keep the shop in Cleveland active and staffed while machine-tool orders declined. As business improved into the next century and the market for Warner & Swasey machine tools became solidified and dependable, telescope manufacturing was gradually phased out, even as a promotional activity.[9] Nonetheless, had telescope manufacture not been a profit-making venture in the 1880s and 1890s, it is quite clear that Warner & Swasey would not have developed that line of work at all.

The case of the contract to remount the Naval Observatory refractor shows the economic and political value of science for two technologists.

The work of building the mounting for the Lick telescope had been awarded to Warner & Swasey in part on the advice of astronomer Edward S. Holden, a former assistant at the U.S. Naval Observatory under Simon Newcomb, who had met Warner in the mid-1870s. Even as the Lick work was just getting under way, Warner & Swasey was bidding on small ancillary instrument work for the Washington, D.C., observatory. Orders for small mounts for photographic instruments and chronographs were received in the late 1880s.[10] By 1888 plans were under way to move the observatory to Georgetown Heights (its present location) and Warner & Swasey was being consulted regarding the possible designs and costs of remounting the large Clark refractor and supplying modern revolving domes and elevating floors for the observatory building.[11] A contract to provide two domes, one 45 feet in diameter and the other 26½ feet, was awarded in December 1890 to the Cleveland firm.[12] By October 1890 the two Cleveland engineers had provided "drawings and specifications" for a proposed elevating floor and were prepared to go to Washington to discuss freely with the navy details for the remounting of the large telescope.[13]

In the spring of 1891 the Navy Department of Equipment, under the direction of its chief, George Dewey, issued specifications and a call for proposals for "Repairing and Remounting Instruments, Etc., at U.S. Naval Observatory."[14] These written specifications called for a large iron column to support the equatorial mounting and a hydraulic ram system to support the elevating floor. Many of the ideas that Warner & Swasey had worked out in the Lick telescope reappear in these government specifications. The firm of Alvan Clark & Sons responded to the printed specifications with firm criticism, suggesting that the newer methods of mounting large telescopes were untried and inferior to the older methods using masonry columns and observing chairs.[15]

As the 15 June 1891 deadline for the proposals came and went, it became evident that the firm of Alvan Clark & Sons had submitted a bid on the work of remounting the telescope.[16] This was surprising to Warner and Swasey, who knew quite well, having essentially coached the navy on the specifications, that the Massachusetts opticians were not fully capable of doing the work as specified. Moreover, the specifications required complete and detailed drawings with the bid proposals. It quickly became apparent that Alvan Clark & Sons was using political maneuvering to gain the navy contract. However, it also became clear that the Clark firm had also engaged the services of Carl Saegmuller, a local Washington, D.C., instrument maker with the Fauth Company and former member of the coast survey. Warner and Swasey felt that their drawings, submitted for consultation to the navy and then for the bid on the work, had been *copied* at least in part by Saegmuller to prepare drawings for Clark to meet the specifications.[17] The Cleveland engineers felt that they were being cheated out of a lucrative contract to which they were entitled, having met all the specifications. As Warner said to Swasey in late June of 1891, they were prepared to "fight the frauds like a Grant if it takes all summer"![18]

The political infighting and memo and letter-writing went on throughout the summer of 1891. By early July, Warner & Swasey had obtained a letter of support from then Ohio Governor William McKinley.[19] The issue was resolved in the fall of 1891 when the technical expertise of the Cleveland firm eventually won out. Review of the Clark bid by William Harkness, Professor of Mathematics with the navy, showed that even with the pirated drawings, the material turned in by the Clark firm with the bid was insufficient to guarantee successful construction. Additionally, the older methods of mounting and operating large telescopes, attested to by the Clark firm, were outmoded and unsuited to the large telescopes of the new, photographically-based astronomy.[20]

In February of 1892 a contract was let to Warner & Swasey to remount the

Fig. 1.   Warner & Swasey mounting for the great refractor of the U.S. Naval Observatory, ca. 1890s. Note the spiral staircase on the south side of the main pier and the rising observatory floor, which permits the use of the relatively simple observing chair. (U.S. Naval Observatory, Washington, D.C.)

Fig. 2. Close-up of the equatorial head of the Warner & Swasey mount for the great refractor of the U.S. Naval Observatory, ca. 1890s. Warner & Swasey clockdrive can be seen in through the glass door in the pier. (U.S. Naval Observatory, Washington, D.C.)

Fig. 3. Eyepiece end of the great refractor of the U.S. Naval Observatory, ca. 1890s. At the eye end the observer could control the motions of the telescope via the knurled handles and control rods, which ran along half the length of the tube to the mechanical gearing in the equatorial head. (U.S. Naval Observatory, Washington, D.C.)

26-inch refractor. The new mounting was installed in June of 1893 and was in full use by December. Although some minor modifications were made over the next several years, the mounting (see figures 1–3) served the observatory largely in this form until a complete modernization in the 1960s.

Even as Warner & Swasey was finishing up the Naval Observatory work, it was also building the mounting for what would be and remains the largest refracting telescope in the world, the 40-inch refractor at the Yerkes Observatory, University of Chicago at Williams Bay, Wisconsin. Its commitment to large instrument building remained undaunted into the first decade of the twentieth century with the work on the 72–inch reflector for the Canadian Astrophysical Observatory,.

Interestingly, and disappointingly, there seems to be little evidence that the instrument work of Warner & Swasey affected the direction of their machine-tool design. No references to such interplay appear in official company accounts of this work. There is no direct evidence in the surviving archival record of such feedback, although historical portraits of the machine-tool industry in American technology, which have put machine tools in the center of technological innovation in the nineteenth and most of the twentieth centuries, would seem to predict it.[21] Since the company proudly used its instrument work as a symbol of its expertise in machine-tool design, it should be expected that it would have pointed out any direct connection. If indeed technological change in machine-tool design was effected through astronomical instrument making, it was so very subtle as to go unnoticed by the practitioners of the work. I suggest that the reason for this is that there was none. Although both Warner and Swasey doggedly pursued excellence in instrument making and produced a revolution in the design and construction of large telescopes, the flow of knowledge was one-way. What came back into technology was not new insight into machine design, but a social stigma born of the higher respect accorded science in America.

While this emulation of science by engineers in the nineteenth century has been often enough suggested as one of the symptoms of the maturation of American engineering, the need for technologically based industry to don the garbs of pure science continues even now. There is a purity of purpose, an altruism of motivation that the term *science* clearly implies to the public, even when science deals with moralistically suspect topics, such as genetic mutations. Technology continues to call itself science and tries to appear to be the same thing. This rhetoric was important in selling machine tools in the 1890s and continues to be important in selling computers in the 1980s. It also continues to be important for engineers to emulate science to maintain the proper social image.

The relationship of technology and science, via instrument making, does not imply or suggest any cognitive or even methodological connection. Rather, the two cultural activities of science and techology are related only in an economic and political way. They interact and draw on one another through outside agencies and agents. If this interaction is viewed, then, as taking place through a cultural filter, then it is not surprising to see very little direct transfer of knowledge from one to the other. Neither really expects such transfer or depends on it. Any acknowledgment of a borrowing of knowledge from the other can in fact be interpreted as political rhetoric designed to guarantee social support and produce appropriate gains—markets, funding, and public image. By viewing the relationship between science and technology as "noncognitive" and looking instead for the broader connections, perhaps historians can begin to perceive better how these two human activities are indeed related.

Note

1. See, for instance, Otto Mayr, "The Science-Technology Relationship as a Historiographic Problem," *Technology & Culture* 17 (1976): 663–72.

2. Nathan Rosenberg, "How Exogenous Is Science?" in Rosenberg, *Inside the Black Box: Technology and Economics* (New York: Cambridge University Press, 1982).

3. For example, Walter Vincenti's articles in which he shows how technology moves along quite nicely without science. See, Vincenti, "Control-Volume Analysis: A Difference in Thinking Between Engineering and Physics," *Technology & Culture* 23 (1982): 145; and "Technological Knowledge without Science: The Innovation of Flush Riveting in American Airplanes, ca. 1930–ca. 1950," *Technology & Culture* 25 (1984): 540.

4. Edward Jay Pershey, "The Early Telescope Work of Warner & Swasey," (Ph.D. diss., Case Western Reserve University, 1982).

5. Admittedly, particularly later in life, Swasey would associate with scientists and allow himself to be classified as such, but by then (the 1920s), "engineering science" had emerged. In the nineteenth century Swasey clearly never saw himself as a scientist nor aspired to that title.

6. For a good overview of the popular support of astronomy in America in the nineteenth century, see Howard S. Miller, *Dollars for Research: Science and Its Patrons in Nineteenth-Century America* (Seattle: University of Washington Press, 1970), especially chapter 2.

7. The Massachusetts optical firm of Alvan Clark & Sons, traditional instrument makers, had ground the lenses for the 36-inch refractor.

8. Various references to the work of Warner & Swasey in building telescopes have often accepted Swasey's comments in the twentieth century that he and his partner had built these instruments at cost or at a loss to aid astronomy. While the work of Warner & Swasey in this century did develop into a quasi-philanthropic advertising scheme, their nineteenth-century work on Lick, the Naval Observatory, and Yerkes clearly made a profit. Figures for the early twentieth-century work on the large reflector for the government of Canada are not yet clear.

9. The retirement of the two founders about 1910 and the turning over of the management of the company to a professional management staff also played a significant role in the decision to phase out instrument making.

10. Various letters from the superintendent of the U.S. Naval Observatory to Warner & Swasey, from February 1886 to May 1886, Warner & Swasey Collection, Box 1, Warner & Swasey Papers, Special Collections, Freiberger Library, Case Western Reserve University, Cleveland (hereafter, W&S Papers).

11. Superintendent R. L. Phythian to Warner & Swasey, 12 April 1888, W&S Papers.

12. Typewritten copy of the contract in the archives of the Naval Observatory, Record Group 78, National Archives (hereafter, NO Archives).

13. Warner & Swasey to Captain T. V. McNair, superintendent of the Naval Observatory, 18 October 1890, NO Archives.

14. Copies of the printed items from the General Printing Office can be found in the W&S Papers.

15. Alvan Clark & Sons to B. F. Tracy, secretary of the navy, 8 June 1891, NO Archives.

16. Warner to Swasey, 16 June 1891, W&S Papers.

17. Telegrams from Warner to Swasey, 23 June 1891, W&S Papers. One suggests that they expose Saegmuller's coast survey discharge and also that the drawings be compared to show the clear forgery. Warner suggests that the form of the spiral staircase on the columns be compared.

18. Ibid.

19. Warner to Swasey, 27 June 1891, contains a transcription of the letter sent by McKinley to Secretary of the Navy Tracy. W&S Papers.

20. Unsigned report to Captain McNair, 30 November 1891. The handwriting in this report matches that of Harkness, according to astronomer Steve Dick of the Naval Observatory, who has some familiarity with its history and some of the documents still stored at the observatory, of which this is one. This is not part of Record Group 78 at the National Archives.

21. The classic reference is, of course, Nathan Rosenberg, "Technological Change in the Machine Tool Industry, 1840–1910," *Journal of Economic History,* 23 (4 December 1963).

The Rise of Industrial Research in Cleveland, 1870–1930

DARWIN H. STAPLETON

The rise and development of industrial research is an important topic in the history of science and technology because it underpins so many of the industries and institutions important to the world economy of our era, notably the electrical and chemical industries. The union of the methods and theories of science with the tools and goals of modern technology may be traced back at least to the seventeenth century in military institutions, and to the eighteenth century in the chemical industry and related fields.[1] But only in the latter half of the nineteenth century did that union bear regular fruit, and scholars generally identify 1870 to 1930 as the era when industrial research took shape.[2]

Studies of industrial research in the United States have focused on the giant corporate laboratories, such as AT&T, General Electric, Eastman-Kodak, and General Motors, or on Thomas Edison's Menlo Park and West Orange establishments.[3] Even surveys of the field tend to take an institutional approach.[4] This paper, in contrast, will consider industrial research as a phenomenon of a particular urban-industrial region (Cleveland, Ohio) and as the product of a network of individuals and institutions. While many of the commonly identified patterns of American industrial research are easily discovered in the Cleveland experience, this approach highlights less frequently considered aspects, such as researchers' relationships to company officers and to higher education.

During the era when industrial research in America took shape, Cleveland grew from a city of 93,000 people in 1870 to 900,429 in 1930 (when it was ranked as the sixth largest city in the United States).[5] Cleveland's rapid growth is usually attributed to its combination of excellent rail and water transportation, and its central location for assembling the ingredients of the iron and steel industry. Yet some portion of the city's rapid industrialization must be credited to the early creation of an industrial research tradition, which brought Cleveland to the forefront of the electrical and chemical industries, manufacturing sectors in which it had no initial geographic or commercial advantage.

The rapid development of the electrical and chemical industries shows clearly in the statistics of the federal manufacturing census. The case of the electrical industry is particularly dramatic. Between 1890 and the late 1920s the value of electrical products in Cleveland jumped more than a hundredfold, and by 1927 they were the fifth-largest category of manufactured goods in the city.[6] An admittedly partisan review of Cleveland's industries in 1930 concluded that the city was a world leader in the manufacture of electrical motors, electrical railway cars, electrical home appliances, arc-welding equipment, and lighting fixtures and bulbs. In chemicals Cleveland was called "the greatest paint, varnish and lacquer manufacturing center in the country," and it had a diverse production of industrial chemicals. Combined with the Akron district, it comprised the greatest concentration of rubber manufacturing in the world.[7]

Cleveland's first steps into industrial research were taken during the Edison era, when independent professional inventors and consultants were leaders in many sectors of American industry. Outstanding examples of Cleveland's professional inventors and consultants include Elisha Gray, Charles Brush, and Elmer Sperry in the electrical industry, and Herman Frasch and a group of college professors in the chemical industry. Examination of their careers demonstrates the role of the region in their achievements. The outstanding case is that of Charles Brush, but the others were also major figures in the heroic age of industrial research.

Elisha Gray, for example, may have been the earliest professional inventor to set up shop in Cleveland. He had attended Oberlin, Ohio, preparatory school and Oberlin College (1857–61) and was inspired by the teaching of Professor Charles Churchill to study electricity. In 1867 Gray filed a patent application for a new telegraphic relay.

Telegraphy was the primary segment of the nascent electrical industry, so Gray was well-advised to apply his skills there. Moreover, since Western Union had consolidated telegraph lines in the United States in the late 1850s, one of the company's centers of business had been in nearby Cleveland. Accordingly Gray went there to show his new relay to Jeptha Wade and General Anson Stager, president and superintendent of Western Union respectively.[8] Wade had moved to Cleveland in 1856 and became a major commercial and industrial leader in the city.[9]

Not only did Wade and Stager approve Gray's device and adopt it for their company's use, but they were so impressed with him that they began a regular relationship. For example, they invested in a new telegraphic printer designed by Gray when it was still a mental image—not even a model. In 1869 Gray moved to Cleveland and established a partnership with Enos Barton, who had been employed by Western Union as an examiner and tester of telegraphic inventions that came to the company's attention. For the next two years the Cleveland partnership grew on the strength of Gray's inventive skill and the patronage of Western Union. Shortly after Gray and Barton moved to Chicago in 1871 their company was transformed into Western Electric, the manufacturing and development arm of the telegraph company.[10]

Like Gray, Charles F. Brush began his career as a professional electrical inventor in association with the telegraph equipment industry in Cleveland. Yet his research and development took a different direction. Brush was born on a farm in Euclid, Ohio, and was educated at Shaw Academy and then Cleveland's Central High School. At the latter institution he showed a deep interest in the chemical and physical laboratory, perhaps stimulated by his teachers or preconditioned by his father's technical expertise in woolen manufacturing.

Brush continued to develop his scientific and technical interests by attending the University of Michigan from 1867 to 1869, receiving a degree in mining engineering. He then returned to Cleveland and set up a business as an analytic and consulting chemist, no doubt an occupation that made good use of his academic training. Since Cleveland was rapidly becoming a major center of the iron and steel industry and its businessmen were investing heavily in mines on the upper Great Lakes, Brush probably expected to provide advice on metallurgical and mineral samples. That business proved modest, however, and in 1873 Brush attempted to make more direct use of his knowledge by forming a partnership with Charles Bingham to market ore and pig iron. Through that enterprise Brush must have come to the attention of many figures in Cleveland's business community.[11]

While Brush's public activities at this time were associated with mining and metallurgy, it appears that his passion was for electricity. Brush later claimed that from his boyhood in the early 1860s he had experimented with batteries, induction coils and carbon arcs; and he probably kept up with changes in the electrical industry through his engineering education and subsequent private reading and experimentation.[12]

Thus when Brush met George W. Stockly, vice-president and general manager of the Telegraph Supply Company of Cleveland, he was able to share a persuasive vision of new directions for electricity. As early as 1873 Stockley asked Brush to serve as a consultant to the company (probably in a role similar to Barton's with Western Union) and soon came to respect Brush's ability. In 1876 Telegraph Supply signed its first research and development contract with Brush.[13]

The Telegraph Supply Company was founded in 1872 by G. B. Hicks, inventor of a telegraphic device, and it manufactured electric bells, fire and burglar alarms, and batteries, as well as telegraphic instruments. The firm's president was Mortimer D. Leggett, former U.S. Commissioner of Patents, who now headed a group of patent attorneys. Although the company was small, it had officers who knew the state of the industry and workmen with practical knowledge of the technology.[14]

Brush was interested in developing dynamos that would be more powerful and more reliable sources of electricity than existing storage batteries (although he foresaw an electrical system composed of both elements). Moreover, he visualized their use in electroplating and arc lighting, recognizing that these uses would determine the necessary characteristics of his power sources. Brush's research therefore considered the production, distribution and consumption of electricity, much as did Thomas Edison's research when he focused his attention on incandescent lighting two years later.[15]

Brush's work began with the dynamo. At first his research facilities were makeshift arrangements in his own apartment and on his parent's farm, but after his experimental dynamo built in the summer of 1876 was successful, he was given a laboratory at the Telegraph Supply Company's shop. There he developed his dynamo and experimented with arc-lighting devices. Brush's crucial arc-lighting inventions were a process for making cheap, high-quality arc carbons; copper coatings for carbons; an automatic feeding device to regulate the length of the electric arc between carbons, and a shunt to control the effect of a burned-out carbon in a circuit.[16]

Brush's research methodology was generally like that of other professional inventors of his time. He kept notes to verify his experiments and ideas, a procedure undoubtedly encouraged by Leggett, who knew that patent disputes often hinged on the evidence of laboratory records.[17] He initiated contacts with university professors interested in electrical engineering and sold or donated dynamos and other apparatus to them, receiving critical feedback on their performance. One of his correspondents was Professor George F. Barker of the University of Pennsylvania, who shortly received similar attention from Edison.[18] Brush also mixed his research methodology and theoretical knowledge with a knack for assembling his own experimental devices and working models, so that he had an immediate understanding of the practical successes and failures of his inventions.[19]

One area in which Brush seems significantly different than other inventors of his stature was his insistence on working alone. Throughout his career he seldom had an assistant in his laboratory, and when he was not present, the door to it was locked. His approach to inventing was in part an extension of Brush's lifestyle which, though not reclusive, was certainly very private. A contemporary stated that Brush was "in no sense a popular man; nor can it be said that he . . . ever sought or won a personal following."[20] But Brush's style also harked back to an earlier era when an unsophisticated patent law and legal system and an easily entered marketplace made pirating of inventions common and lucrative.

Brush's style of research and invention stands in marked contrast to electrical inventors like Edison, Elihu Thomson, and Elmer Sperry, who gathered around them talented assistants to develop, modify, and critique their ideas. Brush's method of research limited his inventive period to those six or seven exciting years of the late 1870s and early 1880s when the electrical industry broke loose from the battery-telegraph paradigm and moved into the era of the central-station dynamo supplying electric light and power. After solving the initial problems imaginatively, Brush was unable to bring sufficient resources to bear on the myriad secondary problems that the new technology encountered. The Brush system's promising combination of dynamos and batteries for generation was, for example, an early attempt to economize on the expense of dynamo operation, but received little of Brush's attention after its initial formulation. The Brush system was an early casualty of the electrical industry.[21]

Brush's arc-lighting system was nonetheless the pioneering central-station electrical technology in the United States. Its prize-winning demonstration at the Franklin Institute in 1878 and successful installation on Public Square in Cleveland in 1879 brought world-wide publicity to Brush, and a flood of orders (to the Telegraph Supply Company) for arc-lighting systems for street lighting. With its business revolutionized, the company was reorganized (and apparently refinanced) by a group of Cleveland capitalists and its name changed to the Brush Electric Company.[22]

In 1880 and 1881 a new manufacturing plant was constructed with a special laboratory for Brush. There he continued his largely solitary research, completing his work on storage batteries, for which he was granted important patents in 1881, and on various elements of the Brush arc-lighting system. But by the mid-1880s the heady days of major inventions and rapid entrepreneurial success were past.[23] Brush was already in the initial years of a fifteen-year span when he influenced and encouraged several other electrical innovators.

The clearest line of Brush influence runs through the founding of the National Carbon Company, an ancestor company of Union Carbide. Brush was a pioneer manufacturer of arc-light carbons, developing in 1877 and 1878 a process for copper-coating the carbons, simultaneously strengthening them and making them better conductors.

In 1881 one of Brush's foremen, William H. Bolton, formed a company to manufacture carbons by the Brush process and in close association with the Brush Electric Company. In 1886 (probably because of serious price competition in the industry) the company was reorganized as the National Carbon Company, with new investors and management. The new president was Washington H. Lawrence, one of the original partners in the Telegraph Supply Company and a co-experimenter with Brush on some elements of the carbon-arc problem.

While Bolton apparently neglected to improve his product in response to the rapidly growing arc-lighting industry, Lawrence had absorbed fully Brush's outlook, and he made research and development a keystone of the company's strategy. When National Carbon built a new manufacturing works in 1892–94 a laboratory was included. Subsequently the company branched out into the manufacture of carbon items for batteries, telephones, and dynamos.[24]

Significantly, under Lawrence's tenure as president (1886–1900), research followed Brush's private style. In 1889 the company hired Clarence M. Barber to conduct research on the arc-carbon process, especially copper-plating. He carried on his work at the company laboratory "to which, by consent of the officers of the company, only Mr. Lawrence, Mr. Barber and such workmen as were employed in the room had access."[25]

We know less about the research style of other enterprises stimulated by Brush. Two very successful Cleveland electrical firms, Brown Hoist (mechanical ship unloaders, established 1881) and Lincoln Electric (arc welders, 1895) were founded by former Brush employees but may not have carried on regular programs of research in their early years.[26] On the other hand, pioneering development of an electrolytic furnace by the Cowles brothers and early work on the Bentley-Knight electric streetcar (the first commercially feasible electric streetcar in the United States), were conducted at the Brush Company's shops from 1883 to 1885.[27] It seems highly likely that Brush was a consultant to both enterprises.

In 1889 Brush sold his stock in the Brush Electric Company to the Thomson-Houston Company and retired from the electrical business. He retained a deep interest in electricity and turned his energies to the construction of a huge windmill to generate electricity for his Cleveland residence. In 1895 his address to a meeting of electric manufacturers in Cleveland was regarded as a rare and special occasion.[28]

Brush's achievements led directly to Elmer Sperry's decision to move to Cleveland. A syndicate of Clevelanders who had invested in Brush and National Carbon took an interest in Sperry's invention of arc-lighting and electric streetcar devices at his shop in Chicago.[29] In 1890, perhaps in recognition that Brush was no longer at the forefront of electrical technology and that the electric streetcar industry seemed on the verge of explosive growth, the Cleveland syndicate financed Sperry's streetcar work in return for patent rights. By 1892 his development work seemed promising enough to go into production, and the Thomson-Houston Company bought the rights from the Cleveland syndicate and set up the Sperry Electric Railway Company. The Company's headquarters were at the Brush Electric plant in Cleveland. Within a few months Thomson-Houston became part of the new General Electric Corporation, and in 1894 Washington H. Lawrence became president of the Sperry Company (concurrent with his presidencies of National Carbon and Brush Electric).[30]

Sperry lived in Cleveland from 1893 to 1905, working in part as a consultant to the company that bore his name as well as to General Electric. Unlike Brush, he chose to remain in the consultant role or to develop patentable inventions for sale, rather than become enmeshed in a bureaucracy. He was constantly searching for fields of electrical technology that were on the verge of birth and growth, where he could solve bottleneck problems.

Cleveland was a fertile field for Sperry. When his attentions turned to automobiles in 1895, he found that his Cleveland syndicate was willing to support him. Like other early automobile entrepreneurs in Cleveland (such as Winton, White, and Baker) he apparently found no difficulty having bodies, wheels and accessories made in the burgeoning industrial center, and he concentrated on the power plant and transmission. At first he worked with gasoline motors, but then returned to his first love, electricity.

Sperry took space in the Brush plant, hired a former Brush superintendant to direct the work, and built six electric vehicles. The Sperry automobile was a technical success, and its patent rights were purchased by the Cleveland Machine Screw Company. The automobile enterprise gave Sperry the expertise to develop small gasoline engines and to patent several aspects of electric storage batteries. Each of those avenues led to the creation of a company to exploit the Sperry innovations.

Sperry's last years in Cleveland were spent developing electrochemical cells for producing a variety of industrial chemicals. Although he worked largely in cooperation with an inventor based in Washington, D.C., at a crucial point Sperry obtained assistance from F. J. Burwell, a chemist at National Carbon in Cleveland. Eventually Sperry's electrochemical patent rights were sold to the founder of Hooker Chemical, and a plant was built at Niagara Falls.[31]

Sperry's departure from Cleveland in 1905 symbolized the end of the era of the professional inventor and beginning of the era of institutional research. While the professional inventor set his own research agendas and generally had direct monetary investment in his patents and manufacturing enterprises, the new breed of researcher was salaried and accepted the direction of an employer.

That shift was foreshadowed in Cleveland by the consulting work of professors at Western Reserve University and Case School of Applied Science. In the 1880s Edward Morley of Western Reserve found his chemical talents in demand by gas, petroleum, and iron companies. Charles Mabery of Case worked with the Cowles brothers on their electrical furnace about 1886 (after their experiments at Brush Electric), then turned to research on petroleum. Case professor Albert Smith trained student Herbert H. Dow, who founded Dow Chemical in 1897, and Smith remained Dow's technical adviser for years. Smith and several other Case faculty carried out important research on electrochemical problems for Dow, and in return he gave funds to furnish an "electro-metallurgical laboratory" at the school in 1906.

Case attempted to institutionalize such consulting relationships in 1910 and 1911 when it established an industrial fellowship program, permitting sustained corporate research in its laboratories. W. A. Harshaw of Harshaw Chemical (Cleveland) created the first fellowship that supported a researcher (expected to be a recent Case graduate) for a year and stipulated that both the school and the fellow would receive payment from the donor for any useful discoveries.[32] This pattern of consultation and research for industry continued, and Case has been listed retrospectively as one of the nineteen American institutions of higher learning that regularly conducted industrial research prior to 1930.[33]

It was the corporate creation of positions and laboratories for industrial research that clearly marked the foundation of institutionalized research. The

Cleveland chemical industry seems to have been a leader in seeking out trained researchers to hire for staff positions. For example, the Sherwin-Williams Company (paint) employed its first chemist, Percy Neyman (a graduate of MIT), in 1884. Eight years later it added E. C. Holton (also an MIT chemistry graduate) to the staff.[34] It seems likely that other large chemical concerns in Cleveland, such as Glidden (paint and varnish), Harshaw Chemical, and Grasselli Chemical, also had chemists by 1900.

General histories of the petroleum industry have often stated that Standard Oil began the first sustained research program in petroleum chemistry by hiring Herman Frasch in the 1880s to work on the problem of sulfur in Ohio and Indiana crude oil.[35] The story is somewhat more complex. Frasch acquired a German education in the pharmaceutical chemistry, emigrated to the United States in 1868, and began work as a pharmacist and chemical consultant in Philadelphia. After developing a new process for making wax from petroleum, he was hired in 1877 by Merriam and Morgan, a Cleveland refining company that was part of the Rockefeller combine.

In Cleveland Frasch continued to work on petroleum refining technology, but he also did research on the manufacture of white lead, alkali, salt, and soda. After an interval of two years in the Canadian oil fields around London, Ontario, Frasch returned to Cleveland in 1886 to work for Standard Oil on desulfurization. In a laboratory and refinery he established on Kinsman Road, he came up with an effective refining process and patented it in 1887. Frasch then assembled a team of professional chemists to solve the problems inevitably encountered in scaling up laboratory apparatus for industrial production. One member of his staff was William H. Burton, a graduate of Western Reserve University with a Ph.D. from Johns Hopkins. Burton subsequently set up a laboratory at Standard Oil's Whiting, Indiana, refinery in 1890, the first permanently staffed research center in the petroleum industry. In 1888 Frasch left Cleveland to work in the Louisiana sulfur industry, and his Cleveland laboratory staff apparently broke up.[36]

The first permanent industrial research laboratories were established in Cleveland about 1900. Perhaps because Washington Lawrence's death in 1900 left the company without an advocate for personal research in the Brush mode, National Carbon created a formal laboratory shortly thereafter. Early research there specialized in carbon brushes for motors and in dry cells. National Carbon hired a series of university-trained chemists, but for years retained a policy of secrecy reminiscent of the Brush/Lawrence era.

One staff member recalled the 1910s as a time when "there still existed rules and regulations prohibiting technical men talking with one another about their work. Each technical man was supplied with the names of other men in the organization with whom he was allowed to talk." These arrangements were resented by the academically trained men who made up more and more of the research staff at National Carbon—a staff that already numbered 91 in 1910.[37] Yet National Carbon's management found it difficult to relinquish the direction of research to these scientific and technical upstarts. A memorandum written by a staff chemist in 1915 centered on the problem:

> We feel continually that we are not taken into the confidence of the management, that we are asked to spend our time on trivial matters when it could be better spent on more important ones, that our advice is not asked for nor desired on questions we should be able to consider intelligently, that some in management look on the laboratory as an expensive (and even somewhat dangerous) toy and that we, both collectively and as individuals, can't be trusted out of the sight of our superiors.[38]

To whatever extent this compaint was valid, the laboratory remained a cornerstone of National Carbon's strategy of technical development. A study of American research laboratories in 1917 listed it as one of the twenty prominent laboratories in the nation. Its research program was then focused on the improvement of arc-carbons, on lighting technology generally, and on quality control of the company's products.[39]

The largest of the permanent industrial research laboratories in Cleveland was Nela Park. The National Electric Lamp Association (NELA) was created in 1901 to stabilize the production of lightbulbs by the independent producers in competition with General Electric. GE took most of the new association's stock, but initially allowed the association to go its own way.

Within two years NELA purchased the former Brush Electric works in Cleveland and established headquarters there, including a development laboratory. The laboratory was put on a regular basis by Edward P. Hyde, appointed director in 1908 after receiving a Ph.D. at Johns Hopkins and working at the Bureau of Standards as a physicist. Hyde's own interests were theoretical, and he lost no time in setting up a purely scientific branch of the laboratory to conduct research on the emission, diffusion, and physiological effects of light. But overall he appears to have followed policies similar to those of Willis Whitney, director of the GE laboratory at Schnectady: these allowed publication of research in purely scientific areas, insisted on the keeping of regular records, and emphasized, in the end, applied research.[40]

Hyde was unhappy with the location of the laboratory in an industrial district where "smoke, gaseous fumes, [and] mechanical and electrical disturbances" intruded on experiments. NELA's leaders accepted his views and in 1910 selected a new site in the adjacent suburb of East Cleveland. Land purchases from 1911 to 1913 totaled nearly 90 acres, and an architect drew up plans for a research park with a grouping of classical buildings in a wooded setting. In the meantime GE had acquired all of the NELA stock, but it chose to complete the plans and designated Nela Park the National Quality Lamp Division of General Electric. It became second only to GE's laboratories at Schenectady in the company's research program, with a budget 60 percent as large as Schenectady's in 1919 and 75 percent of Schenectady's in 1933.[41]

In the first twenty years after its founding, Nela Park assumed the form of the modern research laboratory, with efforts divided into three fields: basic research into phenomena associated with the electrical industry; applied research on the boundary of theoretical knowledge and technology; and developmental work focusing on the scaling up of new processes or trouble-shooting current production problems. In the area of basic research, NELA was able to recruit academically trained physicists, physiologists, and psychologists to do research on topics as wide-ranging as the physiology of the senses, infared absorption spectra, and the phosphorescence of the firefly. Laboratory director Hyde encouraged publication, and in 1913 NELA began issuing an annual abstract of his staff's journal articles. A few of the staff held adjunct teaching appointments at Western Reserve University or Case School of Applied Science. The academic nature of the research was heightened by the creation of the Charles F. Brush Fellowships in physics, which allowed professors on leave from institutions throughout the United States to use the Nela Park laboratories for several months at a time.[42]

The laboratories' applied and developmental work dealt with electric illumination. At the applied level was the work of scientists like Aladar Pacz, who through "almost interminable experiments" developed "a commercially practicable non-sag [tungsten] wire for lightbulb filaments"; and Carl Kenty, who studied "gaseous discharge light sources" such as neon lights.[43] At the developmental level was an extensive glass technology laboratory that employed engineers and experienced glassblowers, an engineering department that studied the performance and characteristics of production-model incandescent lamps, and a lamp development laboratory devoted to solving both current production problems and those of scaling up. By 1930 Nela Park employed over 400, about 90% of whom worked in the applied and development fields.[44]

As these sketches of National Carbon and NELA indicate, Cleveland's industrial research tradition remained strong in the first three decades of the twentieth century. The list of industrial research laboratories collected by the National Research Council in 1920, 1927, and 1930 show that the number of laboratories was growing rapidly and that Cleveland was one of the nation's centers of research, by 1930 ranking as the fifth largest urban concentration (with 43 laboratories). This was slightly ahead of its ranking as a manufacturing center.[45]

By the 1920s a number of major Cleveland corporations had established laboratories of ten to fifty employees, well below the size of Nela Park and National Carbon, but still representing major commitments of capital. The largest of these units typically conducted some research into new products as well as into the means of improving their products. Grasselli Chemical Company, for example, was fundamentally a producer of acids and heavy chemicals, but also developed a business in pigments and rubber accelerators used in the vulcanization of rubber. Even Glidden's laboratory, with a staff of about 15 chemists, physicists, and engineers, could have a program of research as varied as synthetic gums, treated oils, varnishes, paints, enamels, stains, dry colors, insecticides, and nitrocellulose lacquers.[46]

Smaller corporate laboratories usually specialized in quality control and product design. The Cleveland Graphite Bronze Company—makers of bushings, bearings, and washers—had a staff of six studying the alloys used by the company, as well as new bearing designs. Lincoln Electric Company's nine researchers tested various metals for their arc-welding properties and continually improved the arc-welding electrodes and welders that were the company's product. The Industrial Rayon Corporation worked on improving the properties of its viscose rayon, but also sought to control the odors and fumes that escaped during manufacture.

Cleveland also had a number of consulting laboratories. The largest was the J. H. Herron Company, located in Cleveland's business district. It had a staff of 23, including chemists, engineers, and a physicist. The company reported to the National Research Council that one-third of its research involved "alloys, paints, heat insulating materials, cements, plasters, concrete, welding fluxes and concrete surfacing materials."[47] James H. Herron was a respected figure in Cleveland's technical community, having served as president of the Cleveland Engineering Society in 1917–18, and as general chairman of the committee that organized the society's fiftieth anniversary in 1930.[48]

Overall, Cleveland's laboratories reflected both the categories of industry that dominated the city and the national patterns of industrial research. For example, the city's 43 laboratories of 1930 were predominantly product research and development centers, with 23 in that category, while eleven combined research and development with basic research, and the remaining seven were largely testing and quality-control facilities. By industry, the largest group (12) was in the chemical and paint industry, six laboratories were metallurgical, and four each were in the electrical and machinery industries. Nationally, chemical, electrical and metallurgical firms were the earliest and greatest supporters of industrial research.[49] Perhaps historians should expect Cleveland to reflect these patterns, because it helped create them: indeed, Cleveland was an early leader in the rise of industrial research in America.

*   *   *

This study of the rise of industrial research in Cleveland points to at least two observations that surveys of American industrial research and studies of large corporate laboratories have not made. First, it was important in Cleveland for industrial research to have advocates or patrons who understood that it is a process with a substantial period between investment and payoff. In the early period (1870–1900) these individuals were either the inventor-researchers themselves or top officers in a company. They personally directed the course of research and thus carried the burden of creating new products and markets on their own shoulders. Without their personal leadership and initiative, research programs could falter. Later advocates or patrons (after 1900) were corporate leaders working through a bureaucracy. They institutionalized research to the point that it had a life of its own, and they relied upon trained specialists to carry out research that could determine the corporation's future. As the National Carbon laboratory engineer's letter (cited above) demonstrates, some corporate officers may have resented the creation of independent laboratories.[50]

A second observation is that virtually every aspect of industrial research in Cleveland was set in a national or even international context. As I have pointed out elsewhere, the industrialization of Cleveland was a phenomenon that drew heavily on streams of immigrants and technological knowledge from other areas.[51] In terms of industrial research, an obvious route of outside influence on Cleveland was through higher education. From Elisha Gray and Charles Brush to the staff of Nela Park, people with academic experience in science and engineering were usually critical in successful research programs. While Western Reserve University and Case School of Applied Science supplied some of those people, the local demand considerably exceeded local supply. Without any attempt to be exhaustive, I have noted in this essay men who came to Cleveland with training at Oberlin, Michigan, Johns Hopkins, and MIT.

Further research is needed to confirm the validity of these observations. But, as this study suggests, our understanding of industrial research in America will be considerably enriched by examining what happened in significant regional centers like Cleveland.

Notes

1. Anthony F. C. Wallace, *The Social Context of Innovation: Bureaucrats, Families, and Heroes in the Early Industrial Revolution, as Foreseen in Bacon's New Atlantis* (Princeton, N.J.: Princeton University Press, 1982), 21–61; Robert E. Schofield, "Josiah Wedgwood and the Technology of Glass Manufacturing," *Technology and Culture* 3 (1962): 285–97; Robert E. Schofield, "Josiah Wedgwood, Industrial Chemist," *Chymia* 5 (1959): 180–92; Robert E. Schofield, "Wedgwood the Technician," in *The Ninth Wedgwood Seminar, 1964* (New York: Metropolitan Museum of Art, 1971), 130–35.

2. John Rae, "The Application of Science to Industry," in Alexandra Oleson and John Voss, eds., *The Organization of Knowledge in Modern America, 1860–1920* (Baltimore: Johns Hopkins University Press, 1979), 249–268; Kendall Birr, "Industrial Research Laboratories," in Nathan Reingold, ed., *The Sciences in the American Context: New Perspectives* (Washington: Smithsonian Institution Press, 1979), 193–207, especially 194; W. David Lewis, "Industrial Research and Development," in Melvin Kranzberg and Carroll W. Pursell, Jr., eds., *Technology in Western Civilization* (New York: Oxford University Press, 1967), 2: 615–34; Kendall A. Birr, "Science in American Industry," in David D. Van Tassel and Michael G. Hall, eds., *Science and Society in the United States* (Homewood, Ill.: Dorsey Press, 1966), 43–62, 66–80.

3. Leonard S. Reich, "Industrial Research and the Pursuit of Corporate Security: the Early Years of Bell Labs," *Business History Review* 44 (1980): 504–29; Leonard S. Reich, *The Making of American Industrial Research: Science and Business at GE and Bell, 1876–1926* (Cambridge: Cambridge University Press, 1985); Lillian Hoddeson, "The Emergence of Basic Research in the Bell Telephone System, 1875–1915," *Technology and Culture* 22 (1981): 512–44; Reese V. Jenkins, *Images & Enterprise: Technology and the American Photographic Industry, 1839–1925* (Baltimore: Johns Hopkins University Press, 1975), 179–87, 300–318; Kendall Birr, *Pioneering in Industrial Research: The Story of the General Electric Laboratory* (Washington: Public Affairs Press, 1957); Stuart W. Leslie, *Boss Kettering* (New York: Columbia University Press, 1983);

George Wise, "A New Role for Professional Scientists in Industry: Industrial Research at General Electric, 1900–1916," *Technology and Culture* 21 (1980): 408–29; Robert Conot, *A Streak of Luck* (New York: Seaview Books, 1979); Matthew Josephson, *Edison* (New York: McGraw-Hill, 1959); Robert Friedel and Paul Israel, with Bernard S. Finn, *Edison's Electric Light: Biography of an Invention* (New Brunswick, N.J.: Rutgers University Press, 1986); David F. Noble, *America by Design: Science, Technology and the Rise of Corporate Capitalism* (New York: Alfred A. Knopf, 1977); David A. Hounshell and John K. Smith, *Science and Corporate Strategy: Du Pont R&D, 1902–1980* (Cambridge: Cambridge University Press, 1988).

4. E.g., Birr, "Industrial Research Laboratories."

5. William G. Rose, *Cleveland: The Making of a City* (Cleveland: World Publishing, 1950), 361, 697, 873; David D. Van Tassel and John J. Grabowski, eds., *The Encyclopedia of Cleveland History* (Bloomington, Ind.: Indiana University Press, 1987), xxxiii.

6. Department of the Interior, *Compendium of the Eleventh Census: 1890* (Washington, D.C., Government Printing Office, 1894), 782–83; Department of Commerce, Bureau of the Census, *Biennial Census of Manufactures, 1927* (Washington, D.C. Government Printing Office, 1927), 1432.

7. *The Golden Anniversary Book of the Cleveland Engineering Society* (Cleveland: Cleveland Engineering Society, 1930), 24, 27–28, 41, 49–50; see also van Tassel and Grabowski, eds., *Encyclopedia of Cleveland History*, 175–77, 369–71.

8. David A. Hounshell, "Elisha Gray and the Telephone: On the Disadvantages of Being an Expert," *Technology and Culture* 16 (1975): 134–36.

9. Rose, *Cleveland*, 261, 277, 322, 351, 352.

10. Hounshell, "Elisha Gray," 136–38.

11. Harry J. Eisenman III, "Charles F. Brush: Pioneer Innovator in Electrical Technology." (Ph.D. diss., Case Institute of Technology, 1967), 11–26; Mel Gorman, "Charles F. Brush and the First Public Electric Street Lighting System in America," *Ohio Historical Quarterly* 70 (1961): 129–32; *Electrical Engineer* 19 (20 February 1895): 152. Eisenman's dissertation is the only biography of Brush to exploit the Brush Papers, now held in the Special Collections of Freiberger Library, Case Western Reserve University, Cleveland, Ohio.

12. Eisenman. "Charles F. Brush," 12–15, 17–19, 24, 26.

13. Ibid., 26–28, 95–96; agreement, Charles F. Brush with Telegraph Supply Co., 7 June 1876, box 20, Brush Papers.

14. Eisenman, "Charles F. Brush," 43–44, 93–95; William R. Sutton, "Herman Frasch." (Ph.D. diss., Louisiana State University, 1984), 45; Harold C. Passer, *The Electrical Manufacturers, 1875–1900* (Cambridge: Harvard University Press, 1953), 14.

15. Eisenman, "Charles F. Brush," 27; Charles F. Brush, "The Arc Light," 1905, box 8, Brush Papers; Brush to J. W. Langley, 23 December 1876, box 4, Brush Papers; Thomas Parke Hughes, "Thomas Alva Edison and the Rise of Electricity," in Carroll W. Pursell, ed., *Technology in America: A History of Individuals and Ideas* (Cambridge: MIT Press, 1981), 121–23. The writer of a standard history of the electrical industry commented: "[Brush] combined in one person the commercial ability to envision where arc-lighting could be used successfully and the technical ability to invent an arc-lighting system that fitted his commercial vision." Passer, *The Electrical Manufacturers*, 67.

16. Gorman, "Charles F. Brush," 132–37; Passer, *The Electrical Manufacturers*, 14–18; C. T. Richmond, "History of the Carbon Industry in the United States," in *Proceedings of the Fifth Annual Salesmen's Convention* (Cleveland: National Carbon Co., 1911), 2–4; Eisenman, "Charles F. Brush," 59–69.

17. Some of Brush's notebooks and memoranda survive in the Brush Papers.

18. Brush to J. W. Langley, 23 December 1876; Brush to George F. Barker, 9 February 1877 and 29 June 1877; Brush to C. F. Brackett, 1 October 1877: all in box 4, Brush Papers; Edward W. Morley to father, 27 April 1883, Morley Papers, Case Western Reserve University Archives, Cleveland, Ohio; David A. Hounshell, "Edison and the Pure Science Ideal in 19th-Century America," *Science* 207 (8 February 1980): 612–15.

19. Eisenman, "Charles F. Brush," 43–47, 102–103.

20. Ibid., 46, 48, 82, 102, 154, 173; *Electrical Engineer* 19 (20 February 1895): 153.

21. Hughes, "Thomas Alva Edison," 119–22; Thomas Parke Hughes, *Elmer Sperry: Inventor and Engineer* (Baltimore: Johns Hopkins Press, 1971), 206–10; Richard H. Shallenberg, *Bottled Energy: Chemical Engineering and the Evolution of Chemical Energy Storage* (Philadelphia:

American Philosophical Society, 1982), 166–120. My comment on the limitations of Brush's research strategy is substantially the judgment of Brush's biographer Eisenman, "Charles F. Brush," 112–13.

22. Passer, *The Electrical Manufacturers*, 15–16, 18, 20.

23. Eisenman, "Charles F. Brush," 101–4; *The Industries of Cleveland* (Cleveland: Elstner Publishing Co., 1888), 202; Shallenberg, *Bottled Energy*, 53–56, 102, 201.

24. Richmond, "History of the Carbon Industry," 2–4; Passer, *The Electrical Manufacturers*, 58–60; *Electrical Engineer* 19 (20 February 1895): 155–57; *History of the Carbon Products Division, Union Carbide* (New York: Union Carbide, 1976), 26; copy of docket, Cuyahoga County Court of Common Pleas, 16 June 1886, historical files, Parma (Ohio) Technical Information Center, Union Carbide; *Electrical World* 15 (5 January 1895): 8. In 1885 Case Professor Albert A. Michelson twice visited the Bolton factory in conjunction with student research: 20 April and 4 May, William Meriam diary, Case Western Reserve University Archives.

25. *Clarence M. Barber vs. The National Carbon Company, et at.: Brief for the Complaintant, Circuit Court of the United States, Northern District of Ohio, Eastern Division*, ca. 1895, 8, 23–26, 77.

26. Rose, *Cleveland*, 437, 577–78; Daniel M. Bluestone, ed., *Cleveland: An Inventory of Historic Engineering and Industrial Sites* (Washington: Historic American Engineering Record, 1978), 11; *New York Times*, 15 January 1984.

27. Alfred Cowles, *The True Story of Aluminum* (Chicago: Henry Regnery Company, 1958), 32–34, 40; Passer, *The Electrical Manufacturers*, 225–26.

28. Eisenman, "Charles F. Brush," 118–20; *Electrical Engineer* 10 (24 December 1890): 646–47, 19 (20 February 1895): 166.

29. This included Myron T. Herrick, James Parmalee, and Webb C. Hayes. See T. Bentley Mott, *Myron T. Herrick, Friend of France: An Autobiographical Biography* (Garden City, N.Y.: Doubleday, Doran & Co., 1929), 92–93; Hughes, *Elmer Sperry*, 70.

30. Hughes, *Elmer Sperry*, 70–73; *Electrical Engineer* 19 (20 February 1895): 157.

31. Hughes, *Elmer Sperry*, 70–98, 328–31.

32. Edward Morley to father, 26 December 1883, 29 March 1884, 27 September 1885, 13 March 1887, 17 April 1887, 15 December 1887, 29 March 1889, Morley papers; Clarence H. Cramer, *Case Institute of Technology: A Centennial History, 1880–1980* (Cleveland: Case Western Reserve University, 1980), 25–26, 111–112; Trustees' Minutes, Case School of Applied Science, 15 May 1906, 10 January 1910, 3 October 1910, 20 February 1911, Case Western Reserve University Archives; Murray Campbell and Harrison Hatton, *Herbert H. Dow: Pioneer in Creative Chemistry* (New York: Appleton-Century-Crofts, 1951), 9, 10, 34, 38, 67, 68, 83. The beginning of the industrial fellowship system in the United States is discussed in Noble, *America by Design*, 122–124.

33. Harold Vagtborg, *Research and American Industrial Development: A Bicentennial Look at the Contributions of Applied R & D* (New York: Pergamon Press, 1976), 252–53.

34. Williams Haynes, *American Chemical Industry* (New York: Van Nostrand, 1949), 6: 385–86.

35. E.g., Ibid., 1: 250, 266; Rae, "The Application of Science to Industry," 261. See also Charles G. Moseley, "The Capitalist, the Chemist, and Lima Sour Crude Oil," *Journal of Chemical Education* 56 (1979): 657–58.

36. Sutton, "Herman Frasch," chapters 1–4; Charles G. Moseley, "Chemistry and the First Great Gasoline Shortage," *Journal of Chemical Education* 57 (1980): 288–89.

37. H. D. Batchelor to A. V. Wilker, 30 April 1938, historical files, Parma (Ohio) Technical Information Center; Edward Church Smith, "Reminiscences of Forty Years in the Battery Business," 28 June 1945, historical files, Parma (Ohio) Technical Information Center.

38. "What is Wrong with the Laboratory?" attached to D. W. Ordway to M. W. Allen, 1 December 1915, historical files, Parma (Ohio) Technical Information Center.

39. A. P. M. Fleming, *Industrial Research in the United States of America* (London: Department of Scientific and Industrial Research, 1917), 11–12.

40. Ibid., 13–14; Hollis L. Townsend, *A History of Nela Park, 1911–1957* (Cleveland: Plummer, [ca. 1959]), 7–12; "Edward Pechin Hyde," J. McKeen Cattel, ed., *American Men of Science: A Biographical Directory* (New York: Science Press, 1906); *Who's Who in America*. Vol. 12 (1922–1923), p. 1639; Wise, "A New Role for Professional Scientists," 422, 426–27, 429.

41. Townsend, *History of Nela Park*, 11–18; Birr, *Pioneering in Industrial Research*, 83.

42. Fleming, *Industrial Research,* 13–14; *Bulletin of the National Research Council* 1 (1919–21): 85–86, 60 (1927): 57, 81 (1931): 80–81; *Abstract-Bulletin of the Physical Laboratory of the National Electric Lamp Association* (1913 +); "E.Q. Adams," "Bentley Tiffany Barnes," "L. J. Buttolf," "Percy Wells Cobb," "Mary Adelaide Easley," "William Elmer Forsythe," "James DeGraff Tear," J. McKeen Cattell and Jacques Cattell, eds. *American Men of Science,* 5th ed. (New York: Science Press, 1933); E. L. Nichols, "Ernest Fox Nichols," *Biographical Memoirs, National Academy of Sciences* 12 (1929): 117–20.

43. Birr, *Pioneering in Industrial Research,* 43; *Nature* 215 (26 August 1967): 1009.

44. Fleming, *Industrial Research, 13–14; Bulletin of the Natonal Research Council* 1 (1919–21): 85–86; 60 (1927): 57; 81 (1931): 80–81.

45. Compiled from the *Bulletin of the National Research Council* 1 (1919–21), 60 (1927), 81 (1931); *Biennial Census of Manufactures, 1931* (Washington, D.C.: Government Printing Office, 1935), 34. The *Bulletin* indicates a rapid growth in the number of research laboratories in Cleveland, from six in 1920 to 43 in 1930. While the 1920s were a period of rapid growth in the number of American industrial research laboratories, the 1920 list was the NRC's first, and the *Bulletin's* editors noted its inadequacy. The actual rate of growth of Cleveland's industrial research establishments cannot be determined from the *Bulletin.* David Charles Mowery, "The Emergence and Growth of Industrial Research in American Manufacturing, 1899–1945." (Ph.D. diss., Stanford University, 1981), 51. For a brief history of industrial research in Cleveland, see *Encyclopedia of Cleveland History,* 953–54.

46. Much of this discussion and what follows is based upon the National Research Council's surveys of industrial research laboratories, cited in note 45. My remarks on the Grasselli Chemical Company's laboratory are based largely upon notes supplied to me by David A. Hounshell, who studied the condition of the Grasselli laboratory up to 1928, when it was acquired by Du Pont. (See note 3.)

47. *Bulletin of the National Research Council* 60 (1927): 54.

48. *The Golden Anniversary Book of the Cleveland Engineering Society,* 111, 115.

49. Mowery, "The Emergence and Growth of Industrial Research," 54, 104, 215.

50. Reich, in *The Making of American Industrial Research,* 255, points out that the value of industrial research was not always clear to corporate managers in the early twentieth century.

51. Darwin H. Stapleton, "The City Industrious: How Technology Transformed Cleveland," in Thomas F. Campbell and Edward M. Miggins, eds., *The Birth of Modern Cleveland, 1865–1930* (Cleveland: Case Western Reserve Historical Society, 1988), 71–95.

Part IV
Science, Technology, and Culture

From Local to National: Time Standardization as a Reflection of American Culture

ORVILLE BUTLER

Historians have portrayed the period of American history bracketed by the Civil War and World War I as a time when American cultural emphasis shifted from the local community to the national society. Most writers have traced this change through developments in the business community.[1] Although the transition from local to national business markets and concomitant change in corporate structure that characterize post–Civil War American business communities sometimes reflect, or even presage, trends in American culture, portraying the broader society through those developments can be misleading. Unlike the 1920s when President Coolidge asserted that the business of America was business, America in the late nineteenth century contained significant elements that were hostile to many business interests.[2] A few writers have found a transition from a local to a national perspective in American thought.[3] They have found issues that, after the Civil War, were initially perceived as local issues to be solved by local communities but by the beginning of the twentieth century had become national concerns.

In this paper I shall, as a historian, define culture as the characteristic features of a particular stage in the advancement of a civilization. One important feature in any culture is the manner in which the society is organized. The adoption and changing nature of time standards in America provide insights into the transition in American culture from an emphasis on the local community to the concept of a national society.

Time standardization may have been essential to developments in the business community, but it affected all aspects of American society. Businessmen used time to regulate the operation of their factories; contracts and insurance policies usually began and ended at noon; judges and elected officials generally began and ended their terms of office at midnight. Midnight and noon varied from community to community, and the development of common business practices remained difficult. The adoption of a system of time also affected everyday life. Schools began and ended their day by a set schedule; the weekly Sunday church service began, and sometimes ended, at a prearranged time. Even farm wives scheduled their daily activities around the "noon" lunch prepared for her husband and his laborers. The social notes and invitations of upper-class women reflected the importance of time in their daily lives.[4]

While timekeeping in the early nineteenth century, for the most part, was not precise, it emphasized the role of the local community. By the mid-nineteenth century many smaller communities began adopting the time of regional business centers. The development of a system of time standards by the railroads in 1883 did not necessitate their acceptance by society. Some business communities did not completely accept railroad time until the 1890s. Acceptance by the general public took even longer. Advocates of time standards in the 1870s presumed that the average person would continue to use local time. The standards would affect only commercial interests. For the remainder of the nineteenth century, court cases proved that assumption correct. Even attempts to legalize time standards focused on local or, at most, state legislation. National legislation, although occasionally proposed in the nineteenth century, did not receive serious consideration before World War I. The acceptance of standard time during World War I reflects the shift in American culture from a local to a national emphasis. While the push for time standards by scientific and commercial interests after the Civil War mirrors their broader interests, the slower adoption of standard time by popular society, as well as the lateness in the shift from local to national legislation of time standards, indicates that the development of a national perspective took longer than an analysis of business interests suggests.

The lack of uniform time standards prior to the Civil War reflects the dominance of the local community in antebellum American culture. Intercommunity travel was slow and arduous. Most travelers were required to wait at an inn for the next coach. Those with their own means of conveyance used sun time. While a local merchant's business contacts might extend beyond the community and lecturers and traveling ministers brought some semblance of the larger world into the community, the local village remained the focus of social and business concerns. With limited and generally slow interaction with the outside world, most people needed no more accurate measurement of time than was provided by sunrise, noon, and sunset.

The earliest American time standards, in fact, served the navigational needs of shipping interests. Navigators required accurate timepieces to determine longitude. A ship's chronometer would be set to the local time of Greenwich or, in some cases, Paris. The navigator could then determine his longitude by comparing the sun's position in the sky with the theoretical position it would have at the "prime meridian." Small variations in time could result in significant errors, so the accuracy of the chronometer would be checked at every available facility. By the mid-nineteenth century, packet ships operating on a fixed schedule between Europe and America, made accurate time measurements even more crucial.[5]

Railroad networks that developed in the 1840s and 1850s created another demand for precise time standards. Prior to the Civil War these networks served the hinterlands of business centers. The implementation of train schedules necessitated the observance of a common time, and most railroads adopted the time of the business center they served. As a result business communities also adopted "railroad" time.[6] American observatories, seeking a source of income, sold accurate time signals to the railroads for distribution along their lines.[7]

With the exception of New York and Washington, D.C., antebellum time standards primarily served railroad and shipping concerns. The Naval Observatory time ball, dropped by hand, lacked sufficient accuracy to be used by navigators in the 1840s, and lack of visibility from the river made its use as a source for ships' time unlikely.[8] In New York, Mayor Fernando Wood ordered the dissemination of public time in 1857 by the ringing of fire bells each evening at nine o'clock.[9] But in most cases time standards aided commercial interests in unifying business centers with their hinterlands served by a railroad network.

No proposals for a nationwide system of uniform time standards were put forward before the Civil War. The confusing variety of times one had to follow while traveling between commercial centers was resolved by comparing the sundry time with that of a well-known city. For instance, *Dinsmore's American Railroad and Steam Navigation Guide and Route-Book* for 1857 compared the times of principle cities at noon in Washington, D.C.[10] Travelers could convert their schedules to Washington time and then to the time of their local community. Even when Congress established the prime meridian of Washington, D.C., for geographic and astronomical purposes in 1851, it failed to specify its use for time regulation. Congress also deferred to the Greenwich meridian for navigation, thereby eliminating any inferred time keeping status to the Washington meridian.

By the Civil War, there were a variety of regional and local time standards. Most, however, influenced only commercial interests and travelers using the railroads. A few people, no doubt, indirectly adopted a local standard time under the influence of their business obligations. However, these local and regional standards rarely deviated significantly from the "sun time" by which most people appeared content to live. Even after the development of standard time proposals in the 1870s and 1880s few proponents believed they would affect the day-to-day activities of the average citizen. They assumed that beyond business and railroad concerns, people would continue to live by sun or local time. Even the American scientific community, which provided some of the strongest support for time standards after the Civil War, found it difficult to break with local time.

When General A. J. Myer organized the telegraphic service of the United States Weather Bureau in 1870, he required two sets of records for the tridaily weather maps: one kept according to the local time of the observer, the other according to Washington time. In 1873 the "Signal Service," as Myer's agency came to be known, utilized a similar plan for weather observations made by its international observers. By 1878 simultaneous records were required on all American naval vessels.[11]

Cleveland Abbe's research on the aurora borealis in 1874 forcefully exhibited the time problem to the scientific community. Various observers had recorded their data according to "local time," "railroad time," or "standard time." Abbe could not decode these various standards, especially when several railroads, each using different times, passed near an observing station. He turned to the American Metrological Society, suggesting that it study the time problem. The society's president, professor F. A. P. Barnard, in turn, appointed Abbe chairman of its committee on time standards.[12] The American Society of Civil Engineers and the American Association for the Advancement of Science formed similar committees about the same time.

While scientists usually advocated a national or international time, they could not agree on a prime meridian. In 1882 the Signal Service proposed sending standard time signals over its telegraph system, but could not get astronomers to agree on the standard. Naval Observatory astronomers claimed the meridian of the observatory in Washington, D.C., should be the prime meridian for the country; but other astronomers, who financed their research by selling time to local communities and railroad networks, objected to the Naval Observatory's preeminence, while all opposed the use of New York or one of the other commercial centers.[13] F. A. P. Barnard and the director of Russia's Pulkowa Observatory, Wilhelm Struve, suggested a meridian passing through the Bering Strait to avoid national prejudices. Time for the entire world would be measured from that point.[14]

Several scientists, including Benjamin Pierce, suggested dividing the world into twenty-four time zones and setting a prime meridian that would provide universal time for scientists and telegraphic networks. The twenty-four hour zones would differ from each other by exact hours. The American Society of Civil Engineers embraced these proposals at their annual meeting in 1882 and lobbied Congress for an international conference to resolve the problem of a prime meridian.[15]

The Washington International Meridian Conference held in 1884 brought together scientific representatives from more than twenty-five countries. In spite of objections from the French, the conference called for the adoption of the Greenwich meridian. It also proposed the unification of the astronomical and nautical days, the adoption of a universal day for scientific purposes, and called for further research leading to the eventual decimalization of time.[16]

While some proposals from the scientific community, for instance the prime meridian at Greenwich, were eventually adopted, most were ignored by the railroads when they established their time standards. Scientists continued to call for a universal day, the decimalization of time, and twenty-four hour notation throughout the remainder of the nineteenth century.[17] These proposals reflect the broader interests in the scientific community. While scientists during the post-Civil War period sought to standardize most measurements affecting scientific work, their international approach made little sense to business or to members of a community whose world contacts were even more limited than their interactions with other North American communities.[18] Most business and popular concerns of the 1880s were not yet of a national breadth. They could hardly be expected to support time standards requiring an international perspective.

Popular standard-time proposals in the 1870s and 1880s emphasized the local component of time standards. William Frederick Allen, editor of the *Official Railway Guide,* pointed out: "We are all of us more or less imbued with a feeling of local pride, and if the meridian time of the 'Hub of the Universe' is the standard by which the trains on our particular road are run, we feel like holding to it."[19] Only by reconciling standard time with the times by which local communities carried out their daily activities could reformers overcome objections to their proposals.

Science reported with approval a proposal to represent Barnard's "cosmic time" using roman figures while protraying local time in Arabic figures. Admiral John Rodgers, director of the Naval Observatory, suggested that a time based on the meridian of Washington, D.C., could be reconciled with local time by using two sets of hands such as traveling salesmen used in New England to keep track of New York and Boston times. Such clocks would use gilt colored hands for national time and black hands for local time.[20]

Other emphasized the evolution of regional business centers and the time standards of their hinterlands. Edward Holden, director of the University of Wisconsin's Washburn Observatory, suggested a series of "regional time zones" based on the times of the nation's largest business centers. New England would adopt New York's time, while other east coast states would use that of Washington, D.C. New Orleans would provide the standard for southern states while the Midwest would adhere to Chicago's time. Other standards could be developed in the west as business centers consolidated there. With these standards, Holden wrote, everyone would "be supplied with the time he requires most often, and the growth of local standard times will be on a solid basis of use, and not a forced one of an artificial system."[21]

A few advocated a national time based on a meridian in the Mississippi valley. Since nearly fifty percent of the American population lived between seventy five and one hundred five degrees west of Greenwich, proponents argued that such a national time would result in comparatively minor disruptions in day-to-day activities.[22] Others urged the adoption of the time of New York, Philadelphia, Pittsburgh, or Chicago. Most such advocates sought to gain support for their city in commercial conflicts with competing business centers. New York and Philadelphia competed for the business of the mid-Atlantic states, while Chicago and Pittsburgh contended with St. Louis, Columbus, and Cincinnati in their respective attempts to build postwar business empires in the Midwest. To have the time for the entire country based on their meridians would have increased their influence in the expanding national market place.[23]

While reformers sought a means of unifying time after the Civil War, they recognized the necessity of adapting those standards to the local needs of affected communities. The broader national interests of twentieth century America had not yet developed, even in much of the business community. Without support from local and regional interests, or at least the absence of animosity, standard time could not succeed.

However, the spread of a market economy tied to the railroads and a communication network developed through the telegraph provided the major nonlocal influences on community life after the Civil War. By the end of the war the integration of railroad networks began to serve transportation needs between business centers. The development of time zones did not keep up with the expanding railroad network.[24] During the 1870s the Chicago *Tribune* reported thirty-eight official time zones in Wisconsin, twenty-seven each in Michigan and Illinois, and twenty-three in Indiana.[25] While the Railroads did not use every local standard, as late as 1881 the *Railway Review* complained: "Every place of any size professes to fix its own meridian, and smaller places follow the nearest large city. Over a hundred of these irregular standards have received some such recognition."[26]

The time problem became more complicated at the more than three hundred points where two or more railroads connected. Each railroad used its own time, which often differed from that of the connecting city. As a result businessmen in these cities functioned by multiple times. Chicago operated on three conflicting times, while Pittsburgh had six different standards by which the traveler or businessman had to arrange his schedule.[27]

As the first national business in the United States, railroads became the focus of many early time standard proposals.[28] However, most railroad managers in the 1870s saw the coordination of time tables as more important than the development of standards. In 1872 several major northern railroads formed the Time-Table Convention to coordinate schedules on east–west lines. By 1875 the Time-Table Convention came to be known as the General Time Convention. The integration of north-south route schedules resulted in the similar development of the Southern Time Convention. However it was not until 1881, when several scientific societies urged the convention's consideration of their plans, that the convention seriously contemplated time standards. Even then, it referred the problem to Allen, who was to report his proposal at the following meeting.[29]

Like other advocates of time standards, Allen attempted to correlate his standards with existing regional time zones. A survey of times used by railroads in the eastern half of North America revealed two groups roughly centered around the meridians seventy-five and ninety degrees west of Greenwich. While the standards he proposed would be based on these meridians, they would not arbitrarily be fifteen degrees wide. Since the eastern and western boundaries of existing regional zones represented the edges of railroad networks as well as those of business centers and their hinterlands, they would, so far as possible, remain the boundaries of the new time zones.[30]

In accepting Allen's proposals, the General Time Convention emphasized the consolidation of regional time zones into larger standards rather than the development of new ones. Railroads using the regional standards of Boston, New York, Philadelphia, Baltimore, Toronto, Hamilton, or Washington, D.C., would adopt the time of the seventy-fifth meridian. Those following the times of Columbus, Atlanta, Cincinnati, Louisville, Indianapolis, Chicago, St. Louis, Jefferson City, St. Paul, or Kansas City would now set their time by that of the ninetieth meridian. Farther west, the new standards would center on the 105th and 120th meridians west of Greenwich.[31]

Meanwhile, H. S. Haines, general manager of the Savannah, Florida & Western Railroad, proposed the Southern Time Convention adopt a series of regional zones based on the times of Washington, D.C., Atlanta, and New Orleans,[32] whose standards dominated the timekeeping of southern commercial interests. After the General Time Convention finalized its approval of Allen's plan in early October, Haines withdrew his proposal from the Southern Time Convention. On 18 October 1883, it voted to accept Allen's standards.[33]

A month later on November 18, railroads covering 78,000 miles of track converted to Allen's standard time. Seventy of the nation's one hundred largest cities went along with the railroads. Allen and other supporters of standard time later wrote that the acceptance of the standards was universal and almost immediate.[34] However, evidence suggests the contrary, that the adoption of railroad time standards even by the business community was not universal prior to the 1890s.

Commercial interests actively opposed the new time standards when they appeared to threaten regional interests. Chicago almost destroyed the new plan when the city council, upset that "Chicago time" had not been adopted, refused, until the last moment, to abide by railroad time. The Illinois Central Railroad conditioned its acceptance of Allen's plan on that of the city. Newspapers asserted that no railroad with a large business in Chicago or its suburbs would adopt standard time if the city did not.[35] Only after Western Electric, which provided time signals to the city, announced that it would conform to the new standards did the city council agree to go along.[36]

Regional conflict between St. Louis and Chicago played an important role in Chicago's hesitation in adopting standard time. Several newspaper accounts had portrayed the new standards as the times of Philadelpia, St. Louis, and Denver.[37] St. Louis provided Chicago's major competition for midwestern commerce, and regional pride affected Chicago's unwillingness to adopt "St. Louis time."

Western railroads and communities hesitated in adopting the new standards for similar reasons. Laramie and Salt Lake City apparently delayed adoption of railroad time because of the emphasis given to "Denver time" where Denver competed for their hinterlands. Laramie, for instance, did not adopt standard time until the territorial legislature, which the railroads allegedly controlled, forced the change on 20 February 1884.[38]

Many cities along the new time-zone borders feared the division of their hinterlands by the new standards. After adopting railroad time, Louisville reverted to sun time to serve local banking and insurance interests. Detroit, Michigan; Wheeling, West Virginia; and Pittsburgh, Pennsylvania all remained on local time. At least parts of their hinterlands—Port Huron, Michigan; Courtwright and Windsor, Ontario; Benwood, West Virginia; and Washington, Pennsylvania—followed their lead. Columbus and Cincinnati both refused to adopt the new standard and several Ohio communities that had previously followed Columbus time reverted to local time.[39]

Many of the business centers, between eastern and central time zones, had provided regional time standards to surrounding communities. The adoption of railroad standards meant a significant change for these cities, whose local time varied from the standard by nearly thirty minutes, and denied them a powerful cohesive force over their hinterlands. Some no doubt feared the new standards would split the business of the surrounding communities between their competitors to the east and west. Local concerns, then, influenced the acceptance of standard time by business and community interests in the 1880s.

Eleven states enacted standard time legislation in the nineteenth century.[40] Most contained well-developed railroad networks and may have passed the laws in response to railroad pressure. However, state legislation did not guarantee local observance. Allen convinced Belfast, Maine, to adopt standard time shortly before the state imposed it in 1887, but other towns in the state—Bath and Bangor—voted to return to local time in spite of state law.[41] Detroit alternated between standard and sun time well into the twentieth century, in spite of Michigan state law.[42] That cities in Maine and industrial centers as big as Detroit legislated local time in opposition to state laws suggests that even in legal matters, local interests in the 1880s still maintained a powerful influence.

By the 1890s most major commercial centers had adopted railroad standard time. In 1887 Allen coaxed Pittsburgh into dropping local time and by 1889 he had developed support for railroad time among superintendents of education in several Ohio communities.[43] Only twenty–five of the more than two hundred cities larger than ten thousand remained on local time by 1890.[44] Both Columbus and Cincinnati finally adopted standard time and by 1893 the state of Ohio forced fifteen more cities to abide by railroad standards.[45] Outside of Ohio, Detroit; Louisville, Kentucky; Savannah, Georgia; and Augusta, Maine, remained on local time.[46] Louisville banking institutions used local time as late as 1905.[47] Detroit, while still vacillating between local and standard time after 1900, had become the laughing stock of the region.[48] By the mid–1890s such cities were exceptions rather than the rule. All major commercial railroads operated by standard time by 1890 and the growing importance of national markets made local time inconvenient to commercial interests.

The general acceptance of railroad standard time by city government and commercial interests by the mid-1890s reflects the growing national concerns of the business sector. The conflicts over standard time in the decade after 1883 mirrored both the battles between competing business centers and demonstrated the gulf between business interests and those of the local communities that found little need for time standards. Local society did not develop a national view that incorporated time standards until after the turn of the century. One such local concern involving time lay in determining time in legal matters.

Court cases determined legal time in states that did not legislate standard time. In December 1884 Massachusetts Supreme Court Justice Oliver Wendell Holmes, determined that legal time would be that time in general use in a community.[49] Since courts subsequently adopted Holmes's criteria, they provide a measure of the popular acceptance of standard time prior to national legislation. If that measure is accurate, standard time was not generally accepted prior to the twentieth century.

In 1889 the Georgia Supreme Court forced lower courts to operate by local rather than railroad time. The governor had ordered clocks set to standard time with the railroads in 1883,[50] and a judge had subsequently run his court by railroad standards. When this resulted in a Sunday judgment by local time, the state supreme court ruled that the judge could not substitute railroad time for the standard "recognized by the statutes of the State as well as the general law and usage of the country. . . . " The court acknowledged that cities and towns along the railroads used standard time. However, communities distant from the railroads would not usually adopt standard time. To expect everyone to adopt railroad standards, the court decided, would introduce uncertainty and confusion in the administration of the law; especially when sun time, used by a majority, was readily available.[51] The following year the Nebraska Supreme Court declared that it would presume common time—meaning local time—when no evidence indicated another standard was in general use.[52] In 1895 the Texas Court of Criminal Appeals ruled that Texas courts must operate by the sun time of their community.[53]

As late as 1899 the Iowa Supreme Court defended the legality of local time. The court upheld a jury in Creston that determined that even though the railroad, the schools, and businessmen in Creston followed standard time, the majority of the residents continued to follow sun time. Employing Holmes's criteria, the court concluded that local time remained the legal time in Iowa.[54] The explicit distinction between time used by the commercial sector and that used by most Americans indicates the separation of business interests from the rest of society. By the 1890s the business community had begun to emphasize its national characteristics. If time usage can be taken as a measure, most members of society continued to focus on local or, at best, regional communities.

Between 1900 and the beginning of World War I, only two additional states—Minnesota in 1901 and South Dakota in 1909—passed standard time laws.[55] But courts, using Holmes's criteria of general usage, increasingly accepted standard time. The Minnesota Supreme Court in 1989 ruled that railroad standard time had long been the sole standard for the state.[56] In 1905 the Kentucky Court of Appeals found popular acceptance of standard time in spite of the fact that banks in Louisville continued to operate by sun time. Society, the court asserted, had gradually abandoned sun time over the previous ten to fifteen years.[57] The North Dakota Supreme Court declared standard time to be the legal time of that state in 1906. Earlier rulings by Georgia, Nebraska, and Iowa courts in favor of local time were not relevant, the court decided, since by Holmes's criteria, standard time had not been adopted in those states at the time of those rulings.[58] Subsequent court cases determined that most people used standard time in Virginia (1907) and Utah (1911).[59] The Utah Supreme Court went so far as to suggest that the federal government had officially approved although not legally adopted standard time. Only Texas courts, relying on precedence rather than usage, continued to rule against standard time.[60]

Throughout the nineteenth century Congress considered time standards to be the responsibility of state and local government. In 1880 Representative J. Floyd King of Louisiana failed to get support for his proposal to place time balls in all ports of entry and major commercial cities.[61] Two years later King proposed placing time balls in all cities with a population larger than 15,000.[62] Naval Observatory astronomers, seeking to establish a national time, influenced the Commerce committee to report a substitute bill that would transmit Naval Observatory time to those cities. Opponents blocked a vote on the bill.[63] Congress chose to ignore the proposals of the Washington International Meridian Conference of 1884, even after President Cleveland requested their approval in 1888.[64] New York Representative Roswell P. Flower's attempt to legalize railroad standard time in 1891 died in committee, as did a bill introduced the following term by Roswell's successor, Joseph Little, and New York Senator Frank Hiscock.[65]

While Congress refused to legislate national time standards, it did move to resolve confusion over District of Columbia time. Just before the railroads instituted their new standards in 1883, Attorney General Benjamin Brewster ruled that no government institution could adopt them until so authorized by Congress.[66] The secretary of war used this decision to force the Signal Service to return to using local times; a practice it had abandoned in the 1870s.[67] The decision resulted in a dual time standard for Washington, where government offices operated by local time while businesses adopted the new railroad standards.[68]

Early in 1884, Senator Joseph Hawley of Connecticut hoped to remove the confusion by introducing legislation to adopt eastern standard time in the District of Columbia. Kentucky representative John D. White, who earlier supported the Naval Observatory's bid for a national time, proposed an amendment making eastern standard time the legal time for the entire country. "Every civilized country on the globe except the United States," he argued, had "a national standard of time." When his amendment was ruled out of order, White complained that his opponents lacked a sufficiently "national" perspective.[69] In spite of attempts by White and others to legislate a national time, most considered the matter a concern of the local community or the state.

While state and local governments remained the arbiters of legal time into the twentieth century, the federal government's involvement in time standards increased. In 1901 the Naval Observatory established a clock room with precise atmospheric controls, allowing for more exact timekeeping.[70] The chief of the Navy's Bureau of Equipment declared that Navy observatories in Washington, D.C., and Mare Island, California, provided the only sources for government standard time.[71] By 1905 Naval Observatory astronomers were again calling for federal legislation to "fix a legal standard of time everywhere" based on the Greenwich meridian.[72] However, it was not until 1909 that Massachusetts congressmen, Senator Henry Cabot Lodge and Representative Andrew Peters, brought standard time legislation before Congress again. Both bills died in committee.[73] Congress remained reluctant to usurp the states' constitutional prerogative to regulate time. Congressional failure to adopt time standards did not stop debate elsewhere. Several cities, including Detroit and Cleveland, which had refused to adopt standard time in the 1880s, now declared that their interests were more closely allied with those in the eastern time zone.[74] Detroit voted to abide by eastern standard time in 1908.[75] Other cities soon followed. Columbus, Ohio, formed a committee to advocate the adoption of eastern standard time in major cities as far west as Chicago.[76] Jacksonville, Florida, began agitating the General Time Committee of the American Railway Association to place their city in the eastern zone.[77]

Proposals proliferated for a single national, or variations on railroad standard time. While all the suggested standards had been presented in the 1870s and 1880s, new arguments in their favor emphasized their utility in unifying the country rather than relating the standards to local time. Supporters of a national time declared that it would eliminate the problems of midwestern communities forced to use both eastern and central standard times.[78] Advocates of adjustments to railroad time zones urged the federal government to legislate standards.[79]

With America's increasing involvement in Europe prior to World War I, proponents of daylight saving also called for federal legislation. Yet bills to establish a commission to study time standardization still failed to pass committee in 1916.[80] In the following term, astronomer Harold Jacoby urged Congress use its power to fix standards of weights and measures to regulate time. Senator Frank B. Kellogg of Minnesota suggested that time standards came under provisions regulating interstate commerce. Such legislation, he presumed, would be generally adopted by the public.[81] Standard time still failed to come to a vote before 1918.

The Standard Time and Daylight Saving Bill of 1918 reflects the culmination of a shift in American culture from a local to a national outlook. Like the railroad standards of the nineteenth century, it presumed only to regulate the business sector.[82] However, as the debate over the daylight saving provisions reveals, this assumption was no longer valid. The argument that the law regulated only interstate commerce carried little weight. Farmers outside of the South hired laborers who were often paid by the hour rather than by the day. Under daylight saving they expected to begin work before the fields dried each morning and to quit before dark. Housewives rose before the sun to get their husbands to work and their children to school. Workers complained of traveling to work in the dark and going to bed before sundown to obtain adequate rest. Utility companies objected to the effect daylight saving had on the consumption of electricity. All these complaints presumed that a change in time standards would affect people's everyday lives.[83]

After Congress repealed daylight saving, several communities and states instituted daylight saving acts. The Massachusetts legislature, for instance, enacted daylight saving in 1920.[84] New Hampshire, on the other hand, unified state timekeeping by legislating standard time in 1921.[85] Courts allowed daylight saving differing from standard time by one hour on condition that such standards would be limited to times not regulated by federal legislation.[86] However, the courts limited the debate to standard and daylight saving time. Even the Texas Supreme Court, which had ruled in favor of sun time as late as 1916, adopted standard time following national legislation. Although Congress had not technically legislated the standard of time "for all persons subect to the jurisdiction of the United States," the Texas court ruled that had been its practical effect.[87]

The development and eventual acceptance of time standards between the Civil War and World War I provide insights into the changing structure of American culture and suggest directions for further study. Time standards in the 1870s and 1880s remained a matter of local concern. They provided a cohesive force on the hinterlands of regional business centers and to a lesser extent on local and regional society. The formal acceptance of railroad standards by the 1890s mirrored business's growing national concerns. However, the divergence between "railroad standard time" and the "local time" used by many, if not most, Americans in the nineteenth century is reminiscent of the conflicts between populists and some elements of big business during that same period. Significantly many of those conflicts began to recede about the same time the general public came to accept time standards. Court cases suggest that prior to 1900 local time dominated general time keeping in America, while after the turn of the century its use was increasingly limited to those out of touch with the mainstream of American life.

The cultural shift from local to national concerns demonstrated in the acceptance of standard time can also be found in the shifting responsibility for time legislation. Standards could be unified only through national legislation; yet few people in the nineteenth century, even among the advocates of time standards, could escape their cultural constraints to advocate national legislation. Such legislation would make sense only from a cultural viewpoint allowing the perception of time as common to the entire country. The development and acceptance of standard time, then, reflected a change in the structure of American culture from one that was local and community-based to one that was increasingly national and even to which business conformed as well as most localities.

Notes

1. The literature emphasizing the business aspect of this transformation in American culture is enormous. Some works, arbitrarily chosen, include Frederick Lewis Allen, *The Big Change: America Transforms Itself 1900–1950* (New York: Harper & Row, 1952); Howard Mumford Jones, *The Age of Energy: Varieties of American Experience, 1865–1915* (New York: Viking Press, 1973); and Walter LaFeber and Richard Polenberg, *The American Century: A History of the United States Since the 1890s,* 2d ed. (New York: John Wiley and Sons, 1979).

2. The Grange, the Populist Movement, and the Social Gospel, for instance, were influential in significant sectors of American culture. On these movements, see Dennis Sven Nordin, *Rich Harvest: A History of the Grange, 1867–1900* (Jackson: University of Mississippi Press, 1974); John Donald Hicks, *The Populist Revolt, A History of the Farmer's Alliance and the People's Party* (Minneapolis: University of Minnesota Press, 1931); Stanley B. Parsons, *The Populist Context: Rural Versus Urban Power on a Great Plains Frontier* (Westport, Conn.: Greenwood Press, 1973); Bruce Palmer, *"Man Over Money": The Southern Populist Critique of American Capitalism* (Chapel Hill: University of North Carolina Press, 1980); and Norman Pollock, *The Populist Response to Industrial America: Midwestern Populist Thought* (Cambridge: Harvard University Press, 1962). Little has been written specifically about the response of the Social Gospel to the rise of big business. Some aspects are discussed in Thomas M. Jocklin's, "The Civic Awakening: Social Christianity and the Usable Past," *Mid-America* 65 (1982): 3–19.

3. Henry D. Shapiro, *Appalachia on Our Mind: The Southern Mountains and Mountaineers in the American Consciousness, 1870–1920* (Chapel Hill: University of North Carolina Press, 1978), and Alan I Marcus, "Disease Prevention in America: From a Local to a National Outlook, 1880–1910," *Bulletin for the History of Medicine* 59 (1979): 184–203.

4. While time does not play a central role in any of the following articles, they all suggest its importance for mid-nineteenth century American Women. John Mark Faragher, "The Midwestern Farming Family, 1850," in Linda K. Kerber and Jane Dehort Mathews, eds., *Women's America: Refocusing the Past* (Oxford: Oxford University Press, 1982), 115; Gary Paulsen, "Pulling Together," in *Farm: A History and Celebration of the American Farmer* (Englewood Cliffs, N.J.: Prentice Hall, 1977), 83–92; and Elizabeth Hapsten, *Read This Only to Yourself: The Private Writings of Midwestern Women 1880–1910* (Bloomington: Indiana University Press, 1982).

5. On the development of shipping schedules, see Robert G. Albion, *Square Riggers on Schedule: The New York Sailing Packets to England, France and Cotton Ports* (Princeton: Princeton University Press, 1938). Time balls were erected at Washington, D.C. (ca. 1846), San Francisco (1852), and New York (1856) prior to the Civil War. Compare Ian R. Bartky and Steven J. Dick, "The First North American Time Ball," *Journal for the History of Astronomy* 13 (1982): 50–54; *San Francisco's Telegraphy Hill* (Berkeley: Howell-North, 1972), 31–32; *New York Times,* 14 February 1856, p. 8; 3 September 1856, p. 4.

6. George Rogers Taylor and Irene D. Neu, *The American Railroad Network 1861–1890* (Cambridge: Harvard University Press, 1956). This monograph contains an excellent discussion of railroad networks before the Civil War and the integration of those networks by 1890. Taylor and Neu, however, emphasize railroad gauge standardization and only briefly mention time standards.

7. The Cincinnati Observatory established a time service in 1840 for railroads serving that city, while the Allegheny Observatory began selling time to area railroads in 1855. The Goodsell Observatory of Carleton College, Northfield, Minnesota, provided the most widespread time service, sending time signals along seven thousand miles of railroad track when it opened in October 1877. Compare "Standard Time in America," *Science* 22 (1905): 315–18; and Delavan L. Leonard, *The History of Carleton College* (Chicago: Fleming H. Revell Company, 1904), 203–4.

8. Bartky and Dick, "The First North American Time Ball," 51.

9. *New York Times,* 25 January 1857, 6.

10. Dinsmore's table is reproduced in Carlton J. Corliss, *The Day of Two Noons,* 7th ed. (Washington, D.C.: Association for American Railroads, 1953), 7. It appears likely that the traveler converted the various time to the time of his residence or that of a nearby business center before beginning his trip.

11. "Standard Time in America," 316.

12. Ibid.

13. *New York Times,* 1 January 1882, p. 3. Ian R. Bartky asserts that Samuel P. Langley received an income of $66,000 from time signal sales while director of the Allegheny Observatory. Since observatories were hard pressed for research finances, they would naturally be reluctant to give up a major source of income. Bartky, "Naval Observatory Time Dissemination Before the Wireless," *Sky with Ocean Joined: Proceedings of the Sesquicentennial Symposia of the U.S. Naval Observatory December 5 and 8, 1980* (Washington, D.C.: U.S. Naval Observatory, 1983), 6.

14. Cited in Edward S. Holden, "Astronomy," *Annual Report of the Smithsonian Institution for 1882* (1883): 323–24.

15. *New York Times,* 19 January 1882, p. 8; 18 May 1882, p. 5; 10 June 1882, p. 2.

16. House of Representatives, *Executive Documents,* 48th Cong., 2d sess., no. 14. See especially 111–113.

17. For example: C. A. Young, "Standard Time for the United States," *Boston Journal of Chemistry and Popular Science Review* 16 (1882): 25; *Proceedings of the American Academy of Arts and Sciences* 19 (1884): 473–77; W. H. M. Christie, "Universal Time," *Popular Science Monthly* 29 (1886): 795–802; and *Scientific American Supplement* 29 (1890): 11700.

18. For a popular view of many scientific time proposals, see the *New York Times,* 22 May 1882, p. 4.

19. "Standard Time," *Railroad Gazette* 15 (1883): 441.

20. *Science* 2 (1883): 697; *Smithsonian Annual Report for 1882* (1883), 321–22.

21. *Smithsonian Annual Report for 1881* (1883): 321–22.

22. *Proceedings of the American Metrological Society* 2 (May 1879): 27.

23. On the competition between New York and Philadelphia, see Robert G. Albion, "New York and Its Rivals," *Journal of Economic and Business History* 3 (1931): 602–29. Taylor and Neu provide an excellent general discussion of intercity rivalry as it relates to railroads in *The American Railway Network.* See pp. 23–29 for their analysis of the New York-Philadelphia rivalry and pp. 32–33 and 40 for the rivalry between Pittsburgh and Chicago and other midwestern cities.

24. *Traveler's Official Guide* for May 1879 listed 407 railroad companies using 77 time standards, while for May 1883 it recorded 316 companies operating by 49 time standards. The larger railroads on the East Coast emphasized the times of Boston, New York, Philadelphia, Baltimore, and Washington, D.C. Inland, they adopted the times of Columbus, Pittsburgh, Atlanta, Louisville, Chicago, and Jefferson City, Missouri. Further west Denver, Laramie, Salt Lake City, San Francisco and Portland set such standards as existed. Compare *Railroad Gazette* 15 (1882): 441.

25. Cited in Corliss, *Day of Two Noons,* 6, and John F. Stover, *American Railroads* (Chicago: University of Chicage Press, 1961), 157–58.

26. "Uniform Standards of Time," *Railway Review* 21 (1881): 378.

27. Stover, *American Railroads,* 157.

28. "Standard Time in America," 316; Charles Ferdinand Dowd, *System of National Time and Its Application, by Means of Hour and Minute Indexes to the National Railway Time Table . . .* (Albany: Weed, Parsons and Company, Printers, 1870); *Railroad Gazette* 1 (1870): 1; "Uniformity in Railroad Time," *De Bow's Review* n.s. 8 (1870): 464–68; "The General Time Convention," *Railway Review* 21 (1881): 600.

29. "The General Time Convention," *Railway Review* 21 (1881): 600.

30. "Standard Time," *Railroad Gazette* 15 (1883): 441.

31. "Standard Time," 241–42; "Standard Railway Time," *Science* 2 (1883): 494–96.

32. "Standard Time for Southern Roads," *Railroad Gazette* 15 (1883): 21.

33. *New York Times,* 18 October 1883, p. 2; W. F. Allen, "A History of the Movement by which the Adoption of Standard Time Was Consummated," *Proceedings of the American Metrological Society* 4 (1884): 38–39.

34. W. F. Allen, "The Reformation in Time-Keeping," *Scientific American Supplement* 18 (1884): 7472–73; W. F. Allen, "A History of the Movement ... ," 48; Charles J. Higginbotham, "The Time System of the United States," *Scientific American Supplement* 79 (1915): 375; "Fortieth Birthday of Standard Time," *Railway Age* 75 (1923): 866; Corliss, *The Day of Two Noons.*

35. *New York Times,* 10 November 1883, p. 3; 11 November 1883, p. 8; 15 November 1883, p. 1.

36. *New York Times,* 18 November 1883, p. 1.

37. The *New York Times,* for instance, asserted that "even the proud city of Chicago must in this matter acknowledge the supremacy of the meridian of St. Louis ... ," 17 October 1883, p. 4; 29 October 1883, p. 4. Compare *New York Tribune,* 17 October 1883, p. 4, and 19 November 1883, p. 1.

38. *Laws of Wyoming, 1877–84,* chap. 95, p. 161.

39. "Report on Standard Time," *American Metrological Society, Proceedings,* 5 (May 1885): 53, 76–82.

40. *Public Acts of Connecticut, 1881,* chap. 21, 11–12; *Public Acts of Connecticut, 1884,* chap. 1, 1; *Laws of New York, 1884,* chap. 14, 22; *Laws of Wyoming, 1877–84, 161; Public Laws of New Jersey, 1884,* 175–76; *Laws of Maryland, 1884,* chap. 433, 578; *Public Acts of Michigan, 1885,* no. 5, 5; *Laws of Wisconsin, 1885,* chap. 216, 189; *Acts and Resolutions of Maine, 1887,* chap. 29, 20–21; *Public Acts of Pennsylvania, 1887,* 21; *Acts and Resolutions of Florida, 1889,* chap. 3916, 154; and *Acts of Ohio, 1893,* 110.

41. William H. Earle, "November 18, 1883: The Day the Noon Showed Up on Time," *Smithsonian* 14 (1983): 193–208, 206; *New York Times,* 8 May 1887, p. 6.

42. Detroit experimented briefly with standard time, then voted to return to local time in 1900. It remained on local time until after 1905. Just when it adopted standard time is unclear, but local time remained an option in the 1908 election. "Detroit's Different Kinds of Time," *Popular Astronomy* 9 (1901): 36–37; Edward Everett Hayden, "The Present Status of the Use of Standard Time," *Naval Observatory Publications,* 2d ser., 4 (1905–6)'; pt. 4, appendix 4, p. 25; *New York Times,* 26 October 1908, p. 8.

43. "Standard Time and Measures," *Science* 9 (1887): 7–8, 8. Not all Ohio communities were sympathetic however. The Bellaire, Ohio, city council placed its school board members under arrest after they refused to operate the school by local time. *New York Times,* 23 February 1889, p. 1.

44. *Railroad Gazette* 21 (1889): 29.

45. Earle, "November 18, 1883 ... ," 204.

46. *Railroad Gazette* 21 (1889): 29.

47. "Rochester German Ins. Co. v. Peaslee Gaulbert Co.," *Kentucky Law Reports* 120 (April term, 1905): 765.

48. When Detroit again returned to local time in 1900, the Midland, Michigan, *Republican* chided: "Poor halting Detroit is having a hard time with its time—more kinds of time than it knows what to do with. Only time can tell what it will do this time," while the Detroit *Journal* sniffed: "Standard time is the legal time of the City and County and will continue to be until the legislature changes the standard." Quoted in "Detroit's Different Kinds of Time," *Popular Astronomy* 9 (1901): 36–37.

49. *New York Times,* 5 November 1883, p. 2; 5 December 1883, p. 4. The *Times* identifies the case as Clapp v. Jenkins. However no such case was cited in *Massachusetts Law Reports* or the *Northeastern Reporter* for 1883 or 1884.

50. *New York Herald,* 18 November 1883, p. 10.

51. "Henderson & Son v. Reynolds," *Georgia Law Reports* 84 (1889): 162–164.

52. "H. A. Searles et. al. v. Albert Averhoff," *Nebraska Law Reports* 28 (January term, 1890): 668–69.

53. "Parker v. State," *Texas Criminal Reports* 35 (February 1895): 13–15.

54. "R. M. Jones v. the German Insurance Company of Freeport, Illinois," *Iowa Supreme Court Reports,* 110 (October term 1899): 76–80.

55. *General Laws of Minnesota* (1901): chap. 15, p. 16.; *Session Laws of South Dakota* (1909): chap. 46, p. 45.

56. "State v. O.P. Johnson," *Minnesota Supreme Court Reports* 74 (December 1898): 381–85. The Minnesota court only considered the business adoption of standard time, making no mention of popular acceptance. The fact that the state legislature felt constrained to enact a standard time law three years later suggests continued usage of local time.

57. "Rochester German Ins. Co. v. Peaslee-Galbert Co.," *Kentucky Law Reports* 120 (April term 1905): 757–63.

58. "Knud Orvik and Ole G. Olson v. John Casselman," *North Dakota Supreme Court Reports* 15 (December 1905): 37.

59. "Globe & Rutgers Fire Ins. Co. of New York v. David Moffat Co.," *Federal Court Reporter* 1554 (April 1907):19–21; "Salt Lake City v. Robinson, Supreme Court of Utah," *Pacific Reporter* 116 (May 1911): 444–47.

60. "Texas Tram & Lumber Company v. L.B. Hightower, District Judge," *Texas Supreme Court Reports* 100 (October 1906): 127, and "Walker v. Terrell, Comptroller (Court of Civil Appeals of Texas)," *Southwestern Reporter* 189: (June 1916): 75–80.

61. *Congressional Record,* 46th Cong. 2d sess., 1880, vol. 10, pt. 1, 462.

62. *Congressional Record,* 47th Cong., 1st sess., 1881, vol. 13, pt. 1, 108.

63. *Congressional Record,* 47th Cong., 1st sess., 1882, vol. 13, pt. 2, 1755; pt. 7, pp. 6578–80.

64. *Congressional Record,* 50th Cong., 1st sess., 1888, vol. 19, pt. 1, 323, 338.

65. *Congressional Record,* 51st Cong., 2d sess., 1891, vol. 22, pt. 2, 1503; *Congressional Record,* 52d Cong., 1st sess., 1892, vol. 23, pt. 1, 522, 573.

66. *Official Opinions of the Attorneys-General of the United States,* 17 (Washington, D.C.: Government Printing Office, 1890), 619–20; *New York Times,* 13 November 1883, pp. 3, 4.

67. "The Signal Service and Standard Time," *Science,* 2 (1883): 755; "Standard Time in America," 317.

68. *New York Times,* 18 November 1883, p. 8.

69. *Congressional Record,* 48th Cong., 1st sess., 1884, vol. 15, pt. 1, 88; pt. 2, 1760–63.

70. Edgar D. Tillyer, "The Clock Vault of the U.S. Naval Observatory," *Popular Astronomy* 18 (1910): 100–106.

71. Letter, Chief of the Bureau of Equipment, Navy Department to R. G. Aitken, quoted in Aitken's "The Sources of Standard Time in the United States," *Popular Astronomy* 10 (1902): 14.

72. Hayden, "The Present Status of the Use of Standard Time," 15.

73. *Congressional Record,* 61st Cong., 1st sess., 1909, vol. 44, pt. 3, 2347, 2427.

74. Myron E. Wells, "Eastern and Central Time Standards in Ohio and Michigan," *Railway Age Gazette* 60 (1916): 201–202; "Juggling with Standard Time," *Railway Age Gazette* 56 (1914): 1147.

75. *New York Times,* 26 October 1908, p. 8.

76. Wells, "Eastern and Central Time Standards ... ," 201–2; "Juggling with Standard Time," 1147; "Standard Time," *Railway Review* 57 (1915): 644.

77. "Standard Time," 644.

78. For example, Jermain G. Porter, "A Simple Plan for Unifying Time in the United States," *Popular Astronomy* 26 (1918): 188–89.

79. "A Timely Adjustment," *Scientific American* 34 (1918): 522; *Railway Review* 63 (1918): 755–56.

80. *Congressional Record,* 64th Cong., 1st sess., 1916, vol. 53, pt. 9, 8644, 8755.

81. *Senate Reports,* 65th Cong., 1st sess., 1917, no. 46, p. 708.

82. "An Act to Save Daylight and to Provide Standard Time for the United States," *Public Documents* (1918), No. 106; *Monthly Weather Review* 46 (1918): 75–76.

83. See the debates in the Senate and House of Representatives, *Congressional Record,* 66th Cong., 1st sess., 1919, vol. 58, pt. 2, 1304–1336; pt. 4, 3504–3510.

84. *Acts of Massachusetts* (1920): chap. 280, pp. 288–89.

85. *New Hampshire Laws* (1921), chap. 15, p. 27.

86. "Massachusetts State Grange v. Benton, Attorney General. Appeal from United States

District Court for the District of Massachusetts,'' *United States Supreme Court Reports* 272 (October term 1926): 525–29. Supreme Court Justice Oliver Wendell Holmes, who as Massachusetts Supreme Court Justice ruled in the first time case after the institution of standard time by the railroads in 1883, wrote the majority opinion, while Justice James Clark McReynolds wrote a separate concurring opinion.

87. ''J. H. McFarlane et al. v. Ross O. Whitney et al.'' *Texas Supreme Court Reports* 134 (February 1940): 401–3. Litigation involved the time of a 1931 trustee's sale.

From Ehrlich to Waksman:
Chemotherapy and
the Seamed Web of the Past

ALAN I MARCUS

The search for chemotherapeutic agents is an oft told tale. Its chroniclers often start with the Greeks but generally identify Paul Ehrlich's announcement of the synthesis of atoxyl in this century's first decade as its initial success. They then continue their narrative to the present day. Among the milestones noted is the discovery of streptomycin in 1943 by Selman A. Waksman and others. Streptomycin proved the first chemotherapeutic substance effective against the tuberculosis mycobacterium, "the white plague of mankind."[1]

That the development of chemotherapeutic agents from Ehrlich to Waksman confirms the old adage that history is a seamless web appears obvious. Both men expressly sought chemical cures and both found them. In that sense, only their success separated Waksman and Ehrlich from their predecessors. Although the web of history may be seamless in that simple-minded longitudinal manner—equating success with understanding hardly reveals anything meaningful about the past—modern historiography has considered it seamless in another sense. Studying diverse topics and time periods, historians such as Alexandre Koyrè, Thomas Kuhn, Edwin Layton, Garland Allen, and Michel Foucault have suggested that at particular times in the past the web has linked scholarly communities together; history's web within a discipline or subject area at a particular time has been found to be cross-sectional. Allen has indicated that late nineteenth-century biologists shared with each other certain notions about the nature of nature and life.[2] Indeed, such shared assumptions made conceivable not only accords but also debates; these jointly held notions made scientific discourse possible.

But Allen has gone further and has sewn the beginnings of a seam in the longitudinal web by demonstrating that biologists after about 1920 shared a new set of notions about nature and life. To Allen, the biologies of the late nineteenth and mid-twentieth centuries were discontinuous, each based upon a radically different series of shared assumptions.[3] When viewed from this perspective, the history of chemotherapy from Ehrlich to Waksman takes on a different light. An examination of their research shows that these two men approached chemotherapeutic studies from dramatically different premises;

they were not members of the same community of life scientists. Ehrlich's work reflected assumptions common among late nineteenth century life scientists, while Waksman's investigations were predicated upon notions that Allen has maintained characterized life scientists in the years after 1920.

Yet the new cross-sectional webs of history have not been cast nearly broadly enough; at any particular time history's cross-sectional webs have enveloped much more than a single discipline or subject area. The web's strands become a bit looser perhaps, but contemporaries in a great many disciplines, subject areas, and places share fundamentally similar assumptions about the organization of reality itself. They agree with each other about reality's organization, not necessarily its constitutent elements— differences in subject areas preclude that sort of agreement; while they concur about the relationships among the parts and between the parts and whole, that consensus only extends to those relationships.

Among those associated with the new historiographic tradition, Foucault stands preeminent in hinting at multidisciplinary, cross-sectional webs connected in this manner, but his own work covers only two particular periods. His several studies on aspects of eighteenth and nineteenth century thought reveal collectively that each century had a different "order of things" and that these orders were manifested in respectively different discourses and forms of public action. His works imply that the nineteenth century broke sharply with the eighteenth and was marked by, for example, a new treatment of the insane and explanation of insanity; the replacement of language, natural history, and the study of wealth by philology, biology, and political economy; a new medical practice and understanding of disease; a new form of punishment and a new perception of criminality; and a new type of school.[4] Taken individually, Foucault's books reaffirm Allen's position; they provide seams in the longitudinal web in several more places. But taken together, they prove much more powerful and help establish the existence of dense multidisciplinary cross-sectional webs. They suggest that while the new nineteenth-century forms of public action each were embraced by a different set of historical figures and were in that sense disconnected from each other, they all emerged from and reflected the same new order of things, the new notion of the organization of reality; the critical bond in Foucault's work is time—not place, individuals, subject area, or discipline.

The identification of time as the crucial feature is attractive. It reduces the tendency to place disciplines and subject areas within social matrices rather than beside them. It treats social matrices as it does disciplines and subject areas, not as some type of readily discernible products of their society but as products of their era. A particular era's notion of the organization of reality manifests itself in disciplines, subject areas, *and* social matrices; Ehrlich's organizational notions were similar to those reflected in late nineteen-century social arrangements as well as the life sciences, and Waksman's echoed those of the twentieth century. Ideas of reality's organization always circumscribe possibilities. They limit what seems worth attempting, whether the creation of an institutional form or the performance of an experiment, and bound the result's terms. They adjudicate appropriateness. Indeed, the histories of technology and science (and history generally) are littered with discoveries and inventions (and events) that seemed "right" at a later or earlier date but "wrong" within their own time. That the converse is also true goes without saying.

This should not be construed to indicate that a change in the notion of reality's organization results in one particular *Weltanschauung* or spirit of the age. Several of these essentially political or philosophical positions come from a new organizational notion and although each certainly has its partisans, the positions are not necessarily political in the sense of partisan politics. Nor is it meant to suggest that there are insignificant political differences among people or that motivation among individuals does not differ. Instead, it is to argue that despite these differences, there is agreement at any particular time about the parameters that govern both the nature of and the debates about a wide range of activities.

The replacement of the conventional web of history with a series of dense, multidisciplinary cross-sectional ones, each of which has seams separating it from other webs, does violence to the established relationship between past and present. They seemed joined only in the most rudimentary fashion or in some phenomenally complex, irreducible manner. This view's compensation is that particular times in the past are united in a way that exposes intriguing historical possibilities; it conveys new meaning to the idea of context and recasts the classic issue of origins and development in terms of the notion of reality's organization that made the origins and development in question appear viable, desirable, "necessary."

The cross-sectional webs of history provide a way of understanding the past and how and when it changes, but they are silent about what caused it to change. As demonstrated by Foucault and Allen, changes in notions of reality's organization rarely happen dramatically; the seams usually accommodate aspects of both the old and new webs. And these changes do not necessarily manifest themselves in the different subject areas and disciplines at precisely the exact same time. In the case of chemotherapy, for instance, several investigators pursued an approach similar to Paul Ehrlich's even as Selman Waksman began his work. Nonetheless, the disjunctions in the past are clear. History's web has seams.

*   *   *

Ehrlich stood as a major figure in microbiology when he turned his full attention to chemotherapy. In a scientific career that had spanned more than two decades, Ehrlich had made contributions in histologic staining and the production and assay of diphtheria antitoxin, developed an explanation for toxin—antitoxin reactions, and articulated a general theory of immunology. In each of these early studies he had incorporated notions similar to those he would employ in his chemotherapeutic investigations. To Ehrlich, biology was in essence structural; the process of life could be explained and understood in chemical and geometric, or more properly perhaps, physical terms. In each investigation, Ehrlich had assumed that the parts under study were ordered, discrete, and distinct, yet fitted together in a specific manner so as to compose a whole; every part had a certain place in the whole and the whole itself seemed nothing more than the sum of its parts.[5]

These notions led Ehrlich to consider health in a particular fashion. Health was a condition of static equilibrium, in which each body part—cell, tissue, or organ—performed a particular function and in which no foreign substances disrupted any part's activities. Disease, on the other hand, was the improper functioning of a single part, a condition that focused almost exclusively on that part, while the severity of the disease depended upon which part functioned improperly and to what extent. Cure was the restoration of that part through chemical or physical means so as to return the body to its customary state.

Perhaps an analogy would prove helpful. Health was like a rectangular jigsaw puzzle composed of all white pieces. Each piece had a different shape—an integrity of its own—but when all were placed in the proper order, a white rectangle would emerge. Disease was analogous to a piece being altered so as not to fit in its proper position, a situation that left the puzzle incomplete. Cure was effected when the offending piece was amended again to assume exactly its rightful spot.

An example from Ehrlich's early work makes the structural argument clear. His theory of immunology presumed that pathogens or pathogenic substances were chemically structured in such a way that they attached or portions of them attached to specific groups of cells within the body and impaired the bound cells' normal function. The body responded to this pathogenic invasion by producing massive numbers of antibodies. Each of these new and identical antibodies possessed a similar configuration or site, a chemical side chain, and these sites were identical to those of the cells under attack. The rest was simply a matter of statistics. The preponderance of antibodies enhanced the chances for the maintenance of regular function among menaced, but still healthy cells. Instead of uniting with the invaders' original targets, pathogens encountered the antibodies' structurally similar side chains and, due to the greater numbers of these antibodies, generally linked with them. This rendered the bound pathogens harmless to vital cells. The intruders fit with the antibodies, noted Ehrlich remembering the phrases of Louis Pastuer and Emil Fischer, "as male and female screw or as lock and key."[6]

Ehrlich carried with him the lock-and-key metaphor when he began his chemotherapeutic work in earnest. The possibility of chemotherapy had been established earlier: investigators long had known that different chemical substances had differing affinities for various living cells. Ehrlich himself had employed that knowledge in the late 1870s to develop histological staining techniques. His understanding of the staining phenomenon was suggestive. He argued that a stain's expression of color resulted from a "chromophore group," a structural appendage to the basic staining molecule. But Ehrlich regarded the acceptance of stain in a cell "as a chemical combination" and maintained that the possibilities and degrees of union between any particular stain and cell probably depended on the three-dimensional configurations of aromatic hydrocarbons in both the stain and the cell; staining was a consequence of geometric shape. Also early in his career, Ehrlich hypothesized that specific chemotherapeutic agents could be produced to fight specific diseases; he had postulated that artificial substances could be synthesized that, because of their structure, would have great affinity and toxicity for particular pathogens. These chemical structures would not join readily with host tissues and therefore would be virtually harmless to the host. To Ehrlich, these chemical agents seemed "charmed bullets." Investigators needed only to learn "to shoot . . . through the variation of chemical substances."[7]

Ehrlich's chemotherapeutic speculations remained unconfirmed until 1902 when he initiated a search for a chemical agent to cure sleeping sickness, a disease caused by a protozoan from the trypanosome family. He did not randomly create compounds, but drew from his histologic staining experience and selected the aniline dyes, which were accepted readily into many tissues. Ehrlich started shooting by synthesizing derivatives of this compound. He took the basic molecule and added or substituted functional groups so as to vary slightly its configuration. In effect, he sought to alter the compound's structure so that it would possess receptors compatible to the pathogen's sites; the new substance would act like an antibody and bind itself to the trypanosomes.

The aniline dyes proved a propitious choice, and in 1903 Ehrlich's laboratory found trypan red effective in vitro on the trypanosomes of sleeping sickness. Subsequent in vivo tests suggested that the substance possessed little toxicity for host cells. He immediately extended his efforts and sought to synthesize a substance to combat other types of trypanosomes. This work culminated in 1904 in the discovery of atoxyl.[8]

During the next several years, Ehrlich determined the drug's chemical structure, toxicity, and effect on various trypanosomes. He then turned his attention to the recently discovered spirochete of syphilis. As did many of his contemporaries, Ehrlich suspected that spirochetes and trypanosomes were closely related—that they had similar structures—and armed with that assumption, tried atoxyl on syphilis. That the drug showed no potency on the spirochete did not dissuade him, and he set his laboratory to synthesizing a number of structurally similar chemical substances. In 1909 he met with success. His 606th effort, arsphenamine or more commonly salvarsan, demonstrated chemotherapeutic properties against syphilis and other spirochetal diseases, such as yaws and relapsing fever.[9]

Ehrlich was not yet satisfied. He wanted an agent even more effective against spirochetes and tackled that problem for the next five years. His laboratory's 914th creation, neo-salvarsan, met his expectations. It usually required only one dose to rid patients of the pathogen and proved considerably less toxic than salvarsan to the host.[10]

The announcement of neo-salvarsan marked the culmination of Ehrlich's chemotherapeutic career. He died two years later. His endeavors had been restricted to the consideration of disease and health and the relationship among cells, pathogens, and chemotherapeutic agents. Despite this relatively narrow agenda, the assumptions Ehrlich employed about the organization of reality—the assumptions that led him to proceed in the manner that he did and explain his efforts as he did—were symptomatic of the late nineteenth-century life sciences and social thought generally. To Ehrlich as to others, the system—the whole—was static, composed of discrete, fundamentally distinct parts, each of which had a proper place, role, or function. Put another way, the system was divisible into its parts; each part could be isolated and studied and would continue to maintain its integrity. But what made the whole a system was that each part needed to fulfill its specific function or role or be situated in its appropriate position for the whole to be complete, for the system to achieve peak efficiency. In that sense, the whole was simply additive; the system was nothing more than the sum of its properly constituted and properly placed operating parts.

This view of reality's organization fueled a generation of thinkers. While any survey of its extent must remain woefully incomplete, a few examples help to establish the idea's pervasiveness. For instance, animal nutritionists maintained that the animal body comprised a system and that its continued health, development, and ability to work demanded a proper diet. They argued, furthermore, that each component of a proper diet performed a particular function in the animal body, and they set their task as identifying a proper diet's components and their functions. From that assumption came the basic ration work at Weende, E. B. Rosa's and W. O. Atwater's respiratory calorimetry studies, T. B. Osborne's and L. B. Mendel's identification of peptides and amino acids, the detection of a preventative factor for polyneuritis, E. V. McCollum's and M. Davis's discovery of fatsoluble *a* and watersoluble *b*, and E. B. Hart's work with trace minerals.[11] It also found expression in the efforts of late nineteenth-century electrical theorists and entrepreneurs. Indeed, the arc lighting systems of Charles Brush and Paul Jablochkoff are well known.[12] But in the past several decades, historians have focused on Thomas A. Edison as the electrical system builder *par excellence*. In particular, they have celebrated his genius in the conception of a central power station, dynamos, feeder lines, incandescent bulbs, and the like as parts of an electrical generation and distribution system. In this view, Edison's subsequent efforts with electrical motors, kinescopes, and phonographs seem attempts to complete the system by fashioning a reliable consumer product. Without that part of the system drawing power regularly, the entire enterprise would fail; only components would remain. The static notion of a system of discrete and ordered parts apparently also guided Edison in other ways. The telegraph system served as the model for many of his inventions. To Edison, the stock ticker was the printing telegraph; the telephone the speaking telegraph; and the phonograph the talking telegraph recorder. Each employed devices or principles, such as the continuous feed, that Edison had used in his earliest telegraphic dabblings. And each was a telegraph system designed to perform a different function; the order of the parts was changed or new parts necessary to complete the new system were added.[13]

The period's social thought reflected the same organization of reality. For example, Frederick Jackson Turner's frontier thesis was predicated upon the notion that America was a whole composed of several "self-conscious sections," each of which played a specific role in the formation of the nation and its institutions. And Turner concentrated on the frontier precisely because he worried that its disappearance would transform the nation; it would remove a vital cog in the system and as a consequence, alter the system.[14] In a similar manner, the emergence of late nineteenth-century nationalism was a tacit acknowledgement that nations were ordered entities composed of various separable interests. These interests, whether classes, ethnic groups, professions, states, races, principalities, or the like, each needed to operate to their capabilities and in their "proper" way for nations to progress. Indeed, those characteristics that seemed to define any particular national entity served as the social glue and organizing principle for that national system.[15]

These notions of reality's organization also manifested themselves in the efforts of those bacteriologists who sought chemotherapeutic cures in the wake of Ehrlich's announcement of neo-salvarsan. In fact, the apparent success of neo-salvarsan touched off a quest for agents of the type that Ehrlich had uncovered that continued for more than two decades. During this period, bacteriologists tested gold, silver, and sodium salts for chemotherapeutic properties and persisted in modifying the aniline dyes, finally discovering sulfonilamide and other sulfa drugs in the 1930s. Though often disagreeing over their theories' fine points, these investigators all concluded that the structural manipulation of chemical compounds would yield new substances that, because of the configuration, would bind themselves to specific pathogens and free the body from disease.[16]

That researchers continued in the 1920s and 1930s to pursue chemotherapeutic agents in a manner consonant with principles espoused by Ehrlich was deceptive, merely an indication of his success. Well before the advent of the sulfa drugs, the importance of the relationships that led Ehrlich and others to structural chemotherapy—the relationships among the parts and the whole— had waned. A new notion of reality's organization had begun to replace them. This transformation produced a new language to describe the new organization of reality and that language concerned itself with dynamics, not static relationships. The new language was abundantly clear in the sciences.

Concepts, such as ecological niche, homeostasis, the indeterminancy principle, aerodynamics, supraorganicism, interdependence, and the like came into vogue. The examples were as obvious and the case as persuasive in social theory and thought. Keynesian economics, A. O. Lovejoy's unit-ideas, Roderick D. McKenzie's metropolitan communities, Rupert Vance's regional cultures, and Horace Kallen's cultural pluralism each testified to the crystallization of a new set of assumptions of reality's organization. In this twentieth-century view, the parts of the whole seemed neither discrete nor ordered, while the whole itself appeared greater or at least different than the sum of its parts. In this new scheme, the parts received definition not in and of themselves but as part of the whole because of their relationship to the other parts.[17]

In the new organization of reality, the chemical and physical structuralism of the late nineteenth-century seemed outmoded. This earlier approach appeared too simplistic; it failed to take into account the new, intense inter-relationships and interdependencies among parts. Dynamism—the emphasis on the whole as in a state of dynamic equilibrium—made the question of each part's position, function, or role anachronistic. It was the whole that really mattered; relationships, such as those postulated by Ehrlich and others in the late nineteenth century, seemed no longer to represent reality's organization.

Turning to analogy once again, the dynamic new reality's organization approximated a well-cooked stew. None of the ingredients retained the integrity, texture, or taste that it had prior to preparation, and the concoction as a whole tasted different—in many cases better—than the ingredients did separately. Yet the omission, substitution, or reduction of a single ingredient changed the nature and taste of the dish; it created a new stew. Length of cooking and temperature also figured in. These two factors changed the stew's taste; they altered its dynamics.

That view of reality's organization became popular and accepted in the twentieth century. It differed markedly from its predecessor. In chemo-therapy, the earlier notion led to Ehrlich's conception of magic bullets. Twentieth-century assumptions made it possible for Selman A. Waksman and others to think of chemotherapy in a different manner. These assumptions would yield a new chemotherapy, the chemotherapy of antibiotics.

Waksman's career provided a useful example of the distinction between the late nineteenth-century organization of reality and that which followed. He worked in both periods, and his studies changed dramatically over time. Waksman came to America from Russia in 1910 and studied at Rutgers, with Jacob Lipman, and at University of California, with T. Brailsford Robertson. His association with Robertson was especially revealing. Robertson had been a cherished student of Jacques Loeb, whose volume *The Mechanistic Conception of Life* stood as testament to and a model of the late nineteenth-century notion of reality's organization.[18] After Waksman completed his studies, he accepted a post at Rutgers as a soil bacteriologist. His early

investigations there reflected the reality in which he was trained; his microbiology of the soil was predominantly anatomical and any description of
phenomena were reduced to simple chemical reactions. Indeed, Waksman's
research consisted of classifying the actinomycetes—a group of microorganisms naturally occurring in soils and resembling both bacteria and
fungi—according to structure, color, smell, and growth rate. He also sought
to find the place of actinomycetes in the nitrogen cycle and examined the
microbially induced chemical reactions that produced humus and citric,
fumaric, and sulfuric acids. Each investigation presumed that the chemical
reactions of soil microbes might be studied independently, that these reactions
need not be explored in reference to other facets of soil life.[19]

The first hint that Waksman had begun to modify his assumptions appeared
in 1923. As part of his work at the Rutgers Agricultural Experiment Station,
Waksman undertook a series of investigations on the effect of heat, moisture,
caustic lime, and antiseptics on microorganisms in the soil. He concluded that
these agents and conditions effected the soil microbe population differentially;
some reduced the numbers of one type of microbial life but boosted another.
These experiments soon led him to postulate that outside influences caused a
shift in the "microbiological equilibrium in the soil." Each influence would
be conducive to some microbes and antagonistic to others, but that
equilibrium—stasis—would establish itself at a new point so long as no new
agency entered the equation. Waksman termed this relative reshuffling in the
face of external forces "unstable equilibrium, . . . not static but dynamic," and
maintained that its basis lay in the "interrelationships of microorganisms in
the soil."[20]

To Waksman, soil microorganisms now seemed to "carry on their activities,
not as individual forms or even as groups, but as a soil population." Rather
than consider the soil as the sum of simple separable chemical reactions among
microorganisms, he started to treat it as if it were the product of an intensely
interconnected series of complex microbial interactions. In other words, he
began to examine the process of soil building and decay, to assume that it was
a single continually occurring phenomena, and to assert that its dynamics
could not be understood without the consideration of all microbial interactions. Waksman no longer deemed it sufficient to investigate the formation
of a particular chemical by a microbe in the soil without exploring the effects
of that chemical on the various types of soil microorganisms. To do less was
to ignore the complexity and dynamism of the soil and to offer a false and
incomplete portrait.[21]

Waksman embarked upon two courses of action, both based on this new
view of the soil. The first dealt with method. Waksman emerged as an early
critic of the Remy-Lohnis solution method for the analysis of soil microorganisms and perhaps the staunchest champion of Sergei Winogradsky's
"direct method" of investigation. The Remy-Lohnis method took suspensions
of soil microbes, spread them on cultures standardized by substances and

conditions, and noted the chemical reactions. Waksman maintained that this procedure bore virtually no resemblance to what occurred in the soil and, as a result, could reveal little if anything about soil life. He termed it "soil bacteriology" and argued that it was purely artificial. The standardization of cultures and conditions permitted only one equilibrium; it favored some microbes and hampered others. More to the point, he claimed that chemical reactions on artificial media could hardly be representative of soil life. The soil was an intensely complicated microbiological medium and its processes could only be understood as a whole; isolation of individual reactions in the fashion of Remy-Lohnis seemed inadequate.[22]

Those considerations led Waksman to embrace Winogradsky's method. Indeed, their intellectual kinship soon translated into a close personal and professional relationship. Waksman visited him in Germany twice in 1924 and several times thereafter. Together they sought to develop what they termed the "independent science of soil microbiology." They hoped to transform the study of microorganisms of the soil, soil bacteriology, into the study of microorganisms in the soil, the new science of the soil microbiology. Winogradsky's method was crucial to that transformation. Both men maintained that an independent science required "its own special methods." The direct method they advocated was fully in line with their new notions. It rested on the premise that the only way to understand life processes in the soil was to look at the soil; soil was to serve as the culture medium. To be sure, they recognized the necessity of sometimes supplementing soil study with "auxiliary cultures," particularly when investigating a special modification of the soil population. Nonetheless, their direct method employed the soil as medium whenever possible.[23]

Waksman's second initiative was more striking. He repudiated the thrust of his earlier humus work and undertook a sustained attack on the concept of humus itself. His argument echoed that for the direct method. He noted that considerable study had gone into the humus question, but he found it all flawed. Humus was a subject of "great complexity," and research in that area suffered from "faulty methods of investigation." The problem lay with the notion itself. Investigators treated humus as if it was a discrete substance, the sum of specific chemical reactions; they sought to "determine its chemical nature." Such a view was wrong. Waksman asserted that "when two different soils containing the same amounts of organic matter are compared," it can be "readily recognized" that "there is a difference in the quality of the soil organic matter and in the manner in which it decomposes." Indeed, he maintained that the composition of a soil's organic matter—its "humus"— depended "entirely on the organisms taking part in the decomposition and upon the environmental conditions influencing these processes." As a consequence, "humus" differed in composition from place to place; it was impossible to offer a single chemical structure. Waksman urged his misled colleagues to abandon their search and to focus instead on the "genesis" of soil organic matter. In short, he wanted them to investigate process; he advised them to replace soil bacteriology with soil microbiology.[24]

By no means was Waksman the only investigator to adopt this new approach to the soil. In addition to Winogradsky, a number of other scientists also pressed this view[25] In effect, they reconsidered old facts in a new way; these men and women merely asked new and different questions of soil life. That was no small matter. The questioning was a manifestation of a new organization of reality. And that new organization would ultimately transform the field.

Waksman, then, was moving into areas only beginning to be charted. A former student, René J. Dubos, provided him the outlines of a map for one particular route. It would play a significant role in Waksman's decision of a decade later to pursue chemotherapeutic agents. Working for Oswald T. Avery at the Rockefeller Institution for Medical Research, Dubos was given the task in 1927 of searching for a chemical substance that would destroy the polysaccharide coating of the deadly Pneumococcus III bacterium. Such an enzyme appeared necessary to dissolve the capsular material and to permit the body's phagocytes access to devour the microbe. This protein had eluded Avery and his coworkers. They had unsuccessfully subjected the coating to a great many known enzymes.[26]

In essence, Avery and his associates had approached the coating problem in much the same manner as those chemists seeking chemotherapeutic cures after Ehrlich. Dubos took the radically different perspective of a new soil microbiologist; he focused on process. Dubos was unconcerned about the anticipated enzyme's structure, composition, or mechanism. What was crucial was that he knew it existed and where it could be found. The dynamic process of soil building and decay held the key. Soil microbes, either individually or collectively, must secrete a substance capable of lyzing the polysaccharide. The proof of that contention seemed unassailable. Without such an agent in the soil, mountains of the coating would exist in nature. That none did indicated that soil microorganisms produced something that broke down the capsule[27]

Dubos's research program followed this set of assumptions. He recreated "natural" systems, ones based on process, in the lab. He sterilized several different types of culture media and varied the alkalinity, acidity, temperature, and oxygen level of each plate. He then added the polysaccharide covering as the only carbon source. He next poured on unsterilized solutions of soil and sewage sludge. The sludge and soil teemed with microorganisms. To survive in these specialized environments, the microbes had to utilize the sole carbon source and that meant dissolving the polysaccharide. Once a microorganism survived, all that remained was to isolate the substance that lyzed the coating.[28]

Within two years Dubos had identified a microbe that produced the polysaccharide-destroying substance. The separation of that substance took a bit longer. Dubos discovered that the microorganism extruded the desired enzyme only when grown in an environment in which the polysaccharide was

the sole food source.[29] His understanding of this phenomenon was il-
lustrative and reinforced his notion of the dynamic equilibrium of soil life. He
maintained that many soil microbes had the capacity to manufacture two
enzyme types. Microbes always produced the first type, constituent enzymes,
but they only made the second type, adaptive enzymes, to survive in
a new environment. Once that environmental stimulus was removed, the
microorganism ceased adaptive enzyme production.[30]

As his former mentor, Waksman knew of Dubos's Rockefeller Institution
work. He did not miss its significance. In 1932, a year after the appearance of
Dubos's first polysaccharide-related paper, Waksman secured a large National
Research Council grant to study the fate of pathogenic bacteria, particularly
acid-fast bacteria—Pneumococcus III is an acid-fast bacterium—in the soil.
He published his results four years later as a three-part study entitled
"Associative and Antagonistic Effects of Microorganisms." Waksman
discussed the relationships among microbes that would later culminate in the
discovery of streptomycin. Most of Waksman's labors occurred in the
laboratory, not the field. Nonetheless, the environments he created were
"natural"; they reflected the dynamism accorded twentieth-century systems
and were interpreted in that manner. For example, Waksman determined that
the life expectancy of the typhoid bacillus decreased dramatically if a great
number of other microorganisms were added to the medium. He found that
the bacilli could live up to 51 days in sterilized chemically polluted water, but
if placed in sewage sludge—a fine microbial medium—99 percent of the
bacteria died within six hours. He also explored the live cycle of E. Coli in the
soil. He plated this potential pathogen on an agar medium and discovered that
many actinomycetes emitted a substance that would lyze the bacterium. From
these and other similar experiments, Waksman concluded that disease-
producing microbes generally lived but a short time in the soil and that their
destruction was due to specific substances manufactured by soil-inhabiting
microorganisms.[31]

Waksman again was not the first to examine the dynamics of soil life and
its consequences for pathogenic microorganisms. Three investigators at the
Pasteur Institute in Brussels—André Gratia, Sara Dath, and Bernice
Rhodes—had demonstrated a decade earlier that actinomycetes were capable
of lyzing dead staphlycocci bacteria, while a Russian group in the early 1930s
published a survey of the antimicrobial antagonisms.[32] Nor would Waksman
be the first to apply this understanding to the quest for chemotherapeutic substances. Once again that honor went to Dubos. In 1937 Dubos took the crucial
next step and reoriented and extended his research. This reconceptualization
was in the realm of goals, not in notions of the nature of soil life. Dubos had
conceived of his work with the polysaccharide coating of Pneumococcus III
as a search for a vaccine, a search for a substance to protect those exposed
from developing the disease. Chemotherapeutic studies required a different
outlook, one based on cure—not immunity or prevention.[33]

It took Dubos nearly a decade to make that vital shift. But once he accomplished it, he found that his earlier assumptions and labors put him in good stead. Most of his Pneumococcus III efforts seemed directly transferable to his new problem. He merely substituted Gram-positive pathogens (staphylococcus, streptococcus, and the like) for the polysaccharide as the exclusive food source in his cultures and unleashed the soil and sludge microorganisms. By 1939, he had isolated a soil microbe—Bacillus Brevis— that produced a pathogen-digesting substance. His discovery, Tyrothricin, was the first antibiotic agent derived from the soil.[34]

Dubos had kept Waksman abreast of his work. And even before the discovery of Tyrothricin, Waksman had begun to venture into antibiotic research. The New Jersey chemical company, Merck, had provided an additional impetus. It agreed to support up to five graduate students yearly to pursue soil microbe-produced chemotherapeutic substances and to fund in vivo toxicity tests on any substance uncovered. In return Merck expected the rights to all patentable processes.

This agreement was signed in early 1939 and gave Waksman the ability to proceed on a scale unavailable to Dubos.[35] He adopted the principles initiated by his former student, including the exclusive focus on Gram-positive pathogens, but used the opportunity provided by the Merck connection to vary more widely the conditions in which he grew his cultures. Waksman also decided to concentrate his search on the soil microbes he knew best, the actinomycetes. He introduced three tests to screen his soil actinomycetes for antibiotic activity. Each owed its origin to the assumption that soil life was inextricably complex and that this complexity manifested itself in microbial associations and antagonisms. Waksman's first innovation, the crowded-plate technique, took a culture of pathogens as its starting point. He then spread a soil suspension on top the culture and examined the plate for areas where actinomycetes had influenced pathogenic growth. His spray-plate technique reversed the process. He first plated actinomycetes in culture, then sprayed on pathogens, and finally noted which soil cultures remained pure. His third new test, the cross-streak method, combined elements of the first two. He alternated the plating of several parallel rows of pathogens and actinomycetes and looked for "zones of inhibition," sites where pathogenic growth was inhibited. In each instance that Waksman found inhibitory activity, he made a pure culture of the actinomycete and then checked to see if it would flourish with pathogens as its only food source.[36]

Waksman achieved initial success almost immediately. Within nine months he had isolated his first antibiotic substance, Actinomycin. The drug proved not the panacea it appeared, however. In vivo tests with mice suggested that Actinomycin was far too toxic for human use.[37] Waksman persisted in his research but a development outside his lab soon encouraged him to reexamine his focus. He learned in 1941 of the efficacy of penicillin against a broad spectrum of Gram-positive bacteria and recognized that a second antibiotic agent effective in fighting similar bacteria would be deemed redundant. That assessment led him to turn his attention to Gram-negative pathogens; he simply substituted E. Coli wherever he had employed previously streptococci or staphylococci.[38] Success again apparently came quickly but it too proved illusionary. His laboratory discovered Streptothricin in 1942. Preliminary tests indicated that it was effective against a wide range of Gram-negative bacteria and nontoxic. More extensive work revealed otherwise. Streptothricin had a delayed toxicity; experimental animals given the drug would seem to recover from disease only to die suddenly some days later.[39]

Not until 1943 did Waksman's laboratory isolate the substance for which he would received the Nobel prize, Streptomycin. This chemotherapeutic agent attacked most Gram-negative bacteria and held little toxicity for hosts.[40] More important, subsequent tests at Merck and, particularly, the Mayo Clinic established Streptomycin's utility in fighting tuberculosis (the white plague of mankind) and other acid-fast bacteria.[41] Its genesis was similar to that of the other substances identified by Waksman and his associates. All these results stemmed from the specific notion of the dynamics of soil life; Waksman incorporated techniques consonant with that notion. Waksman had set the methods of antibiotic research and his announcement of Streptomycin provided a powerful stimulus to other investigators. In that sense, Waksman's influence was comparable to Ehrlich's some thirty years earlier. Both had set an agenda for chemotherapeutic studies.

There the similarity ended. Waksman and Ehrlich pursued chemotherapeutic agents from different perspectives. These differences were not a matter of interest or specialty. Each conceived of reality's organization in a quite different manner. Ehrlich's work was based on the supposition that living systems were composed of discrete, distinct parts. As a consequence, he emphasized the principles that bound the discordant parts together, structure and function. Waksman's efforts were predicated upon a much more complicated system, one in which the parts received definition to a degree from their relationship to other parts. That assessment led him to stress processes, not principles.

Neither Ehrlich's static system nor Waksman's dynamic one separated them from their contemporaries. To be sure, Waksman and Ehrlich were unusual in the subjects they studied and in their expertise; few were as interested in chemotherapy or as proficient in soil microbiology or immunology. But the assumptions about reality's organization that each applied in his specialized work were shared by his compatriots in time. Those commonalities were not restricted to fellow life scientists, but pervaded virtually every sphere of thought. Those of the late nineteenth century accentuated order, fit, hierarchies, structure, position, and/or function precisely because the parts of their systems were capable of integrity when separated. Those of the twentieth century could afford no such conceit. Their parts lost their special meanings when isolated; their parts' particularistic meanings came only from context.

Notes

1. See, for example, Paul De Kruif, *Microbe Hunters* (New York: Harcourt, Brace and Company, 1926); Samuel Epstein and Beryl Williams, *Miracles from Microbes* (New Brunswick: Rutgers University Press, 1946); Boris Sokoloff, *The Miracle Drugs* (Chicago: Ziff-Davis Publishing Co., 1946); Helmuth M. Bottcher, *Wonder Drugs—A History of Antiobiotics,* translated by Einhart Kawerau (Philadelphia: J. B. Lippincott, 1964); Harry F. Dowling, "Comparisons and Contrasts Between The Early Arsphenamine and Early Antiobiotic Periods," *Bulletin of the History of Medicine,* 47 (1973): 236–49, W. D. Foster, *A History of Medical Bacteriology and Immunology* (London: Heinemann Medical Books, 1970), 196–217; Hubert A. Lechevalier and Morris Solotorovsky, *Three Centuries of Microbiology* (New York: McGraw-Hill, 1965), 429–92; and John C. Sheehan, *The Enchanted Ring: The Untold Story of Penicillin* (New York: Murray Printing Co., 1982).

2. On this point, see, for instance, Alexandre Koyrè, *From the Closed World to the Infinite Universe* (Baltimore: Johns Hopkins Press, 1957); Thomas S. Kuhn, *The Structure of Scientific Revolutions,* 2d ed. (Chicago: University of Chicago Press, 1970); Edwin Layton, "Mirror-Image Twins: The Communities of Science and Technology," in George H. Daniels, ed., *Nineteenth-Century American Science. A Reappraisal,* (Evanston: Northwestern University Press, 1972), 210–20; Garland Allen, *Life Sciences In the Twentieth Century* (New York: John Wiley and Sons, 1975), for late nineteenth-century life science, see especially 211–39, 73–81; and Michel Foucault, *The Order of Things, An Archaeology of the Human Sciences,* (New York: Random House, 1970).

3. Allen, *Life Sciences* 94–111, 113ff.

4. Michel Foucault, *The Order of Things; Madness and Civilization: A History Insanity in the Age of Reason* (New York: Vintage, 1973); *The Birth of the Clinic: An Archaeology of Medical Perception* (New York: Vintage, 1975); and *Discipline and Punish: The Birth of the Prison* (New York: Vintage, 1979). Also of utility is his *Archaeology of Knowledge* (New York: Vintage, 1972).

5. Historians tend to view Ehrlich's work in parts, as it related to subject areas. See, for instance, Martha Marquardt, *Paul Ehrlich* (New York: Henry Schumann, 1951); John Parascandola, "The Theoretical Basis of Paul Ehrlich's Chemotherapy," *Journal of the History of Medicine and Allied Sciences* 36 (1981): 19–43; and John Parascandola and Ronald Jasensky, "Origins of the Receptor Theory of Drug Action," *Bulletin of the History of Medicine* 48 (1974): 199–220. Many of Ehrlich's scientific papers have been collected and published in F. Himmelweit, ed. and trans., *The Collected Papers of Paul Ehrlich,* 3 vols. (London: Pergamon Press, 1956–57). All citations to Ehrlich's publications (*Papers*) are from those volumes. On Ehrlich's structural approach, see, for example, "Contributions to the Theory and Practice of Histological Staining," *Papers,* 1: 65–98; "The Assay of the Activity of Diphtheria-Curative Serum and Its Theoretical Basis," *Papers* 2: 107–25; "Observations upon the Constitution of the Diphtheria Toxin," *Papers* 2: 134–42; "On Immunity with special reference to Cell Life. Croonian Lecture," *Papers* 2: 178–95; and "The Requirement of the Organism for Oxygen," *Papers,* 1: 433–96.

6. *Papers,* "The Assay," 2: 14; and "On Immunity," especially 2: 185.

7. *Papers,* "Contributions," especially 1: 67–69, 73–76; P. Guttman and P. Ehrlich, "On the Action of Methylene Blue on Malaria," *Papers,* 3: 15–20; and Paul Ehrlich, "Address Delivered at the Dedication of the Georg Speyer House," *Papers,* 3: 56–61.

8. Paul Ehrlich and F. Sachs, "Die Darstellung von Triphenylmethanfarbstoffen aus Brommagnewiumdimethylanilin als Vorlesungsversuch," *Papers,* 3: 21–23; and Paul Ehrlich and K. Shiga, "Farbentherapeutische Versuche bei Trypanosomenerkrankung," *Papers,* 3: 24–37.

9. Paul Ehrlich, "Chemotherapeutische Trypanosemen-Studien," *Papers,* 3: 81–105; Paul Ehrlich, "Die Behandlung der Syphilis mit dem Ehrlichschen Praparat 606," *Papers,* 3: 240–46; and Paul Ehrlich and S. Hata, *Die Experimentelle Chemotherapie der Spirillosen,* (Berlin: Springer, 1910).

10. Paul Ehrlich, "Schlußbemerkungen zu Abhandlungen über Salvarsan III," *Papers,* 3: 457–84; and Paul Ehrlich, "Chemotherapy," *Papers,* 3: 505–18.

11. W. O. Atwater, "Science Applied to Farming," *American Agriculturist* 34 (1875): 91–92, 130, 175–76; W. O. Atwater and F. G. Benedict, *A Respiration Calorimeter with Appliances for the Direct Determination of Oxygen,* Publication no. 42 (Washington, D.C.: Carnegie Institution of Washington, 1905); Thomas B. Osborne, *The Vegetable Proteins* (London: Longmans, Green, and Co., 1912); C. Funk and E. A. Cooper, "Experiments on the Causation of Beri-beri," *Lancet* 11 (1911): 1266–69; E. V. McCollum and M. Davis, "The Essential Factors in the Diet during Growth," *Journal of Biological Chemistry 23* (1915): 231–36; and E. B. Hart and H. Steenbock, "Iron in Nutrition," *Journal of Biological Chemistry,* 76 (1928): 325–58.

12. See Henry J. Eisenman II, "Charles Francis Brush: Pioneer Innovator in Electrical Technology," (Ph. D. diss., Case Western Reserve University, 1967); Malcolm MacLaren, *The Rise of the Electrical Lighting Industry During the Nineteenth Century* (Princeton: Princeton University Press, 1943), 68–71; John Winthrop Hammond, *Men and Volts: The Story of General Electric* (Philadelphia: J. B. Lippincott, 1941), passim; and Harold C. Passer, *The Electrical Manufacturers, 1875–1900,* (Cambridge: Harvard University Press, 1953), passim.

13. See Thomas P. Hughes, *Networks of Power. Electification in Western Society, 1880–1930,* (Baltimore: Johns Hopkins University Press, 1983), 18–46; Thomas P. Hughes, "The Electrification of America: The System Builders," *Technology and Culture* 20 (1979): 124–61; Passer, *The Electrical Manufacturers,* passim; and Matthew Josephson, *Edison: A Biography,* (New York: McGraw-Hill, 1959).

14. Frederick Jackson Turner, "The Significance of the Frontier in American History," *Annual Report of the American Historical Association* (1894): 199–227. Also see Frederick Jackson Turner, "Contributions of the West to American Democracy," *Atlantic Monthly* 91 (1903): 83–96.

15. See Robert H. Wiebe, *The Search for Order, 1877–1920* (New York: Hill and Wang, 1967); Sidney Fine, *Laissez Faire and the General-Welfare State* (Ann Arbor: University of Michigan Press, 1956); George Mosse, *The Crisis of German Ideology* (New York: Grosset and Dunlap, 1964); and Carlton J. H. Hayes, *A Generation of Materialism, 1871–1900,* (New York: Harper & Row, 1941), 196–327. Cities were conceived in a similar manner. See Alan I Marcus, "The City as a Social System: The Importance of Ideas," *American Quarterly* 37 (1985): 332–45.

16. This point is well-documented in the secondary literature. See Foster, *Medical Bacteriology,* 203–17.

17. John Maynard Keynes, *The General Theory of Employment, Interest, and Money* (New York: Harcourt, Brace, 1936); Arthur O. Lovejoy, *The Great Chain of Being: A Study of the History of An Idea* (Cambridge: Harvard University Press, 1936); R. D. McKenzie, *The Metropolitan Community* (1933; New York: Russell and Russell, 1967); Rupert B. Vance, *Human Factors In Cotton Culture: A Study in the Social Geography of the American South* (Chapel Hill: University of North Carolina Press, 1929); Horace Kallen, *Culture and Democracy in the United States: Studies In the Group Psychology of the American Peoples,* (New York: Boni and Liveright, 1924). Walter B. Cannon was one of the life scientists who sought to apply his biological speculations to society. See his *The Wisdom of the Body* (New York: W. W. Norton, 1932), 287–306.

18. Selman A. Waksman, *My Life With Microbes* (New York: Simon and Schuster, 1954), 60–101; and Jacques Loeb, *The Mechanistic Conception of Life* (Chicago: University of Chicago Press, 1912). Also see Loeb's *Forced Movements, Tropisms, and Animal Conduct* (Philadelphia: J. B. Lippincott, 1918).

19. See, for instance, Selman A. Waksman and Robert E. Curtis, "The Actinomycetes of the Soil," *Soil Science* 1 (1916): 99–134; Selman A. Waksman, "The Importance of Mold Action in the Soil," *Soil Science* 6 (1918): 137–55; Selman A. Waksman and J. S. Joffe, "The Oxidation of Sulfur By Microorganisms," *Proceedings of the Society For Experimental Biology* 18 (1921): 1–3; S. A. Waksman and E. B. Fred, "A Tentative Outline of the Plate Method For Determining the Number of Microorganisms in the Soil," *Soil Science* 14 (1922): 27–28; Selman A. Waksman, Clara H. Wark, Jacob Joffe, and Robert L. Starkey, "Oxidation of Sulfur by Microorganisms in Black Alkali Soils," *Journal of Agricultural Research* 24 (1923): 297–305; Selman A. Waksman and S. Lomanitz, "Contributions to the Chemistry of Decomposition of Proteins and Amino Acids by Various Groups of Microorganisms," *Journal of Agricultural Research* 30 (1925): 263–81; Selman A. Waksman, "Cellulose and its Decomposition in the Soil by Microorganisms,"

Proceedings of the International Society of Soil Science 2 (1926): 293–304; and Selman A. Waksman, "The Micro–Biological Complexities of the Soil and Soil Deterioration," *Journal of the American Society of Agronomy* 18 (1926): 137–42.

20. Selman A. Waksman and Robert L. Starkey, "Partial Sterilization of the Soil, Microbiological Activities and Soil Fertility," *Soil* science 16 (1923): 137–56, 247–68, 343–57; Selman A. Waksman and Robert L. Starkey, "Influence of Organic Matter upon the Development of Fungi, Actinomycetes and Bacteria in the Soil," *Soil Science* 17 (1924): 373–78; and Selman A. Waksman, *Principles of Soil Microbiology* (Baltimore: Williams and Wilkins, 1927), 739.

21. Waksman, *Principles,* 642–43, 834–43.

22. Selman A. Waksman, "Soil Microbiology in 1924: An Attempt at an Analysis and a Synthesis," *Soil Science* 19 (1925): 201–46; Selman A. Waksman, "Soil Biology and Biochemistry: Recent Progress," *Soil Science* 25 (1928): 29–36; Sergei N. Winogradsky, "Sur l'étude microscopique du Sol," *Comptes rendus* 179 (1924): 367–67; Sergei N. Winogradsky, "La methode directe dans l'étude microbiologique du sol," *Chimie Et Industrie* 11 (1924): 215–22.

23. Waksman, *Principles,* 834–38, and "Soil Microbiology in 1924," 220–46; Winogradsky, "Sur l'étude," 367–71 and "La Methode," 215–22. Waksman thought so highly of his friend that he penned a brief biography. See Selman A. Waksman, *Sergei N. Winogradsky: His Life and Work* (New Brunswick, N.J.: Rutgers University Press, 1953), especially pp. xi–xiii, 42–48.

24. Selman A. Waksman, "The Origin and Nature of the Soil Organic Matter or Soil 'Humus': I. Introductory and Historical," *Soil Science* 22 (1926): 123–62. Also see Selman A. Waksman, "What is Humus?" *Proceedings of the National Academy of Sciences* 11 (1925): 463–68; Selman A. Waksman, "On the Origin and Nature of the Soil Organic Matter or Soil 'Humus', V. The Role of Microorganisms in the Formation of 'Humus' in the Soil," *Soil Science* 22 (1926): 421–36; Selman A. Waksman, Florence G. Tenney and Kenneth R. Stevens, "The Role of Microorganisms in the Transformation of Organic Matter in Forest Soils," *Ecology* 9 (1928): 126–44; and Selman A. Waksman, *Humus: Origin, Chemical Composition, and Importance in Nature* (Baltimore: Williams and Wilkins, 1938), especially 183–89, 414–16. Waksman also extended his analysis to peat and maintained that it too was not a useful concept. See Selman A. Waksman, "Chemical Composition of Peat and the Role of Microorganisms in its Formation," *American Journal of Science,* 5th series, 19 (1930): 32–54. Also see Selman A. Waksman and Robert L. Starkey, *The Soil and the Microbe* (New York: John Wiley and Sons, 1931).

25. See, for example, M. M. S. DuToit and H. J. Page, "Studies on the Carbon and Nitrogen Cycles in the Soil. III. The Formation of Natural Humic Matter," *Journal of Agricultural Science* 20 (1930): 478–88; H. J. Page, "Studies on the Carbon and Nitrogen Cycles in the Soil. I. Introductory," *Journal of Agricultural Science* 20 (1930): 455–59; H. R. Rosen and L. Shaw, "Studies on Sclerotium Rolfsil, with Special Reference to the Metabolic Interchange between Soil Inhabitants," *Journal of Agricultural Research* 39 (1929): 41–61.

26. Saul Benison, "René Dubos and the Capsular Polysaccharide of Penumococcus: An Oral History Memoir," *Bulletin of the History of Medicine* 50 (1976): 459–77; and René Dubos and Oswald T. Avery, "Decomposition of the Capsular Polysaccharide of Pneumococcus Type III by a Bacterial Enzyme," *Journal of Experimental Medicine* 54 (1931): 51–71.

27. Benison, "René Dubos," 464.

28. Dubos and Avery, "Decomposition," 52–54.

29. Ibid., 54–70; Oswald T. Avery and René Dubos, "The Protective Action of a Specific Enzyme against Type II Pneumococcus Infection in Mice," *Journal of Experimental Medicine* 54 (1931): 73–89; and René Dubos, "Factors affecting the Yield of Specific Enzyme in Cultures of the Bacillus Decomposing the Capsular Polysaccharide of Type III Pneumococcus," *Journal of Experimental Medicine* 55 (1932): 377–91.

30. Dubos, "Factors," 387–91.

31. Selman A. Waksman, "Associative and Antagonistic Effects of Microorganisms," *Soil Science 43* (1937): 51–68, 69–76, 77–91.

32. André Gratia and Bernice Rhodes, "De l'action lytique des Staphlycoques vivants sur les staphlocoques tués," *Comptes Rendus Hebdomadaires des Séances et Mémoires de la Société de Biologie et de ses Filiales* 90 (1924): 640–42; André Gratia and Sara Dath, "Propriétés bacteriolytiques de certaines moissures," Ibid. 91 (1924): 1442–43; André Gratia and Sara Dath, "Moisissures et microbes bacteriophages," Ibid., 92 (1925): 461–62; André Gratia and Sara Dath, "De l'action bacteriolytique des streptothrix," Ibid., 92 (1925): 1125–26; André Gratia and Sara Dath, "A propos de l'action bacteriolytique de streptothrix," Ibid. 93 (1925): 451; André Gratia

and Sara Dath, "Propriétés bacteriolytiques des streptothrix, Ibid. 94 (1926): 1267–68; E. A. Rodionova, "The Influence of Metabolism Products of Micro-Organisms on the development of Others," *Arkhiv Biologicheskikh Nauk,* 30 (1930): 335–44; J. S. Borodulina, "The Mutual Relations of Soil Actinomycetes and B. Mycoides," *Mikrobiologia* 4 (1935): 561–86; and M. I. Nakhimovskaia, "The Antagonism between Actinomycetes and Soil Bacteria," *Mikrobiologia* 6 (1937): 131–57. Also see, for example, G.P.Alivisatos, "Ueber Antagonismos zwischen Pneumokokken und Staphylokokken," *Zentralblatt für Bakteriologie, Parasitenkunde, Infektionskrankheiten und Hygiene Originale,* 94 (1925): 66–73; C. Arnaudi, W. Kopazcewski, and M. Rosnowski, "Les antagonismes physico-chimiques des microbes," *Comptes Rendus Hebdomadaires des Séances de l'Académie des Sciences* 185 (1927): 153–56; and J. M. Lewis, "Bacterial Antagonism with special reference to the Effect of Psudomonas Fluorescens on Spore-Forming Bacteria in Soils," *Journal of Bacteriology* 17 (1929): 89–103.

33. Benison, "René Dubos," 473–75.

34. René J. Dubos, "Studies on a bactericidal Agent extracted from a Soil Bacillus," *Journal of Experimental Medicine* 70 (1939): 1–10, 11–17. Later work showed that Tyrothricin was composed of two substances. See René J. Dubos and Rollin D. Hotchkiss, "The Production of bactericidal Substances by Aerobic Sporulating Bacilli," *Journal of Experimental Medicine* 73 (1941): 629–40.

35. Waksman, "My Life," 203–4; and Merck and Co., *By Their Fruits* (Rahway, N.J.: Merck Sharp and Dohme Research Laboratories, 1963), 10–12.

36. Waksman remembered his innovations in Selman A. Waksman and Albert Schatz, "Soil Enrichment and Development of Antagonistic Microorganisms," *Journal of Bacteriology* 5 (1946): 305–15; Selman A. Waksman and Hubert A. Lechevalier, "The Principle of screening Antibiotic Producing Organisms," *Antibiotics and Chemotherapy,* 1 (1951): 125–32; and Selman A. Waksman, "Searching for new Chemotherapeutic Agents—A Travelogue," *Journal of the Mount Sinai Hospital,* 16 (1950): 267–84. Also see, Selman A. Waksmen, Edith Bugie, and Albert Schatz, "Isolation of Antibiotic Substances from Soil Micro-Organisms, with Special Reference to Streptothricin and Streptomycin," *Proceedings of the Staff Meetings of the Mayo Clinic* 19 (1944): 537–48.

37. Selman A. Waksman and H. Boyd Woodruff, "Actinomyces Antiobioticus, a New Soil Organism Antagonistic to Pathogenic and Non-pathogenic Bacteria," *Journal of Bacteriology* 42 (1941): 231–49, and H. J. Robinson and S. A. Waksman, "Studies on the Toxicity of Actinomycin," *Journal of Pharmacology and Experimental Therapeutics* 74 (1942): 25–32.

38. Selman A. Waksman and H. Boyd Woodruff, "Streptothricin, a New Selective Bacteriostatic and Bactericidal Agent, Particularly Active against Gram-Negative Bacteria," *Proceedings of the Society for Experimental Biology and Medicine* 49 (1942): 207–10.

39. Ibid.; Epstein and Williams, *Miracles,* 134; and H. J. Metzger, S. A. Waksman, and L. H. Pugh, "In Vivo Activity of Streptothricin against Brucella Abortus," *Proceedings of the Society for Experimental Biology and Medicine* 51 (1942): 251–52.

40. Albert Schatz, Edith Bugie, and Selman A. Waksman, "Streptomycin, a Substance Exhibiting Antibiotic Activity against Gram-positive and Gram-negative Bacteria," *Proceedings of the Society for Experimental Biology and Medicine* 55 (1944): 66–69.

41. H. C. Hinshaw and W. H. Feldman, "Streptomycin in Treatment of Clinical Tuberculosis: A Preliminary Report," *Proceedings of the Staff Meetings of the Mayo Clinic* 20 (1945): 313–18; F. A. Figi, H. C. Hinshaw, and W. H. Feldman, "Treatment of Tuberculosis of the Larynx with Streptomycin: Report of a Case," Ibid., 21 (1946): 127–30; and H. C. Hinshaw and W. H. Feldman, "Streptomycin: A Summary of Clinical and Experimental Observations," *Journal of Pediatrics* 28 (1946): 269–74.

One Culture: Creativity in Art, Science, and Technology

HARRY J. EISENMAN

Is it possible that C. P. Snow's articulation of the two-cultures thesis is only twenty-five years old? For many American academicians this intellectual schism seems a permanent fixture in their approaches to and attitudes toward education. They have embraced a two-cultures mentality with their words and their deeds: cries for scientific literacy in their undergraduates, for liberal arts sensitivities among their science and engineering students, for greater public understanding of the technical world, and for less bifurcation of the communication and information processes within the community of scholars all attest to Snow's prescience.

As scholarship within academic communities becomes increasingly specialized, few faculty find the time or inclination to seek an understanding of the other culture. Recent reports urging a renewed commitment to excellence in undergraduate education with emphasis on common courses, shared values, critical thinking, and facility in both the arts and the sciences confirm that academicians have neglected the common bonds of intellectual activity and stressed the divisions and specializations of thinking and learning.[1] Although one finds exceptions to the two-cultures division in certain faculty and students, in curricula designed to "bridge the gap," and in various programs dedicated to enhancing understanding of the humanities or the sciences, one also finds, in an operational sense, a great gap in understanding and communication between those called scientists and those labeled humanists. Snow's two-cultures model seems more deeply rooted in academicians' behavior than many of them would admit.

The embrace of the model has not been universal. Snow himself anticipated potential criticism by suggesting that the schism was widest in Britain and characteristic of the twentieth century.[2] Frank Raymond Leavis, a renowned British literary critic, attacked Snow's faith in science and his optimism for that discipline's abilities to meet mankind's needs.[3] Observers and proponents of the social sciences claim that three cultures rather than two define the intellectual universe. Many historians of science and technology class themselves as members of yet another group, one that bridges the gap between the two cultures. A third-culture posture grew to a multiculture thesis.[4] Along with these many variations on the topic, Snow, in his rebuttal to the critiques of his original two-cultures formulation in the Rede Lectures, reaffirmed his belief in the two-cultures schism.[5]

Although the Snow-Leavis debate is the controversy most familiar to contemporary academicians, the articulation of the two-cultures division really began in the last century. Once science moved from natural philosophy to an independent profession in the early nineteenth century, men of science and their counterparts in the literary world began arguing over the differences of these two cultures. Until that time, no real distinction was made between the artist and the scientist.[6] But by the second half of that century, men such as Herbert Spencer, Thomas Henry Huxley, and Matthew Arnold had entered the arena of debate. Spencer and Huxley advocated the introduction of science into traditional curricula with a shift toward the scientific perspective. Arnold defended a more conventional humanistic emphasis in learning. The debate continued well into the twentieth century with leading figures such as Jacob Bronowski and Bertrand Russell supporting Huxley's position and D. H. Lawrence, E. M. Forster, and Archibald MacLeish siding with Arnold's humanism.[7]

Within the past few decades, however, artists, scientists, and other academicians and intellectuals have attempted to seek a common ground linking the two cultures. The journal *Leonardo,* founded in 1967, focuses on the linkages of the arts, sciences, and technology and publishes many articles treating the two-cultures debate. Many of these works focus on models and the process of creativity as evidence that art and science form one culture.[8] Cognitive psychologists and philosophers of aesthetics have written extensively on the subject in recent years; their work complements the historical studies from the history of science and technology that see commonality, and often identicality, between the artistic and technical cultures.[9]

Yet, historians provide a particularly valuable perspective for a one-culture interpretation. Foremost among those many historians who have demonstrated this one-culture thesis is Cyril Stanley Smith. In his pioneering work, Smith has shown that the division of art, science, and technology into separate activities is a recent development.[10] In their conceptual analysis and work with materials, early technologists and scientists closely resembled artists and often were one in the same. Developments in welding, alloying, shaping, heat treatment, casting, ceramics, glazing, paint pigments, and electroplating were achieved by artisans before specialized engineers used them.[11] Smith cites a wide variety of objects including medieval stained glass, Greek vases, eighteenth-century porcelain, a sixteenth-century breastplate, first-century glass beads, a Shang dynasty (1200 B.C.) bronze vessel, and Egyptian jewelry as examples of this art-science linkage, in which the artisan-scientist and the craftsman-technologist are commonplace.[12] His insights have made academicians keenly aware of the direct connections among art, science, and technology as viewed through history's organizing eye.

Such historical insights provided Robert Schofield with evidence that has linked a worldview to styles in both science and the arts. This evidence reinforces the notion of a one-culture thesis. In seeking a unity of creative process, he drew on examples from several historical eras to define styles common to both science and art for each of the Gothic, Renaissance, baroque, rococo, and romantic periods.[13]

In the Gothic world, Schofield argues, scientists and artists emphasized planimetric space, an attitude that dominated the intellectual milieu of the era. Two-dimensional representation and change of place ruled Ptolemaic-Aristotelean cosmology as it did Medieval painting. When Renaissance thinkers embraced three-dimensional conceptualization, the creative world of scientists and artists reflected that emphasis. Copernican cosmology stressed movement through space so that astronomers worried about volumes and movements within those volumes. Artists, such as Leonardo and Leon Battista Alberti, focused on proportionality, relationships among volumes, and perspective. Harmony and proportion dominated the new polyphonic music and the literature of authors like Shakespeare and Spenser just as they did the model building of scientists like Galileo.[14]

The baroque era exchanged bounded volumes for conceptions of infinity that resulted in a new notion of unbounded space in science and literature. Descartes's universe of relational motion and Milton's unbounded symbols represent this change in intellectual style. In the rococo period, belief in a mechanical universe influenced the wide spectrum of thinking from Newton's clock-like world to Pope's poetry. Likewise, in the romantic era, as the static-mechanical world gave way to the evolutionary-organic state of nature, science and the arts reflected this changed worldview from Darwin to Dickens and Delacroix.[15]

These common threads to the fabric of imagination and creativity suggest that scientists and artists, although different in their ultimate goals, begin with a common ground of understanding upon which they draw in establishing their interpretations of the world. For Schofield, these creators of "cultural artifacts" share an "idea of order" that makes them heirs of that cultural tradition and partners, not adversaries, in the culture of intellectuals.[16]

A similar partnership holds for technology and art. Eugene Ferguson and Brooke Hindle have demonstrated that engineers and artists share much in common. In his important article, "The Mind's Eye," Ferguson analyzes the art-technology connection, especially since the Renaissance.[17] Technology depended upon visual representation in the works of Leonardo, Ramelli, Agricola, and Biringuccio. Try to imagine *De Re Metallica* without illustrations. Leonardo's secret notebooks symbolize his creative imagination made visual. Models, pictorial or tangible objects, have been key factors in the work of engineers from the time of the medieval master craftsman to that of members of eighteenth-and nineteenth-century scientific societies. Using a wide historical sweep, Ferguson cites several successful engineers who were equally successful artists: Leonardo, Robert Fulton, Benjamin Henry Latrobe, Samuel F. B. Morse, and John Rogers.[18]

Their talents reflect what Brooke Hindle has called the capacity for spatial thinking.[19] Using Robert Fulton and the steamboat and Samuel Morse and the telegraph, Hindle demonstrates the importance of drawing in the resolution of design problems of the steamboat and telegraph. Fulton's sketching talent aided him greatly in testing various design solutions, while his sense of overall spatial design was invaluable in the planning of complete machines.[20] Morse's talent lay in his artist-trained mind's ability to cope with "synthetic-spatial" thinking by bringing order out of endless possible relationships between disparate and varying elements of a device.[21] Hindle stresses that Morse avoided knowing too much technical detail about the telegraph and instead concentrated on the essential elements of this electrical device with the goal of designing a practical system.[22]

In applications such as these, standard classifications fail. When does the work of the artist end and that of the engineer begin? Before the nineteenth century, clear distinctions between the two skills were not made; indeed, much of what is now considered technology was termed the mechanical arts. Hindle and Ferguson enhance the notion that skilled artisans and technicians often were the same people.[23] Technologists and artists often shared common talents with their abilities to think in geometric or plastic terms and to project a finished design or object from the disparate elements of an invention or innovation. Conceiving a "working whole" is an essential thought process for creativity, be it in engineering or in art.[24]

As science, technology, and the arts grew into separate professions during the nineteenth and twentieth centuries, their distinctive differences seemed to sharpen rather than to blur. Yet we can find several examples of strong interrelationships within the three categories just as Robert Schofield has done for earlier epochs. Modern artists embraced new scientific ideas and models as well as technological inventions or innovations in their works. Although some connections between modern art and science are tenuous, both fields engaged in "similar interpretations of the *visual significance* of such fundamental concepts as: light, depth, color, space, time, form, ... optical effects, randomness, and redundancy."[25] The Impressionists and neo-Impressionists were chief examples of this linkage. For example, Manet's paintings embrace a "flatness of imagery" characteristic of photography.[26] Color theories influenced Monet, Cezanne, and Delacroix; Seurat's pointillism demonstrated his interest in the science of vision. At the same time that non-Euclidian geometry captured the attention of scientists, the distorted forms and shapes of Manet, Van Gogh, Cezanne, and Picasso entered the art world.[27] The Bauhaus-trained Klee and Kandinsky interpreted gestalt psychology on canvas while an interest in geometric elements—angles, dots, lines—provided a motif for Mondrian, Larry Poons, Frank Stella, David Smith, and Kenneth Noland. The twentieth-century scientific phenomena of randomness, information theory, statistical mechanics, and mathematical models appear in the works of Ellsworth Kelly, Franz Kline, Jackson Pollack, while evidence of "redundancy" fills the canvases of Ad Reinhardt, Robert Irwin and Mark

Rothko.[28] One art critic, Leo Steinberg, argued that modern art relied on science for its images by its borrowings from cloud chambers, X-rays, and underwater and telescopic landscapes.[29]

The world of technology mirrored this parallel vision. For the modern era, the most obvious examples lie in structural engineering. Gustave Eiffel, John Roebling, Robert Maillart, and Christian Menn combined structural theory, engineering know-how, and aesthetic consciousness in their bridge designs. In works such as the Douro Bridge near Oporto, Portugal, and the Rouzat Viaduct over the Sioule River near Gannat, France, Eiffel strove to integrate functional design with "conscious symbolic meaning."[30] His lattice work of iron towers and arches presented an economical, efficient, airy and elegant result merging functionalism with aestheticism.

On the American scene, John Roebling's suspension bridge designs epitomize the pioneering efforts of artists-engineers. Combining beauty and utility in his structures, Roebling placed the visual impact of his bridges on a footing equal to their structural needs. Indeed, the aesthetic sometimes dominated the design considerations; the many stays and cables in the Brooklyn Bridge owe more to appearance than to functional design.[31] Yet the resulting interplay between the massive Gothic towers and the light, curving cables has provided aesthetic interest to generations of observers and artists.[32] Roebling approached bridge design with holistic thinking, visualizing a total structure as a spatial, three-dimensional object. This design talent places Roebling in the company of Robert Fulton, Samuel F. B. Morse, and other artist-engineers whose conceptual abilities rank them in the forefront of creative actors in the enterprise of technology.

In the twentieth century, Robert Maillart and Christian Menn represent the continuation of that tradition of the artist-engineer. Aware of the visual impact of their work, both men realized the twin goals of engineering economy and elegance in bridge design. Maillart achieved his dramatic designs by using the plastic medium of reinforced concrete, a material that allowed him maximum artistic license. His Salginatobel Bridge near Schiers, Switzerland , exemplifies the artistic structure with its low cost, economy of materials, and great aesthetic impact.[33] Of Maillart's Schwandbach Bridge near Hinterfultigen, Switzerland, David Billington has said that this "strikingly thin, fully integrated" deck-stiffened concrete bridge "springs . . . from the imagination of the engineer."[34] This leap of imagination, this artistic talent and insight are traits that mark the common culture of creativity so characteristic of modern society.

In similar fashion, Christian Menn used prestressed concrete to construct his artistic bridges. This building technique allowed him to place aesthetic considerations first in his priorities for the bridge design. Of the strikingly elegant Ganter Bridge above Brig, Switzerland, Billington said, its "deepest meaning lies in its expression of an artistic personality."[35] Menn's ability to transform a basic structural technique into objects of austerity and elegance make it difficult to separate the technical from the creative in his work. Surely, he and Maillart would classify themselves as artists as well as engineers. They recognize no cultural divide.[36]

In the realm of painting and sculpture, postwar American art, especially in three-dimensional form, contains many examples of an art-technology-science combine. The late sculptor David Smith provides excellent examples. An experienced machinist and welder by trade, Smith created metal pieces renowned for their uniqueness and industrial aesthetic.[37] Calling his studio and farm the Terminal Iron Works, Smith set up an artist's machine shop complete with a factory-like stockroom containing vast supplies of ordinary industrial materials such as "bolts, nuts, taps, dies, paints, solvents, acids, protective coatings, oils, grinding wheels, polishing discs, dry pigments, waxes, chemicals, spare machine parts" as well as a variety of metals including sheets of aluminum, bronze, copper, cold and hot rolled steel, and stainless steel.[38] Production of a sculpture replicated machine shop techniques as Smith himself did the grinding, welding, polishing, and painting of his works. An unsuspecting visitor to the Terminal Iron Works would have had difficulty distinguishing this artist's studio from the typical small machine shop so integral to industrial production.

Smith's art was deeply rooted in his industrial background and training. Familiar with metalworking techniques, he felt comfortable creating sculpture from the metals of his time. His creative work would parallel his earlier experience in factories. The process of art and the process of technology were one in the same for Smith. Art became technique and technique became art.

In addition, his pieces reflected the sensations of an industrial age; Smith viewed steel as an appropriate material because it symbolized "structure, movement, progress, suspension ... destruction and brutality."[39] By using industrial and mechanical forms, he celebrated the raw, simple yet powerful effect of commonplace materials.[40] Signature pieces such as the *Cubi* and *Zig* series placed painted and stainless steel in a new, artistic construct by establishing an impressive industrial aesthetic of machine shop vernacular.[41]

Congruent with his art-technology linkage, Smith saw a unity of art and science. Using various images from the print media as sources for new forms, versed in crystal structure and mineralogy, and aware of modern paint chemistry, he drew on scientific ideas and developments for inspiration and implementation. For example, the *Cubi* series seems modeled after a microphotograph of diatoms Smith found in a magazine.[42] Smith constantly searched the technical and scientific literature as sources for his sculptural shapes and forms. Both the materials of technology and the models of science became intricately linked in Smith's pioneering work.

In similar fashion, Robert Morris incorporated the materials of the industrial vernacular into his sculptural pieces. With a series of untitled works, Morris expressed his aesthetic interest in commonplace materials by using aluminum, steel, metal mesh, and felt in various configurations. Simple geometric forms placed in regular fashion, often in the serial imagery of repeating forms or shapes, characterize Morris' emphasis on modern materials.[43]

This embrace of art and technology grew out of Morris's experiences in studying engineering and art, serving in the Army Corps of Engineers, and doing graduate work in art history. Such a background initiated Morris to the world of industrial materials and shapes. However, only when he began to work closely with them did he begin to be influenced by the process of fabrication, the process of materials, and the process of technology and industrialism. Material, methods, and processes all generated a reaction from him so that a work literally unfolded in his mind and in physical fact as he created it. His ability to conceptualize a final product matched that of Robert Fulton or Samuel F. B. Morse in their artistic and inventive endeavors.[44] For Morris, the creative process in art and in technology were identical, were of one culture.

Although David Smith's and Robert Morris's work with industrial art echoes an intuitive grasp of various materials and processes, Kenneth Snelson produced works grounded in a thorough theoretical analysis of shapes, forms, and matter. As a student of Buckminster Fuller at Black Mountain College, Snelson studied compression and tension as architectural and structural phenomena. These studies led to his "tensegrity" concepts based on tension in steel cables and integrity in metal tubes as arranged in a space frame or sculpture. The resultant elegant works, with their hollow tubular units linked by cable, create the sensation of suspended, floating metal tubes in a state of intricate balance, as in Snelson's *Audry I* and *Audry II*. Through his studies of tensegrity, and the resultant continuous-tension and discontinuous-compression structures, Snelson developed an empirical mathematics to calculate the forces in his sculptures. Engineers who mistakenly thought such calculations presented little challenge learned otherwise when they discovered the complexity of these reflexive and interdependent structures. Snelson had created a complex piece of architectural engineering unique enough to patent the "continuous tension, discontinuous compression structures" in 1965 (patent no. 3,169,611). Sculpture based on tensegrity principles resembles structural engineering so much that the two are indistinguishable.[45] Snelson properly claims the titles of engineer, architect, and artist.

Further, Snelson the artist and Snelson the scientist are intertwined. For this artist, science and art are identical. In his research on the structure of the atom, on the theory of tension-compression structures, and on sculpture, Snelson argues that he engages in different facets of the same aesthetic endeavor.[46] The mechanical and the aesthetic are inseparable in space frames, architectural or sculptural.[47] Snelson's attitude and activity mirror that of an engineer. He has said that structure is "involved with forces, the stressing of pieces together, the kind of thing you find in a suspension bridge. . . . It is a definition of what is going on to cause that space to exist."[48] He is as interested in pushing to the limits of a structure as he is in an aesthetic result; often the two coincide.[49] Simultaneously, Snelson must satisfy the aesthetic elements such as scale, size, surface, and proportion with the technical requirements of endurance, stress, and support in his pieces.[50] Doing so requires a deft hand and an analytical mind. Snelson's hallmark sculptures of delicately balanced metal tubes attest to his success as structural engineer, theoretical scientist, and modern sculptor.

Industrial materials, scientific theories, and contemporary sculpture also combine in the works of Donald Judd. Drawing on materials such as stainless steel, plexiglass, galvanized iron, and painted and anodized aluminum, Judd created "cerebral, ultraformalistic" sculptures representing the industrial aesthetic.[51] Geometry and mathematical series fashion much of Judd's pieces such as *Untitled* (1972), which consists of aluminum boxes arranged in geometric progression on a wall. Judd also used arithmetic, inverse natural number, and Fibonacci progressions to design his placement of simple geometric forms. Using modular pieces, Judd attempted to make the mathematical visible. This theoretical base to simple shapes made of bold industrial materials and colors brings science and technology together in the creation and production of art. Judd spent the bulk of his time designing forms on paper; these plans then were sent to a fabricator for production. The plexiglass and stainless steel box, *Untitled* (1966), with its precise edges, smooth shiny surfaces, and nearly perfect geometric form, could just as easily be an industrial product as a piece of modern art. Such activity is little different from the work of a design engineer and indicates that the line between artist and engineer is very thin at best.[52] Only the final purpose of Judd's work defines it as art rather than an industrial product created from a theoretical design.

Facility with technical processes and material science mark the vacuum-formed plastic works of Craig Kauffman. His approach to creating sculpture paralleled that of a research engineer. The dearth of knowledge about vacuum forming led him first to library research work and then to the small factories producing such material. With a very sketchy knowledge about the process gained from his research, Kauffman turned to the actual production site to observe and to discuss the industrial manufacturing of vacuum-formed materials. The more he learned about plastic's properties and its behavior under certain conditions, the more he modified his own approach to the material and to his art.[53]

Increasingly influenced by the process of vacuum forming, Kauffman focused less on object and more on material. He strove to create the most nearly perfect vacuum-formed painted plastic convex shell. In a work such as *Untitled* (1967), Kauffman sprayed paint very carefully on the inner surface of his plexiglass forms.[54] His interest as an artist centered on the material as object and the application of the paint so that a viewer might gain some idea of how the shape was formed and how the paint was applied as he observed the sculpture. This information is as technical as it is artistic and would apply to a materials research laboratory as well as an artist's studio. Indeed, artists like Kauffman focus on the beauty of material, its surface, as well as the optical effects created in the "refined use of industrial materials."[55] He created several pieces of sculpture using the same mold; only the colors of the sprayed paint changed. Such a practice confirms Kauffman's fascination with color, surface, texture, and reflection rather than with an object itself. To satisfy that interest in these new materials, Kauffman had to become experienced in the manufacture and handling of vacuum-formed plastics and aware of the technological advances in painting plexiglass. For him there was little distinction between the creative process of art and the scientific and technical process of material science. The process was more important than object.[56]

The fascination with process and the neglect of object also appeared in the work of John McCracken. Like Kauffman, he emphasized surface, not form. He too versed himself in the technical and scientific foundations of producing finished plastic surfaces so that the creation of art was an interactive process between the worlds of art and of science and technology.

McCracken produced a series of fiberglass-covered plywood planks in different colors. Works such as *The Absolutely Naked Fragrance* (1967) and *Untitled* (1970) are placed against a wall to be judged by the quality of their surfaces, not by their shapes. McCracken's challenge focused on producing a nearly flawless fiberglass surface in both texture and color—not always an easy task with that material. As an artist, he tried to extend the technical possibilities of certain materials beyond those reached in the commercial marketplace.[57] In that endeavor, he and fellow artists such as Craig Kauffman acted as research scientists and development engineers, advancing the understanding and application of paint and plastic in the creation of their sculptures. In many ways, such art is more easily understood by the technical community, which can appreciate the challenges of producing such pure forms and surfaces, than by the traditional art connoisseur. And it clearly blurs the distinction between art and technology.

Through the historical perspective, with examples from artists, scientists, and engineers, and drawing on the work of recent American artists-engineers-scientists, academicians successfully may challenge the persistent two-culture mentality about the intellectual world. Historians have been joined by many cognitive psychologists, psychiatrists, and sociologists of science in arguing a one-culture thesis.[58] Distinguished scholars, such as Cyril Stanley Smith, Robert Schofield, Brooke Hindle, Eugene Ferguson, and Melvin Kranzberg, have examined various aspects of science and technology and found linkages to the world of art, especially in the realm of creativity. These investigations have discovered a common pattern of intellectual activity binding artist, engineer, and scientist into one creative culture. As academicians attempt to enhance our understanding of science and technology, of the intellect, and of creativity and invention; academicians should consider the abundance of tangible objects that surround them, either from the artistic or technical world. This evidence demonstrates a unity of creative human activity. That unity should challenge C. P. Snow's divided world and replace it with a coherent interpretation of the cognitive domain.

Notes

1. "Text of New Report on Excellence in Undergraduate Education," *The Chronicle of Higher Education*, 24 October 1984, 35–49.
2. C. P. Snow, *The Two Cultures and A Second Look* (Cambridge: Cambridge University Press, 1969), 68–69.
3. Frank Raymond Leavis, *Two Cultures? The Significance of C. P. Snow* (New York: Pantheon, 1963).
4. Kenneth E. Boulding, "The 'Two Cultures,' " in Melvin Kranzberg and Carroll Pursell, Jr., eds., *Technology in Western Civilization* (New York: Oxford University Press, 1967), 2: 686–95.
5. C. P. Snow, *The Two Cultures,* 68–100.
6. Arthur Marwick, "Some Conclusions: What Have We Done and Why Have We Tried to Do It?" in Martin Pollock, ed., *Common Denominators in Art and Science* (Aberdeen: Aberdeen University Press, 1983), 187.
7. For more detail on the historiographic battleground, see David K. Cornelius and Edwin St. Vincent, eds., *Cultures in Conflict: Perspectives on the Snow-Leavis Controversy* (Chicago: Scott Foresman & Company, 1964); and Ronald J. Bieniek, "Evolution of the Two Cultures Controversy," *American Journal of Physics* 40 (May 1981): 417–24. I am indebted to my colleague, Dr. Ronald Bieniek of the Department of Physics at the University of Missouri-Rolla for information on the bibliography of this controversy.
8. For example, see Paul C. L. Tang, "On the Similarities between Scientific Discovery and Musical Creativity: A Philosophical Analysis," *Leonardo* 17 (Winter 1985): 261–68; Sheldon Richmond, "The Interaction of Art and Science," *Leonardo* 17 (Spring 1985): 81–86. Richmond argues that the two cultures should be viewed as "functionally interdependent" and that such an interaction leads "to parallel developments in both domains," p. 81; David R. Topper has published a series of excellent bibliographic notes and essays in various *Leonardo* volumes (see vols. 15, 16 and 17) under the title, "Historical Perspectives on the Visual Arts, Science and Technology." For both the similarities and differences between art and science, see "Comment on the Relations of Science and Art" in Thomas S. Kuhn, *The Essential Tension: Selected Studies in Scientific Tradition and Change* (Chicago: University of Chicago Press, 1977.)
9. In addition to Pollock, *Common Denominators,* see Maria Teresa Miliora, "The Creative Attitude," 131–46, and Rebecca McBride DiLiddo, "Scientific Discovery: A Model for Creativity," 108–24, in David G. Tuerck, ed., *Creativity and Liberal Learning: Problems and Possibilities in American Education* (Norwood, N. J.: Ablex Publishing Corporation, 1987); Denis Dutton and Michael Krausz, eds., *The Concept of Creativity in Science and Art* (The Hague: Martinus Nijhoff, 1981). One of the best recent studies is George Levine, ed., *One Culture: Essays in Science and Literature* (Madison: University of Wisconsin Press, 1987). This volume contains an excellent introductory essay by Levine and very useful bibliographies. A large body of work from cognitive psychology treating creativity theories and analysis identifies similarities between scientific and artistic creativity. For example, see K. J. Gilhooly, *Thinking: Directed, Undirected and Creative* (London and New York: Academic Press, 1982); and Teresa M. Amabile, *Social Psychology of Creativity* (New York: Springer-Verlag, 1983). John R. Suler, "Primary Process Thinking and Creativity," *Psychological Bulletin* 88 (1980): 144–65, makes some distinctions between primary and secondary process thinking, the former characteristic of the arts and the latter of science, yet argues that the separation of artistic and scientific creativity is artificial and misleading when linked to these two kinds of thinking. A standard work for cognitive psychologists in this field is M. I. Stein, *Stimulating Creativity,* vols. 1 and 2 (London and New York: Academic Press, 1974, 1975). I am indebted to my colleague in psychology, Professor Ronald T. Kellogg, for his helpful guidance in the literature of creativity theories.

10. Cyril Stanley Smith, "Art, Technology, and Science: Notes on their Historical Interaction," *Technology and Culture* 11 (1970): 533.

11. Cyril Stanley Smith, "On Art, Invention, and Technology," *Technology Review* 78 (1976): 6–7.

12. Cyril Stanley Smith, *From Art to Science: Seventy-Two Objects Illustrating the Nature of Discovery* (Cambridge: MIT Press, 1980), passim.

13. Robert E. Schofield, "The Eye of the Beholder: A Critical Examination of Some Cultural Aspects of Scientific Creativity," *Leonardo* 16 (1983): 133–37.

14. Ibid., 135–36.

15. Ibid., 136.

16. Ibid., 137.

17. Eugene S. Ferguson, "The Mind's Eye: Nonverbal Thought in Technology," *Science* 197 (1977): 827–36.

18. Ibid., 833.

19. Brooke Hindle, *Emulation and Invention* (New York: Norton, 1981), 22.

20. Ibid., 22, 83.

21. Ibid., 93.

22. Ibid.

23. Melvin Kranzberg, "Confrontation or Complementarity?: Perspectives on Technology and the Arts," from Margaret Latimer, Brooke Hindle, and Melvin Kranzberg, eds., *Bridge to the Future: A Centennial Celebration of the Brooklyn Bridge* (New York: New York Academy of Sciences, 1984), 333–34.

24. Edwin T. Layton, Jr., "Technology as Knowledge," *Technology and Culture* 15 (1974): 36, 39.

25. Paul C. Vitz and Arnold B. Glimcher, *Modern Art and Science: The Parallel Analysis of Vision* (New York: Praeger, 1984), 32.

26. Ibid., 50–53.

27. Ibid., 50–53, 77–79, 81–89, 141.

28. Ibid., 194, 229–30, 240–42.

29. Ibid., 255.

30. David P. Billington, "Bridges and the New Art of Structural Engineering," *American Scientist* 72 (1984): 22–24.

31. Mario Salvadori and Christos Tountas, "The Brooklyn Bridge as a Work of Art," in Latimer, Hindle, and Kranzberg, eds., *Bridge to the Future,* 73, 311–12.

32. Ibid., 81.

33. Billington, "Bridges and the New Art," 26–27.

34. Ibid., 27.

35. Ibid., 31.

36. David Billington, "Building Bridges: Perspectives on Recent Engineering," in Latimer, Hindle, and Kranzberg, eds., *Bridge to the Future,* 309–11, 316, 322.

37. William Tucker, "Four Sculptors, Part 4: David Smith," *Studio International* 181 (1971): 28; Cleve Gray, ed., *David Smith on David Smith* (New York: Holt, Reinhart and Winston, 1968), 25.

38. David Smith, "Notes for David Smith Makes a Sculpture," *Art News* 67 (1969): 47. Rosalind E. Krauss, *Terminal Iron Works: The Sculpture of David Smith* (Cambridge: MIT Press, 1971), 6.

39. Gray, *Smith on Smith,* 54.

40. Tucker, "Four Sculptors," 28.

41. Barbara Rose, *American Art Since 1900* (New York: Praeger, 1968), 254.

42. M. Bochner, "Primary Structures," *Arts,* 40 (1966): 34.

43. Martin Friedman, "14 Sculptors: The Industrial Edge," *Art International* 14 (1970): 31–33.

44. Michael Compton and David Sylvester, *Robert Morris* (London: Tate Gallery, 1971), 7, 16, 115.

45. John Coplans, "An Interview with Kenneth Snelson," *Artforum* 5 (1967): 46–49; Stephen A. Kurtz, "Kenneth Snelson: The Elegant Solution," *Art News* 67 (1968): 48–51.

46. Kurtz, "Kenneth Snelson," 48.

47. Ibid., 50.

48. Howard N. Fox, *Kenneth Snelson* (Buffalo, N.Y.: Buffalo Fine Arts Academy, 1981), 11

49. Ibid., 17.

50. Ibid., 13.

51. Friedman, "14 Sculptors," 55.

52. William C. Agee, "Unit, Series, Site: A Judd Lexicon," *Art in America* 63 (1975): 40–49.

53. John Coplans, *Los Angeles Six* (Vancouver, B.C.: Vancouver Art Gallery, 1986), 25. Robert McDonald, *Craig Kauffman: A Comprehensive Survey* (Los Angeles: Fellows of Contemporary Art, 1981), 25.

54. Friedman, "14 Sculptors," 37.

55. Ibid.

56. Jacques Ellul has commented on the increasing importance of process or means in modern art in his "Remarks on Technology and Art," *Social Research* 46 (1979): 812, 826.

57. John Coplans, *Los Angeles Six,* 37.

58. In addition to those sources mentioned in notes 7, 8, and 9, see Roger N. Shepard, "Externalization of Mental Images and the Act of Creation," in B.S. Randhawa and W.E. Coffman, eds., *Visual Learning, Thinking, and Communication* (New York: Academic Press, 1978); Albert Rothenberg, *The Emerging Goddess: The Creative Process in Art, Science and Other Fields* (Chicago: University of Chicago Press, 1979); and, Karin D. Knorr, Roger Krohn, and Richard Whitley, *The Social Process of Scientific Investigation* (Dordrecht, Holland: D. Reidel, 1981). These studies stress the importance of visual conceptualization, of intuition, of artistic talent, and of imagery in the creative process for both artists and scientists.

Natural Science and Visual Art:
Reflections on the Interface

DAVID TOPPER

Ostensibly worlds apart, art and science are often scrutinized for possible connecting links by artists, philosophers, historians, scientists, or engineers.[1] I first began thinking seriously about the problem about two decades ago when I was a graduate student studying history of science with Robert Schofield. I was also teaching a section of a freshman course, an introduction to humanities for science and engineering students, which dealt in part with the art-science relationship. Schofield also taught a section of that course, and, as I recall, he provided the readings in this area; later he taught a separate course on art and science. And just recently he published a provocative article in which he attempts "to demonstrate the essential unity of creative process" in the arts and science in the West since the late Middle Ages.[2]

Meanwhile, I graduated and pursued the issue on my own. After exploring the relationships between art and science in a rudimentary fashion in graduate school, I thought that my subsequent research would expand my understanding of their connections. But my teaching duties forced me to confront methodological blockages that seemed to arrest the flow across the art/science interface. I teach courses in both the history of science and the history of art, so that, except for occasional remarks on specific historical connections between art and science, my teaching is confined to art and science as relatively independent disciplines. Thus, teaching separate courses in which art and science are considered individually has compelled me to confront their differences. Of course, this does not preclude my perception of real historical links, but it does create a context of skepticism regarding simplistic generalizations.

The slightly ambiguous nature of my resulting methodological approach is revealed in this paper. Part 1 begins with an argument for the independence of art and science and concludes with the specification of some criteria within which real historical interaction may be discerned. Then, in Part 2, I shift from a historical to a conceptual approach, and drawing upon the empirical components of art and science, I deduce some conceptual links between them.

1

Some of the best theoretical work in art and science, studied independently, is consistent with the view that art and science are fundamentally autonomous enterprises. In particular, two contemporary theorists come to mind, Thomas S. Kuhn in science history and theory and E. H. Gombrich in art history and theory. Although the work of each has often been interpreted as providing a vehicle for crossing the art-science interface, in fact neither theorist traverses the interface or believes that it should be ignored.

It is true, of course, that Kuhn's adoption of the philosophical position that stresses the theory-dependence of scientific observation has given aid and comfort to theorists hypothesizing a holistic approach to culture. Indeed, his concept of paradigms as "styles" of scientific practices, grounded in historical periods, and continually being replaced by subsequent paradigms—all of which sounds similar to "revolutions" in artistic styles —does appear to give credence to the holistic viewpoint. Nevertheless, at the core of Kuhn's theory is the problem of demarcation; that is, the endeavor to specify criteria for distinguishing "true" science from nonscientific pursuits—and the latter include the arts. To Kuhn, science is a unique activity, and the historian discloses that uniqueness by approaching the subject as a discipline. Scientific groups are more than mere collections of professionals committed to a theory; they are also individuals "trained in a sophisticated body of traditional theory and of instrumental, mathematical, and verbal techniques"; accordingly, the historian, to penetrate to the core of the scientific enterprise, must pursue so-called internal history.[3] Kuhn is emphatic on this point: "When scientific ideas are discussed without reference to the concrete technical problems against which they are forged, what results is a decidedly misleading notion of the way in which scientific theories develop."[4] This disciplinary approach to history delineates the uniqueness of human enterprises, be they art or science, and thus seriously subverts the holistic approach. Indeed, Kuhn once attended a conference where he was confronted with this issue; in his concluding comments he made explicitly clear his view that "science and art are very different enterprises" in terms of their means and ends, in the manner in which their practitioners perceive present and past innovation, and in the audiences to which they are directed.[5] The autonomy of science vis-à-vis art is thus a crucial comparative factor for Kuhn.

E. H. Gombrich's "scientific" approach to art has also been erroneously viewed as a holistic rendering of art and science. Most often cited in this context is his seminal work, *Art and Illusion.*[6] Yet that book, along with Gombrich's other writings on art theory, may be seen as crossing the interface only from a remote tactical viewpoint, not a historical one. That Gombrich draws upon the science of perception, that he approaches art from an analytical-rational perspective, that he believes in "the science of art"[7]—these merely provide a procedural link between art and science. For the aim of art historians should be "to operate on the same level of abstraction and precision as do the other scientists."[8] Thus in this sense, and only in this sense, are art and science linked for Gombrich. But the path from this link to an actual historical interrelationship is fraught with snares.

A striking example is the often-postulated connection between art and science in the first decade of this century, specifically involving Einstein's relativity theory and Picasso's cubism. Neither the fact that Picasso knew nothing of Einstein's 1905 papers in the *Annalen der Physik* when he painted the *Demoiselles d'Avignon* (1906–7) nor that Einstein himself denied any knowledge of cubism at the time he was working on the theory has deterred some theoreticians from linking the two.[9] An illuminating example is the case of Paul M. Laporte, who published a two-part paper (in 1948 and 1949) postulating a connection between Einstein's science and Picasso's art.[10] The details of Laporte's paper are not of importance here for he submits the usual litany of links supposedly revealing the relationship: non-Euclidian geometry and four-dimensional space, observer-object interaction, multifaceted viewpoints of the perception of things, and so forth. The fascinating feature of this case is that Laporte sent a copy of his manuscript to Einstein before submitting it for publication, and in his reply Einstein emphatically stated, "I find your comparison rather unsatisfactory." Although Einstein conceded that there are some similarities between scientific and artistic activity in that they "attempt to assemble from parts a whole" and that "the resulting order" is "of a high order," nonetheless, he felt that on concrete matters they have little in common.[11] Regarding the issue of cubism and relativity, Einstein pointed to Laporte's error in relating the principle of relativity to the multiple viewpoints of cubism. In relativity, any inertial system is sufficient for describing the physical nature of the world. "A multiplicity of systems of coordinates is not needed," writes Einstein. In short, the variant viewpoints of the cubists are in no way related to the emphasis upon invariant forms in relativity. And Einstein concludes his letter thusly: "This new [cubist] artistic 'language' has nothing in common with the Theory of Relativity."[12] Laporte's misconception is understandable: the appellation "relativity" masquerades the absolute nature of the role of invariance in the popular accounts of the theory. What is not as clearly acceptable, however, is Laporte's decision to publish the paper despite Einstein's clear-cut delineation of relativity and cubism. Intellectual misgivings apparently later compelled Laporte to publish Einstein's letter with an explanation.

Another instance in which Einstein divorced relativity and cubism is reported in Wolf Von Eckardt's book on the architect Eric Mendelsohn.[13] Mendelsohn's first major work was the so-called Einstein Tower, an observatory and astrophysical laboratory built in 1920 near Potsdam. The novel structure's major purpose was for testing the gravitational red-shift prediction of Einstein's general theory of relativity. Through the tower, Einstein became acquainted with Mendelsohn. In 1941 Mendelsohn sent Einstein a copy of Sigfried Giedion's book *Space, Time and Architecture*,[14] the title of which was probably based on Eddington's popular book *Space, Time and Gravitation*. In his book Giedion draws a parallel between relativity and modern art and architecture. The idea was "ludicrous" to Mendelsohn, and he sent the appropriate pages of the book to Einstein. The answer came in one of Einstein's doggerel verses: "It's never hard some new thought to declare / If any nonsense one will dare. / But rarely do you find that novel babble / Is at the same time reasonable." And he added, as a postscript, "It is simply bull without any rational basis."[15]

The parallels between art and science drawn by Giedion and Laporte are rooted in the belief in the role of the zeitgeist. This concept from nineteenth-century historiography is often the crucial factor in most holistic theories of history. Many historians, however, view zeitgeist as a deceptive methodological concept. Gombrich, in particular, has been a most vocal opponent of this concept, which he views as fundamentally Hegelian, as applied to cultural history. "It is this belief in the existence of an independent supra-individual collective spirit," he writes, "which seems to me to have blocked the emergence of a true cultural history."[16] Of course, Gombrich does admit that "obviously there is something in the Hegelian intuition that nothing in life is ever isolated, that any event and any creation of a period is connected by a thousand threads with the culture in which it is embedded." Nevertheless, he continues, "it is one thing to see the interconnectedness of things, another to postulate that all aspects of a culture can be traced back to one key cause of which they are the manifestations."[17] This emphasis upon—or, more precisely, this search for—connections is surely compatible with a disciplinary approach to history, such as Kuhn's.

In fact, the art historian Linda D. Henderson has explored the connection between art and science around 1900 and effectively exorcised the zeitgeist. Her research into the concept of the fourth dimension has revealed that the use of this concept by artists and theorists of the cubist and futurist movements can be traced to the many popular accounts of multidimensional geometry in the late nineteenth and early twentieth centuries.[18] These popular books were the major source of such modern concepts as a fourth dimension, but this had nothing to do with the theory of relativity, which had yet to be popularized. Henderson showed that there is a connection between art and science; and the ubiquitous nature of popular science books precludes resorting to a zeitgeist. Henderson's analysis thus reveals tactics for a trustworthy approach to the interface problem as a historical phenomenon.[19] In short, despite their intrinsically autonomous natures, art and science can and do cross the interface.

2

But must the historian's task be confined to seeking specific historical interrelations between art and science? Not at all, for links may also be forged through what I would call a conceptual analysis. This viewpoint shifts the emphasis of study from the essential factors of a subject to the manner in which it is customarily perceived—the social consensus, so to speak. Analyzing art and science in this way casts light upon the interface, without concealing

Rosa Villosa, Pomifera.   *Rosier Velu, Pomifère.*

Fig. 1.   **Pierre-Joseph Redouté (painter), Chapuy (engraver),** *Rosa Villosa, Pomifera* **(Gift of Print Club of Cleveland, Cleveland Museum of Art.)**

the autonomy. This may be exemplified by considering an exhibition assembled several years ago at the Cleveland Museum of Art by Lynette I. Rhodes entitled "Science within Art."[20] Drawn from the permanent collection of the museum, the artifacts displayed were pictures of a botanical, zoological, and anatomical nature—items that, upon initial inspection, appear to exhibit the theme of science within art. The point of the show was to reclassify the art objects, to shift their context from art to science.

Consider some concrete examples. Several botanical illustrations were displayed in the show, in particular some works of Pierre-Joseph Redouté, "the most celebrated flower painter of his day." Although the pictures are owned by the Cleveland Museum of Art, a change of context—perhaps initiated by noticing the Latin nomenclature at the bottom of the picture—may transform the work from art to science. Thus regarding the *Rosa Villosa* (see fig. 1), Rhodes writes that "Redouté faithfully records the life cycle of the apple rose as it develops from a tightly closed bud to a full-blown bloom. A detail of the rose hip, or seed pod, is drawn to the right of the stem."[21] Like illustrations in science textbooks, one can learn about nature from such pictures, just as Redouté, a careful observer, worked from nature. On one occasion, in fact, Redouté was visited by John James Audubon, the celebrated American wildlife painter. Not surprisingly, a picture by Audubon is in the show; it depicts two peregrine falcons with bloodied beaks pecking at their prey. As with the botanical illustrations, so for the zoological ones—Rhodes proposes a shift in context from art to science. How authentic, however, is this transformation? Initial inspection raises some doubts, for as Rhodes notes, Audubon's pictures "are uneven in technical control and scientific accuracy."[22] Does this fact therefore erect a barrier that blocks the shift in context? That is, do Audubon's pictures have no ornithological significance? I think they have. Audubon's famed *Birds of America* (1838) was a means of visually preserving aspects of American birds; indeed, he was displaying their behaviour, habits, and character. Listen to his caption for the American Redstart: "I have looked for several minutes at a time on the ineffectual attacks which this bird makes on wasps. . . . The bird approaches and snaps at them, but in vain. . . . This male bird is reproduced in the plate in this posture."[23] The artist, in other words, was painting something like a psychological portrait of a bird, portraying something of its temperament. Rhodes's inclusion of Audubon in the show is therefore apparently vindicated. Moreover, the transition back to art is easily made; a naturalist who may own an Audubon picture merely needs to hang it over the fireplace.

How far can such contextual transformations be extended? Has Rhodes, in fact, limited her realm of analysis? Was there more "science within art" in the museum than she availed herself of? I submit there was. The Cleveland Museum owns other pictures of flowers and birds (seventeenth-century Dutch still lifes, for example) that depict nature as closely as—perhaps sometimes even better than—many naturalist artists. Was there a logical barrier preventing their inclusion in the show? If not, then why not extend the logic further and include some of the museum's splendid portraits of people, too? Does not Rhodes's selection betray a residue of the conventional division of what constitutes art and science? Perhaps it was the factual-sounding captions accompanying the naturalist pictures that made them easy prey for a change in context, and hence they (rather the other above-mentioned classes of pictures) found their way into the exhibition.

I would even be bold enough to carry this argument further, by asserting that Rhodes should have included a landscape in the show. This may, I suspect, seem strange, but I should like to justify it through the example of John Constable. Recent scholarship has drawn attention to the importance of visual verisimilitude in Constable's rendering of natural phenomena, particularly the sky. The historical development of his landscapes reveals an increasing emphasis upon cloud formations; in particular during the period 1821–22 Constable executed a number of cloud studies (perhap fifty or more). Most were oil sketches, apparently done on the spot, with specific meteorological data written on the back, such as the date, place, time of day, direction of viewpoint, condition of the wind, and other weather-related factors. In a famous letter to his friend, the Reverend John Fisher, he wrote: "I have done a good deal of skying—I am determined to conquer all difficulties and that most arduous one among the rest. . . . The sky is the *source of light* in nature—and governs everything."[24] These "skying" studies were then available to be incorporated, when needed, into "complete" oil paintings, the aim presumably being an accurate depiction of the sky in his art.

In 1937 the meteorologist L. C. W. Bonacina, on the occasion of the centenary of Constable's death, presented a paper before the Royal Meteorological Society of London in which he stressed the visual accuracy of Constable's depiction of the weather. As Bonacina wrote (adding some questionable pedagogy):

> Constable's skies are . . . highly realistic. . . . The dark lines of falling rain occur in the background of many of the pictures, and for this reason alone the study of Constable in the public galleries might form an invaluable introduction to landscape meteorology for city-bred children who are largely cut off from the natural scenery that country children grow up with.[25]

Bonacina also pointed out that many of Constable's skies were less "an exact facsimile of some particular sky" and more of a depiction of a generalized "type." Nevertheless, he did not consider this a criterion for criticism; indeed, Bonacina spoke of the "fidelity to type" and how "valuable" this may be to students of meteorology. In commenting upon two specific works, Bonacina was moved to state: "The spectator might well remark. 'How absolutely true to type, how well I know that kind of "troubled" sky'—a replica of which in nature he has never seen at all."[26]

Drawing upon such distinctions, Kurt Badt, in his book *John Constable's Clouds* (1950),[27] thrust a wedge between types and individuals by stressing Constable's apparent debt to meteorological theory. Essentially, Badt's thesis was that Constable derived his knowledge of clouds less from direct observation and more from the system of cloud classification put forward at the time by Luke Howard in his essay, "On the Modification of Clouds." In short, according to Badt, Constable's art only minimally relied upon direct empirical observation.

Badt's book initiated a debate, the form of which will be familiar to historians of science. At the core of the debate was the question, "What were the relative roles of theory and experience in Constable's art?" The first challenge to Badt was Louis Hawes's article of 1969, "Constable's Sky Sketches."[28] Recognizing Constable's possible knowledge of Howard's cloud categories and adding as well a probable knowledge of Thomas Forster's book of 1813, *Researches about Atmospheric Phenomena* (another important contemporary study of meteorology), Hawes presented a series of convincing arguments minimizing the role of theory in Constable's art and stressing instead "intensive observation and repeated sketching of skies."[29]

Subsequent scholarship has fundamentally supported Hawes's thesis. There have, however, been some qualifications of the details. The most recent work was Paul D. Schweizer's article, "John Constable, Rainbow Science, and English Color Theory,"[30] which revealed Constable's serious study of the scientific principles of the rainbow. Schweizer's analysis shifted the historiographical theme from theory *versus* observation to a more subtle, and I think meaningful, level: it was his contention that Constable believed "that an artist can never correctly paint a rainbow, or for the matter any other natural phenomenon, without first becoming familiar with the scientific principles upon which it is based."[31] Consequently Schweitzer did not denigrate the role of visual observation in Constable's art.

Similarly, John Thornes's 1979 article on "Constable's Clouds,"[32] which clarified the relative roles of Howard and Forster, ultimately concluded that Constable was a keen observer of nature, however much he was familiar with meteorological theory. Like Bonacina, Thornes is a meteorologist. Another scientist, Ronald Rees, published in 1976 "John Constable and the Art of Geography." He underscored Constable's debt to the study of nature. Rees contended, "Constable tried to replace the conceptual approach to the landscape with one based on experience and observation,"[33] and he went on to speak of Constable's "knowledge of the morphology" of the landscape.[34] Thus the empirical component of Constable's art was further supported from the viewpoint of a geographer.

Constable, I'm sure, would be pleased to know that his work has been scrutinized so carefully by scientists. Indeed, in the last years of his life, Constable delivered a series of lectures on the history of landscape painting to scientific and literary societies in Hampstead, Worcester, and London, his aim being to elevate landscape art to the status of a science. To cite one example, in the spring of 1836 he delivered three lectures before the Royal Institution; Constable's son was a student there and the lectures were arranged through Michael Faraday, who became interested in Constable upon hearing of his idea that landscape painting was as much a science as an art. Another lecture, planned for the following summer, was never delivered; Constable died on 30 March 1837.

There is no verbatim transcription of any of the lectures. What remains are some original notes by Constable and detailed notes taken by his friend and biographer, the young American painter Charles Leslie. The essential argument of Constable may be gleaned from these sources.[35] He apparently began each series of lectures with a remark on science and art. At the Royal Institution he said:

> I am here on the behalf of my own profession.... I hope to show ... that it is *scientific* and well as *poetic;* that imagination alone never did, and never can, produce works that are to stand by a comparison with *realities.*[36]

On the same theoretical plane he broached the issue of the artist's dependence upon preconceived forms; Constable called this "mannerism," as opposed to the artist relying upon his own experience. "Manner," he said, "is more or less an imitation of what has been done already.... Nothing but a close and continual observation of nature can protect [painters] ... from the danger of becoming mannerists."[37] Great art, however, is less mannered. Thus, he continued:

> I have endeavored to draw a line between genuine art and mannerism, but even the greatest painters have never been wholly untainted by manner.[38]

Despite this fact of the artistic process, genuine artists attempt to free themselves from the constraints of imitating other artists; the aim of the landscape artist, of course, being to imitate nature. Hence, in perhaps his most often-quoted statement, Constable said:

> Painting is a science, and should be pursued as an inquiry into the laws of nature. Why, then, may not landscape be considered as a branch of natural philosophy, of which pictures are but the experiments?[39]

To emphasize the point, he concluded the lecture by telling the audience:

> In such an age as this, painting should be *understood,* not looked on with blind wonder, nor considered only as a poetic aspiration, but as a pursuit, *legitimate, scientific, and mechanical.*[40]

In conclusion, then—and returning to the issue of Rhodes's exhibition of "science within art"—the example of Constable may be said to provide a

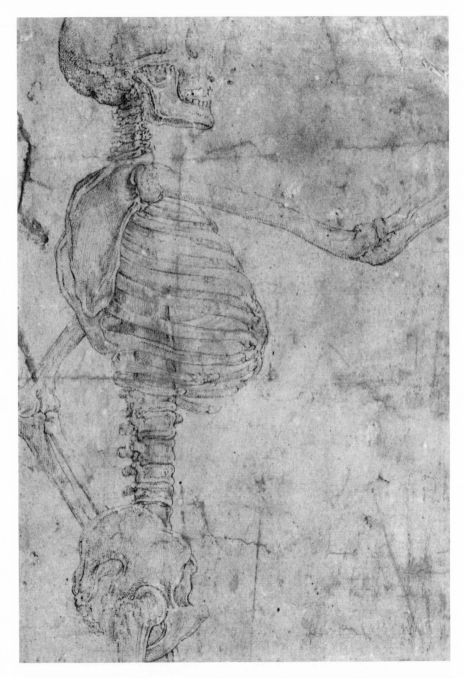

Fig. 2. Giovanni Battista Franco, *Half-Length Skeleton in Profile*. Drawing: pen and brown ink. (Gift of Mr. and Mrs. Claude Cassirer, Cleveland Museum of Art.)

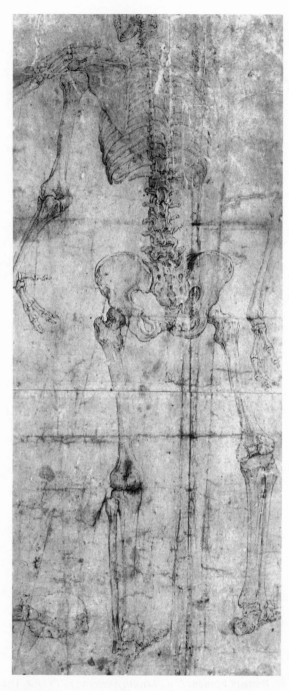

Fig. 3. Giovanni Battista Franco, *Full-Length Skeleton from the Back*. Drawing: pen and brown ink. (Gift of Mr. and Mrs. Claude Cassirer, Cleveland Museum of Art.)

meteorological case, comparable to the previously discussed ornithological examples.

What else may be included? Anatomy, too? Drawings of skeletons, flayed muscles, cross sections of the brain, and so forth seem appropriately placed in an exhibit of "science within art." Surely most visitors to the show viewed the sixteenth-century drawings of skeletons by G. B. Franco (see figs. 2 and 3) either as scientific illustrations or as academic studies for the purpose of learning skeletal structure—in both cases, within an anatomical context. Nevertheless, both interpretations are wrong. Thomas R. Forbes's important note on Franco's drawings has revealed them to be anatomically inaccurate.[41] Specifically, the vertebral column, sternum, ribs, iliac crest, pelvis, and some other parts are incorrectly depicted; indeed, according to Forbes's analysis, they cannot all come from the same body! What then was the purpose of these drawings? Louise S. Richards says: "Franco was not primarily interested (as was Leonardo, for instance) in scientific anatomical studies. . . . Rather his interest was pictorial and visual. His arrangements of the separate bones. . . . in decorative, rhythmic patterns on his paper shows the same delight in pleasing shapes which he doubtless enjoyed in his study of ancient relief sculpture."[42] Apparently Franco found aesthetic pleasure in the ripples of skeletal ribs. Visual verisimilitude was not his goal.

But Rhodes knows this. "Franco," she writes, "was more interested in the artistic idealization of the human body than in scientific accuracy."[43] Why then did she include Franco's work in the show? Apparently for the same reason the pictures by Redouté and Audubon were chosen—their seemingly easy transition from art to science. But, as seen, that ostensible transition has proved to be a decoy, leading not to "science" but elsewhere. A better choice, I have argued, would have been an accurate landscape or still life. By thus at once expanding the realm of subjects beyond conventional classes, yet delimiting the choice by applying the criterion of visual accuracy (the latter, an empirical component, however problematic that may be) Rhodes would have avoided the ensuing difficulties. At least the application of the criterion would have resulted in the correct inclusion of Constable (or the equivalent) and the elimination of Franco.

The example of Franco is a reminder that empiricism is not a necessary factor in art. For example, there are formal similarities between Leonardo's drawing of the flow of the Arno River and Paul Klee's painting, "A Page from the City Records." The drawing, taken from Leonardo's notebook, was made while planning a method of diverting the river's flow; it depicts something like an aerial view. In contrast, from the point of view of the empirical component, the Klee painting is meaningless hieroglyphics. Or said another way: whereas an engineer could learn something about the flow of the Arno River from Leonardo's drawing, an historian could not glean any information about the "City" from Klee's picture. Needless to say, Klee (like Franco) did not intend his art to contain such empirical information. But Leonardo did. Indeed, the empirical component in his art is manifestly seen in the following story told by the eminent scientist and historian of science, Stephen Jay Gould:

> While visiting Florence . . . I became fascinated by the "comparative anatomy" of Gabriel's wings as depicted by the great painters of Italy. The faces of Mary and Gabriel are so beautiful, their gestures often so expressive. Yet the wings, as painted by Fra Angelico or by Martini, seem stiff and lifeless, despite the beauty of their intricate feathering.
>
> But then I saw Leonardo's version. Gabriel's wings are so supple and graceful that I scarcely cared to study his face or note the impact he had upon Mary. And then I recognized the source of the difference. Leonardo, who studied birds and understood the aerodynamics of wings, had painted a working machine on Gabriel's back. His wings are both beautiful and efficient. They have not only the right orientation and camber, but the correct arrangement of feathers as well. Had he been just a bit lighter, Gabriel might have flown without divine guidance. In contrast, the other Gabriels bear flimsy and awkward ornaments that could never work.[44]

Like Leonardo, Michelangelo was an empiricist in art. It is surely common knowledge that Italian Renaissance artists diligently studied human anatomy. But the extent to which Michelangelo went in his quest for anatomical accuracy has only now been revealed through the work of James Elkins. An expert in "medical surface anatomy," Elkins has pursued a study of the external contours of the body's surface—the very subject of a sculptor like Michelangelo. Elkins's thorough study of the works of the sculptor reveals this astonishing fact: although present-day medical surface anatomy nomenclature comprises about 600 terms for describing the specific features of the human body, Elkins was required to increase this to over 800 terms to account for all the details represented by Michelangelo. "Repeatedly in the course of research [into Michelangelo's work]," writes Elkins, "puzzling features eventually proved to be anatomically normal, even if they were so unimportant and obscure that five centuries of Western medicine had not needed to give them names."[45] That Michelangelo was an empiricist with a vengence would presumably be an understatement; that this example indicates that within the field of surface anatomy art preceded science seems to be a historical fact.

Examples of this sort, I think, point to this truism: science, as an autonomous discipline, does not have a monopoly on empiricism. Moreover, is not the theory-dependence of science roughly the inverse measure of its empirical context? The sciences, simply put, form a spectrum: certain fields or subfields (such as geology, meteorology, and areas of anatomy and physiology) have a "high" empirical content, at least higher than those at the other end of the spectrum (such as cosmology and areas of high-energy physics and microbiology), which rely heavily upon heuristics, speculation, or *a priori* assumptions, often of an aesthetic nature. These latter examples point to another truism: art, as an autonomous discipline, does not have a monopoly on aesthetics. In short, the relative empirical and aesthetic factors in science have their corresponding counterparts in art. And it is this conceptual relationship that I think provides a meaningful strategy for crossing the interface between art and science.[46]

Notes

This paper is an expanded version of "Truth and Beauty Revisited: On Classifying Art and Science," delivered at the annual meeting of the American Society of Aesthetics, in Banff, Alberta, Canada, on 29 October 1982.

1. See David Topper and John H. Holloway, "Interrelationships between the Visual Arts, Science and Technology: A Bibliography," *Leonardo* 13 (1980): 29–33, "A bibliographic Up-Date," *Leonardo* 18 (1985): 197–200.

2. Robert E. Schofield, "The Eye of the Beholder: A Critical Examination of Some Cultural Aspects of Scientific Creativity," *Leonardo* 16 (1983): 133–137.

3. Thomas S. Kuhn, *The Essential Tension: Selected Studies in Scientific Tradition and Change* (Chicago and London: University of Chicago Press, 1977), 119. Of course, Kuhn's pioneering work remains *The Structure of Scientific Revolutions* (Chicago and London: University of Chicago Press, 1962).

4. Kuhn, *Essential Tension,* 136.

5. Kuhn, "Comment on the Relations of Science and Art," in *Essential Tension,* 340–51. This article is reprinted from *Comparative Studies in Society and History* 11 (1969): 403–412, where articles by the other participants may be found.

6. E.H. Gombrich, *Art and Illusion: A Study in the Psychology of Pictorial Representation,* 2nd rev. ed. (Princeton: Princeton University Press, 1969).

7. E.H. Gombrich, *Meditations on a Hobby Horse and Other Essays on the Theory of Art* (London: Phaidon, 1963; 2nd ed., 1971), ix. See also his most recent, *The Image and the Eye: Further Studies in the Psychology of Pictorial Representation* (Ithaca: Cornell University Press, 1982).

8. Gombrich, *Meditations,* ix.

9. For a list of relevant articles, see Topper and Holloway, "Interrelationships," 31.

10. Paul M. Laporte, "The Space-Time Concept in the Work of Picasso," *Magazine of Art* 41 (1948): 26–32, 156; and "Cubism and Science," *Journal of Aesthetics and Art Criticism* 7 (1949): 243–56.

11. Einstein, quoted in Paul M. Laporte, "Cubism and Relativity," *Art Journal* 25 (1966): 246.

12. Ibid.

13. Wolf von Eckardt, *Eric Mendelsohn* (New York: George Braziller, 1960).

14. Sigfried Giedion, *Space, Time and Architecture,* 5th rev. ed. (Cambridge: Harvard University Press, 1967), 434–36, for the discussion on cubism and relativity. I have not seen the first edition.

15. Einstein, quoted in von Eckardt, *Mendelsohn,* 14.

16. E. H. Gombrich, *In Search of Cultural History* (Oxford: Clarendon Press, 1969), 36.

17. Ibid., 30.

18. Linda D. Henderson, "A New Facet of Cubism: 'The Fourth Dimension' and 'Non-Euclidean Geometry' Reinterpreted," *Art Quarterly* 34 (1971): 410–33: "The Merging of Time and Space: The Fourth Dimension in Russia from Ouspensky to Malevich," *The Structuralist,* no. 15/16 (1975–76): 97–108; "Italian Futurism and the Fourth Dimension," *Art Journal* 41 (1981): 317–27; and *The Fourth Dimension and Non-Euclidean Geometry in Modern Art* (Princeton: Princeton University Press, 1983). Discussion of this may also be found in Gerald Holton's essay, "Einstein and the Shaping of Our Imagination," chap. 5 in his collected essays, *The Advancement of Science, and Its Burden* (Cambridge: Cambridge University Press, 1986), 108–110.

19. In a series of review articles, I have delineated a number of potentially fruitful areas of research on art and science in various periods of history. See David Topper, "Historical Perspectives on the Arts, Sciences, and Technology," *Leonardo* 15 (1982): 234–37; *Leonardo* 16 (1983): 321–24; *Leonardo* 17 (1984): 124–26, 213–15; *Leonardo* 18 (1985): 50–52, 203–6. I have discussed the specific case of Henderson's methodology in "On a Ghost of Historiography Past," *Leonardo* 21 (1988) 76–78.

20. Lynette I. Rhodes, *Science within Art* (Cleveland: Cleveland Museum of Art, 1980).

21. Ibid., 23–24.

22. Ibid., 29.

23. John James Audubon, *The Original Watercolor Paintings by John James Audubon* (New York: American Heritage Publishers, 1966), plate no. 365.

24. The letter is dated 23 October 1821 and reprinted in R. B. Beckett ed., *The Fishers,* Volume 6 of *John Constable's Correspondence* (Ipswich: Suffolk Records Society, 1968), 76–78.

25. L. C. W. Bonacina, "John Constable's Centenary: His Position as a Painter of Weather," *Quarterly Journal of the Royal Meteorological Society* 63 (1937): 483–90, 484.

26. Ibid., 485–86. Throughout the paper I am avoiding the issue of particulars and types. I should like to point out, however, that I believe the distinction is really one of degree rather than kind, and that most ensuing problems are resolved through semiotic analysis.

27. Kurt Badt, *John Constable's Clouds* (London: Routledge and Kegan Paul, 1950).

28. Louis Hawes, "Constable's Sky Sketches," *Journal of the Warburg and Courtauld Institutes* 32 (1969): 344–65.

29. Ibid., 348.

30. Paul D. Schweizer, "John Constable, Rainbow Science, and English Color Theory," *The Art Bulletin* 64 (1982): 424–45.

31. Ibid., 441.

32. John Thornes, "Constable's Clouds," *Burlington Magazine,* November 1979, 697–704.

33. Ronald Rees, "John Constable and the Art of Geography," *Geographical Review* 66 (1976): 59–72, 61.

34. Ibid., 71.

35. R. B. Beckett, ed., *John Constable's Discourses* (Ipswich: Suffolk Records Society, 1970), 28–74. See also Leslie Parris, Conal Shields, and Ian Fleming-Williams, eds., *John Constable: Further Documents and Correspondence* (Ipswich: Suffolk Records Society and The Tate Gallery, 1975), 5–24.

36. Beckett, *Discourses,* 39.

37. Ibid., 58.

38. Ibid., 68.

39. Ibid., 69. See also Parris, *et. al., Documents,* 20.

40. Beckett, *Discourses,* 69.

41. Louis S. Richards with a Note by Thomas R. Forbes, "Giovanni Battista Franco's Anatomical Drawings in Cleveland," *Journal of the History of Medicine and Allied Sciences* 20 (1965): 406–9.

42. Ibid., 408.

43. Rhodes, *Science,* 15.

44. Stephen Jay Gould, *The Panda's Thumb: More Reflections in Natural History* (New York and London: W. W. Norton, 1982), 306.

45. James Elkins, "Michelangelo and the Human Form: His Knowledge and Use of Anatomy," *Art History* 7 (1984): 176–86.

46. Empiricism, as delineated here, should not be confused with truth. Needless to say, a footnote is no place to digress on "truth," but I would like to mention one facet of this issue, since I may otherwise be misunderstood. A picture's empirical content is not a sufficient condition for its truth content. A momentary reflection on travel posters provides an illustration of this, specifically with regard to what these pictures leave out. Or consider the example of two writers working with the same facts but arriving at different conclusions. A further example: consider the different interpretations of the same hominid fossil by the anthropologists Mary Leakey and Don Johanson. These examples, however, do not entail a relativistic interpretation of truth, for we may still speak absolutely of illusions, errors, and lies. Indeed, that is why I believe the empirical component in art, however minimal it may be within the context of Art, is an important component within the larger context of Life.

Bibliography of the Writings of
Robert E. Schofield

"Founding of the Lunar Society of Birmingham (1760–1780): Organization of Industrial Research in Eighteenth Century England." Ph.D. diss., Harvard University, 1955.

"Did Divis Erect the First European Protective Lightning Rod, and Was His Invention Independent?" *Isis* 43 (1952): 358–64. With I. Bernard Cohen.

"John Wesley and Science in Eighteenth-Century England." *Isis* 44 (1953): 331–40.

"James Watt's letter to Joseph Priestley, 26 April 1783." *Annals of Science* 10 (1954): 294–300.

"Josiah Wedgwood and a Proposed 18th-century Industrial Research Organization." *Isis* 47 (1956): 16–19.

"Membership in the Lunar Society of Birmingham." *Annals of Science* 12 (1956): 118–36.

"The Industrial Orientation of Science in the Lunar Society of Birmingham." *Isis* 48 (1957): 408–15. Reprinted in *Science, Technology and Economic Growth in the Eighteenth Century,* edited by A.E. Musson, 136–97. London: Methuen, 1972. German translation, Hermann Vetter. Frankfurt: Suhrkamp, 1977.

Isaac Newton's Paper and Letters on Natural Philosophy and Related Documents. 2d. ed., 1978. Cambridge: Harvard University Press, 1958. As assistant to I. Bernard Cohen, editor.

"Josiah Wedgwood, industrial Chemist," *Chymia* 5 (1959): 180–92

"The Society of Arts and the Lunar Society of Birmingham, (i) and (ii)." *Journal of the Royal Society of Arts* 107 (1959): 512–14, 668–71.

"Scientific Background of Joseph Priestley," *Annals of Science* 13 (1957, printed 1959): 148–63.

"A New Academic Discipline Is on the Ascendancy," *Kansas Teacher* 67 (1959): 28–30. With James E. Gunn.

"Joseph Priestley's American Education," *Early Dickinsoniana, The Boyd Lee Spahr Lectures 1957–1961.* Carlisle, Pa.: Dickson College, 1961.

"Boscovich and Priestley's Theory of Matter." In *Roger John Boscovich: Studies of his Life and Work on the 250th Anniversary of his Birth,* edited by Lancelot Law Whyte. London: George Allen and Unwin, 1961.

"Josiah Wedgwood and the Technology of Glass manufacturing," *Technology and Culture* 3 (1962): 285–97.

"Histories of Scientific Societies: Needs and Opportunities for Research," *History of Science* 2 (1963): 70–83.

"Electrical Researches of Joseph Priestley," *Actes Xeme Congrès Internationale d'histoire des sciences* (1962), 2, 807–09.

The Lunar Society of Birmingham: A Social History of Provincial Science and Technology in 18th-century England. Oxford: Clarendon Press, 1963.

"The Lunar Society, Science and the Industrial Revolution," *Voice of America Forum Lectures, History of Science,* no. 4 (1964)

"Still More of the Water Controversy." *Chymia* 9 (1964): 71–76.

"Electrical Researches of Joseph Priestley." *Archives Internationales d'Histoire des Sciences* 64 (1963, pub. 1964): 277–86.

"Joseph Priestley, the Theory of Oxidation and the Nature of Matter." *Journal for the History of Ideas* 25 (1964): 285–94

"On the Equilibrium of a Heterogeneous Social System." *Technology and Culture* 6 (1965): 591–95.

Joseph Priestley, *History and Present State of Electricity,* 3d. ed., 1775. New York: Johnson Reprint Corp., Sources of Sciences series no. 18, 1966. Edited with critical introduction.

"Legend and Reality in 18th-century British Natural Philosophy." *Case Graduate Review* 3 (1966): 19–23.

"The Lunar Society of Birmingham: A Bicentennial Appraisal," *Notes and Records of the Royal Society* 21 (1966):, 144–61.

A Scientific Autobiography of Joseph Priestley (1733–1804) Selected Scientific Correspondence . . . with Commentary. Cambridge: MIT Press, 1966.

"Joseph Priestley, Natural Philosopher," *Ambix* 15 (1967): 1–15.

"The Lunar Society and the Industrial Revolution," *University of Birmingham Historical Journal* 11 (1967): 94–111.

Mechanism and Materialism: British Natural Philosophy in an Age of Reason. Princeton: Princeton University Press, 1970.

"An 18th Century Physician in His Library," in *Principles of Medical Librarianship,* edited by Robert C. Chesier. Cleveland: Case-Western Reserve University Press, 1970.

"Wedgwood the technician," in *Ninth Wedgwood International Seminar,* 23–25 April 1964. New York: Metropolitan Museum of Art, 1971, 125–35.

"The Counter-Reformation in 18th-century Science—Last Phase." In *Perspectives in the History of Science and Technology,* edited by Duane H.D. Roller. Norman: University of Oklahoma Press, 1971, 39–54.

"What is Modern in the Eighteenth Century—Not Science." In *The Modernity of the Eighteenth Century, Studies in Eighteenth Century Culture,* edited by Louis T. Milic. Cleveland: Case Western Reserve University Press, 1971, 61–66.

Man and Nature: An Introduction to the Humanities in Science. Cleveland: Program in the History of Science and Technology, Case Western Reserve University, 1971. Second edition, 1974. Edited with Wesley C. Williams.

"A Discourse on the Branches of Natural Philosophy most particularly related to Chemistry [as by Joseph Priestley]." *Journal of Chemical Education* 53 (1976): 409–13.

"An Evolutionary Taxonomy of 18th-century Newtonianisms," *Studies in 18th-Century Culture* 7 (1978): 175–92.

"Joseph Priestley on Sensation and Perception." In *Studies in Perception: Interrelations in the History of Philosophy and Science,* edited by Peter K. Machamer and Robert G. Turnbull. Columbus: Ohio University Press, 1978, 336–54.

"Joseph Priestley and the Physicalist Tradition in British Chemistry." In *Joseph Priestley: Scientist, Theologian and Metaphysician,* edited by Lester Kieft and Bennett R. Willeford Jr. Lewisburg, Pa.: Bucknell University Press, 1980.

Stephen Hales: Scientist and Philanthropist. London: Scholar Press, 1980. With D. G. C. Allan.

"Atomism from Newton to Dalton." *American Journal of Physics* 49 (1981): 211–16.

"The Eye of the Beholder: Cultural Content of Science." *Leonardo* 16 (1983): 133–37.

"Theology, Physics and Metaphysics; Unity and Utility in Joseph Priestley's Thought." *Enlightenment and Dissent* 2 (1983): 69–81.

"The Professional Work of an Amateur Chemist: Joseph Priestley, 1733–1804," in *Oxygen and the Conversion of Future Foodstocks.* Proceedings of the Third BOC Priestley Conference. London: Royal Society of Chemistry, 1983, 410–431.

"Joseph Priestley, British Neoplatonist, and S. T. Coleridge." In *Transformation and Tradition in the Sciences: Essays in Honor of I. B. Cohen,* edited by Everett Mendelssohn. Cambridge: Cambridge University Press, 1984, 237–54.

"Science for the Enlightened Everymen: Charles Willson Peale's Philadelphia Museum," *American Studies* (forthcoming, Fall 1989).

Index